Lecture Notes in Mathematics

Edited by A. Dold, Heidelberg and B. Eckmann, Zürich

373

A. E. R. Woodcock

IBM Th. J. Watson Research Center, Yorktown Heights, NY/USA

T. Poston

Instituto de Matemática Pura e Aplicada, Rio de Janeiro/Brazil

A Geometrical Study of the Elementary Catastrophes

Springer-Verlag

Berlin · Heidelberg · New York 1974

AMS Subject Classifications (1970): 57-02, 57D45, 57D70, 58F10

ISBN 3-540-06681-0 Springer-Verlag Berlin · Heidelberg · New York
ISBN 0-387-06681-0 Springer-Verlag New York · Heidelberg · Berlin

Offsetdruck: Julius Beltz, Hemsbach/Bergstr.
2141/3140-54321

GENERAL INTRODUCTION

The following papers constitute a study of the precise geometrical nature of the Cuspoid Catastrophes, the Elliptic, Hyperbolic and Parabolic Umbilics and the Reduced Double Cusp. Stereographic-pair figures permit a 3-dimensional realization of the Catastrophe Manifolds of the Cuspoid Catastrophes when observed through a stereo viewer. In introducing these precise pictures into the literature, it is hoped that they will aid in the application of Catastrophe Theory to the understanding of complex systems such as those of Biology, Physical Chemistry, Economics and Sociology.

Yorktown Heights, New York
May 1973

A. E. R. Woodcock

CONTENTS

THE GEOMETRY OF THE ELEMENTARY CATASTROPHES (1). THE CUSPOIDS

by

A. E. R. Woodcock
IBM Thomas J. Watson Research Center
Yorktown Heights, N.Y. 10598, U.S.A.

and

T. Poston
Instituto de Mathemática Pura e Aplicada
Rio-de-Janeiro, Brazil

ABSTRACT: The Elementary Catastrophes arise as stable singularities in a system of potentials parameterized by a manifold C (the Control Space) on a manifold X (the Behavior Space) and represented by a smooth map:

$$V: C \times X \to R.$$

The present paper describes the geometry of these singularities for potentials of the type:

$$V = \frac{x^{n+2}}{n+2} + A\frac{x^n}{n} + B\frac{x^{n-1}}{n-1} + \ldots + Rx.$$

and termed the Cuspoids.

This work was begun when the authors were both at the Mathematics Institute, University of Warwick, Coventry CV4 7AL, Warwickshire, England.

INTRODUCTION:

A catastrophe is a singularity in a map that arises stably in the following way: consider a system of potentials parameterized by a manifold C (the control space), on a manifold X (the behavior space) and represented by a smooth map

$$V: C \times X \to R$$

where the potential V_c on X corresponding to a point $c \in C$ is given by

$$V|\{c\} \times X$$

The set $M = \{(c,n) \in C \times X | \nabla_x V(c,n)=0\}$
(reducing to $\{(c,n) | \frac{\partial V}{\partial x} (c,n) = 0\}$ in the case that X is one-dimensional, as in the Zeeman Catastrophe Machine (1,2) and the standard types described in this paper), is generically a manifold by Sard's Theorem, of the same dimension as C. It is known as the catastrophe manifold. A Catastrophe is then a singularity of the Catastrophe Map.

$$\chi: M \to C$$
$$(c,x) \rightsquigarrow c$$

such that a sufficiently small perturbation of the potential does not alter its diffeotype. The greatest interest is in the cases where dim $C \leq 4$, where it is known (3) that the potentials giving rise to stable singularities in this way are open dense in the space of smooth maps $C \times X \to R$ with a suitable topology, irrespective of the dimension of X. Moreover, every such singularity is of the local diffeotype of one or other of a list of algebraic examples known as the seven elementary catastrophes.

Though we are concerned with (locally) maps $R^n \to R^n$, where $n = \dim C$, the definition of the catastrophes makes them distinct from singularities of maps $R^n \to R^n$ stable with respect to perturbations of the map. Neither class in fact contains the other: certain singularities are map-stable, but cannot arise from potentials in the manner described above, while some catastrophes are stable with respect only to perturbations of the potential, not to direct perturbations of the catastrophe map itself. The latter type of stability is, however, important (for example in the classification of light caustics) just as conservative dynamical systems need to be discussed with reference to perturbations only through changes in the Hamiltonian, since considered simply as vector fields they are almost without exception unstable. The intersection of these two stability classes of singularities consist of the Cuspoids, arising as the only catastrophes possible when X is one-dimensional. These have the local diffeotype of the catastrophes arising from potentials of the form:

$$V(c_1 \ldots, c_n, x) = x^{n+2} + c_1 x^n + \ldots + c_n x.$$

where $c_1, \ldots, c_n$ are local parameters on C. (For particular cases it is often more convenient to parameterize C by a, b, c,...rather than $c_1, c_2 \ldots$ etc., the indexed notation being used here only to give the general form). Thus, which of them are possible depends on the dimension of C: there is one more for each added dimension. As we shall see, each involves two one-parameter families of the cuspoid arising in the next dimension down. (Note: it is sometimes convenient to use

$$V(c_1, \ldots, c_n, x) = \frac{x^{n+2}}{n+2} + c_1 \frac{x^n}{n} + \ldots + c_n x,$$ which is

equivalent by a change of coordinates in C to the above, to give a simpler form upon differentiation.)

Our next papers (4,5,6) will discuss the geometry of the non-map-stable elementary catastrophes.

We are concerned, then, with the geometry of the manifolds and projections

$$M = \{(c_1, \ldots c_n, x) \mid (n + 2)x^{n+1} + nc_1 x^{n-1} + \ldots + c_n = 0\} \quad R^n x\ R \ (c,x)$$

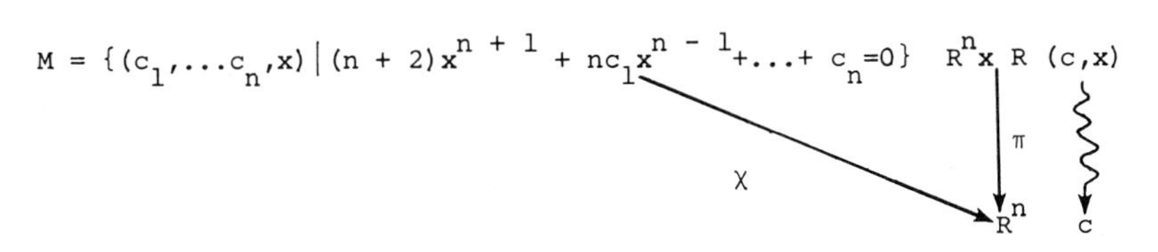

In the case that dim C = 1, we are concerned with cubic potentials of the form: $V(a,x) = \frac{x^3}{3} + ax$, which have graphs changing from those like Fig (1a) for positive a to those like Fig (1b) for negative a. M is the set

$$M = \{ (a,x) \mid x^2 + a = 0\}$$

so that the catastrophe manifold and map are as shown in Fig (1c). This, then, is the <u>Fold Catastrophe</u>, the only one possible when dim C = 1, and is involved in all higher catastrophes.

If dim C = 2, singularities of χ must be differentially conjugate either to a multiple of the fold by a trivial dimension, i.e., a potential of the form

$$R^2 \times R \to R$$

$$(a,b,x) \to \frac{x^3}{3} + ax$$

or to the <u>Cusp Catastrophe</u>, involving quartic potentials:

$$V(a,b,x) = \frac{x^4}{4} + \frac{ax^2}{2} + bx$$

which are therefore as illustrated in Fig (2), having either a single minimum or two minima and a maximum as critical points. (Notice, however, that the potential $-V(a,b,x)$ has the same differential type, but minima and maxima are interchanged. This gives radically different behavior in applications see, for example (1,2). It is known as the dual Cusp, and must be kept in mind along with the standard form when looking for descriptions of potential-minimizing phenomena. (The Fold, however, gives the same behavior when the negative of its potentials are taken; it is self-dual. In general, all the odd-numbered dimension cuspoids are self-dual, while the even-numbered dimension ones are not.) The manifold M, given by:

$$M = \{(a,b,x) \mid x^3 + ax + b = 0\}$$

is now shown in Fig (3), where the fold line is the smooth curve:

$$M \cap \{(a,b,x) \mid \frac{\partial^2 V}{\partial x^2} = 3x^2 + a = 0\}$$

on which the projection also has a singularity, so that its image is the curve:

$$\{(a,b) \mid 4a^3 + 27b^2 = 0\} \subseteq R^2$$

This fairly complete picture (missing only the graphs of V along each (a,b)-fixed vertical line, which would require another dimension for display) is possible in a single diagram only because dim $C \ngtr 2$. In higher dimensions, the limitations of our space-time force a more piece-meal approach. There are essentially two lines of attack: one, carried out very effectively by Godwin (7,8,9), is to examine analytically the bifurcation set in the control space (which, being the image of the set of singular points of χ is given as the set of solutions of the equations:

$$\frac{\partial V}{\partial x} = 0; \; \frac{\partial^2 V}{\partial x^2} = 0.$$

with x eliminated) and to indicate for each component of the complement the number of each type of critical point occurring for c chosen in that component.

Essentially, this is to draw for each catastrophe the analogue of the lower half of Fig (3). However, this technique involves polynomial equations of the order of the following:

$$p = \sqrt[3]{\left\{\left(\frac{v}{8} - \frac{u^2}{32}\right) + \sqrt{\left[\frac{u^4}{16} - \frac{u^2 v}{2} + v^2\right) \cdot \frac{1}{16} + \frac{t^3}{54}\right]}\right\}} + \sqrt[3]{\left\{\frac{v}{8} - \frac{u^2}{32} - \sqrt{\left[\left(\frac{u^4}{16} - \frac{u^2 v}{2} + v^2\right) \cdot \frac{1}{16} + \frac{t^3}{54}\right]}\right\}}.$$

arising from the elimination of x. This causes unnecessarily hard work, and the result does not assist in visualizing the manner in which M actually sits over C - the analogue of the upper half of Fig (3). Our approach is therefore different. The shape drawn at the top of Fig (3) is a ruled surface: each value x_o of x gives an affine equation in (a,b), (taking the x-axis vertical) of the form:

$$\{(a,b,x_o) \mid (2x_o)a + b = -4x_0^3\}$$

If an evenly-spaced representative set of the projections

$$\{(a,b) \mid (2x_o)a + b = -4x_o^3\}$$

of such lines into C is drawn (Fig (4)) the bifurcation set B appears as their envelope. B is then visibly composed of the images of two fold lines, meeting at the higher singularity of the 'pucker point' whose image is the cusp. We have thus arrived at both a simplification of the computation (since drawing straight lines is easier than sketching polynomial curves) and a transfer of some of the visual information of the upper half of

Fig (3) into the lower half. This visual three-dimensionality of Fig (4) is rather striking, is even more pronounced in some of the later figures and becomes very compelling when animated (10) or displayed stereographically (14). Thus we have way of partially reconstructing the three-dimensional nature of these surfaces without resorting to perspective drawing.

(This context incidentally clarifies the classical theory of envelopes, almost incomprehensible when considered in terms of curves in the plane, and often oddly justified (see, for example (11)). The envelope of a family $g(x,y,\theta)$ of curves in R^2, parameterized by θ, is given by the equation:

$$\frac{\partial g}{\partial \theta} = 0$$

simply because it is the image in R^2 of the singularities in the projection to R^2 of the surface:

$$\{(x,y,\theta) \mid g(x,y,\theta) = 0\}$$

in R^3. It comes wholly into this setting if we consider the potential:

$$V(x,y,\theta) = \int_0^\theta g(x,y,t)\,dt.$$

so that we are treating also the generic geometry of envelopes. These considerations are among the historical roots of catastrophe theory; for a far more general discussion of them, see (12).)

When the dimension of the control space is greater than two, it is necessary to work with two dimensional sections of it since it is impossible (as yet) to program the computer to draw planes on three-dimensional space.

Furthermore, it is difficult to assume that the eye would readily perceive the resulting envelope - a surface, now, with singularities. For each cuspoid a single equation, non-linear only in x, defines the manifold M, which is thus ruled in codimension one by a one-dimensional family of hyperplanes, parametrized as a family by x. For a generic plane slice $P \subseteq C$, since for all Catastrophes dim M = dim C a general position argument shows that: $\pi^{-1}(P) \cap M$ is a 2-manifold, ruled by lines (in particular applications these are replaced by contours on M of the potentials, the Bifurcation Set still appearing as the projection of their envelope; see, for example, (1)). The same drawing techniques therefore apply as for Fig (4). For a non-generic slice, the manifold, but not the ruled character of $\chi^{-1}(P) = \pi^{-1}(P) \cap M$ fails; we have, in fact, as P passes through such a slice, a surgery of $\chi^{-1}(P)$ along a straight line. This is visible in Fig (9): any lack of clarity should be resolved by reference to the animated version (10) where the surgery is remarkably clear.

For the fullest mental grip on the geometry of these Catastrophes, the integration of these two-dimensional slices can be aided by perspective drawings of three-dimensional slices through the bifurcation set. For dim C=3, the Swallowtail, only one is involved (Fig. 4); for dim C=4, the Butterfly, Fig. 5 shows a sequence of slices of the form a=constant. For all the Cuspoids such slices by three-dimensional hyperplanes are possible, and the reader should find it helpful with the aid of the computer pictures to draw the resulting shapes. For all the Cuspoids, the most illuminating slices are among those with a among the coordinates held constant with the rich parts of the structure corresponding to negative values of a.

Study of Figs 3, 4, 5 and the computer drawings should make clear the manner in which the Cusp Catastrophe "organizes" two one-parameter families of Folds, the Swallowtail two families of Cusps, the Butterfly two families of Swallowtails (each of the $k<0$ slices in Fig. 5 has two Swallowtail points) and so on: the reader will find it an excellent exercise to prove this analytically. We shall discuss further the way in which a higher catastrophe can act as an organizing center for lower ones, so that in applications its global geometry may be essentially involved through the central singularity may not appear in the displayed behavior, in the context of the Parabolic Umbilic (5).

The figures shown in this paper show that each higher order catastrophe, when plotted on the appropriate plane, generates the relevant lower-order catastrophes:

	Plane				
Higer order Catastrophe	(A - B)	(B - C)	(D - E)	(D - E)	(E - F)
Star:	Cusp	Swallowtail	Butterfly	Wigwam	Star
Wigwam:	Cusp	Swallowtail	Butterfly	Wigwam	-
Butterfly:	Cusp	Swallowtail	Butterfly	-	-
Swallowtail:	Cusp	Swallowtail	-	-	-
Cusp:	Cusp	-	-	-	-

The singularity structure discussed here does not exhaust the significant geometry of the elementary catastrophes: for the geometry of the associated Maxwell set (relevant for applications with significant thermodynamic noise), see (13).

BIBLIOGRAPHY

1. Poston, T. & Woodcock, A. E. R., Zeeman's Catastrophe Machine, Proc. Camb. Phil. Soc., to appear.

2. Zeeman, E. C., Catastrophe Machines, Towards a Theoretical Biology IV, ed, C. H. Waddington.

3. Thom, R., Stabilité Structurelle et Morphogéneŝe, Benjamin-Addison-Wesley. (1973)

4. Woodcock, A. E. R. & Poston, T., The Geometrical Properties of the Hyperbolic & Elliptic Umbilics, (this volume).

5. Woodcock, A. E. R. & Poston, T., The Geometrical Properties of the Parabolic Umbilic, (this volume).

6. Woodcock, A. E. R. & Poston, T., The Geometrical Properties of the Reduced Double Cusp, (this volume).

7. Godwin, N., Ph.D. Thesis, University of Warwick (1970).

8. Godwin, N., Thom's Parabolic Umbilic, I.H.E.S. No. 40, p.117-138, (1972).

9. Godwin, N., The Compactified Parabolic Umbilic, Proc. Lond. Math. Soc., to appear.

10. Woodcock, A. E. R., The Elementary Catastrophes, A Film (to appear).

11. Edwards, An Elementary Treatise on the Differential Calculus, Macmillan, London, 1892.

12. Thom, R., Sur La théorie des enveloppes, Journ. de Math t.XLI-Fasc. 2, 1962.

13. Godwin, N. & Poston, T., On the Maxwell set of catastrophes with finite noise, to appear.

14. Woodcock, A, E. R., Stereographic Reconstructions of the Catastrophe Mfds of the Cuspoid Catastrophes (this volume).

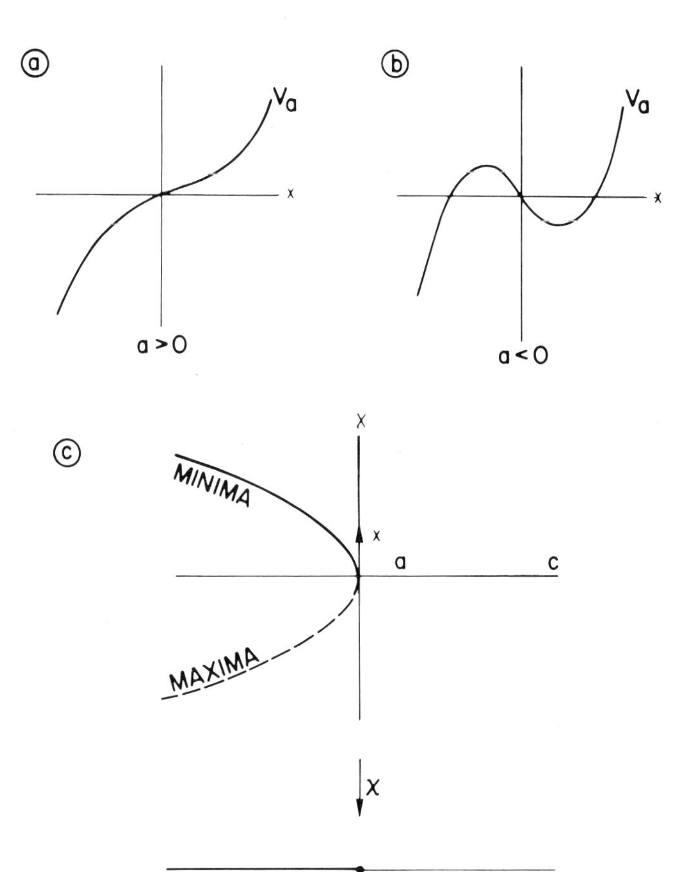
a
b
V_a
x
V_a
x
a > 0
a < 0
c
X
MINIMA
x
a
c
MAXIMA
X
1 MINIMUM, 1 MAXIMUM OVER EACH POINT
NO MINIMA OR MAXIMA
1 UNSTABLE CRITICAL PT. (INFLEXION)

Fig. 1

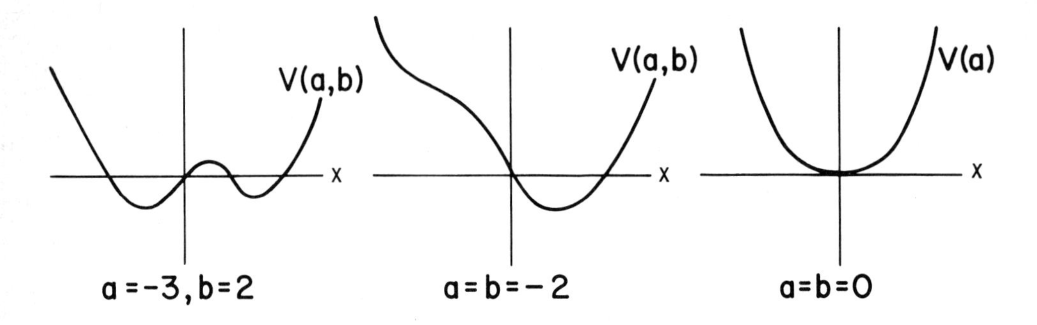

Fig. 2

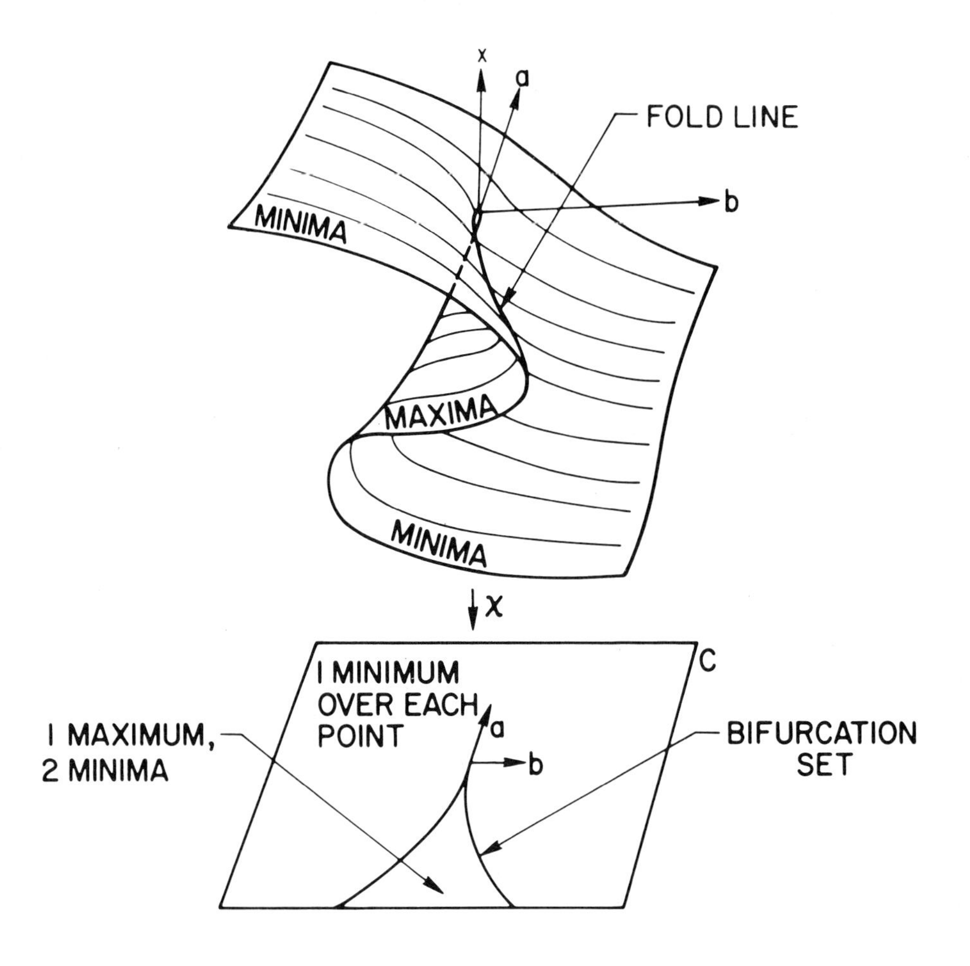
x
a
FOLD LINE
b
MINIMA
MAXIMA
MINIMA
x
C
I MINIMUM
OVER EACH
POINT
a
b
I MAXIMUM,
2 MINIMA
BIFURCATION
SET

Fig. 3

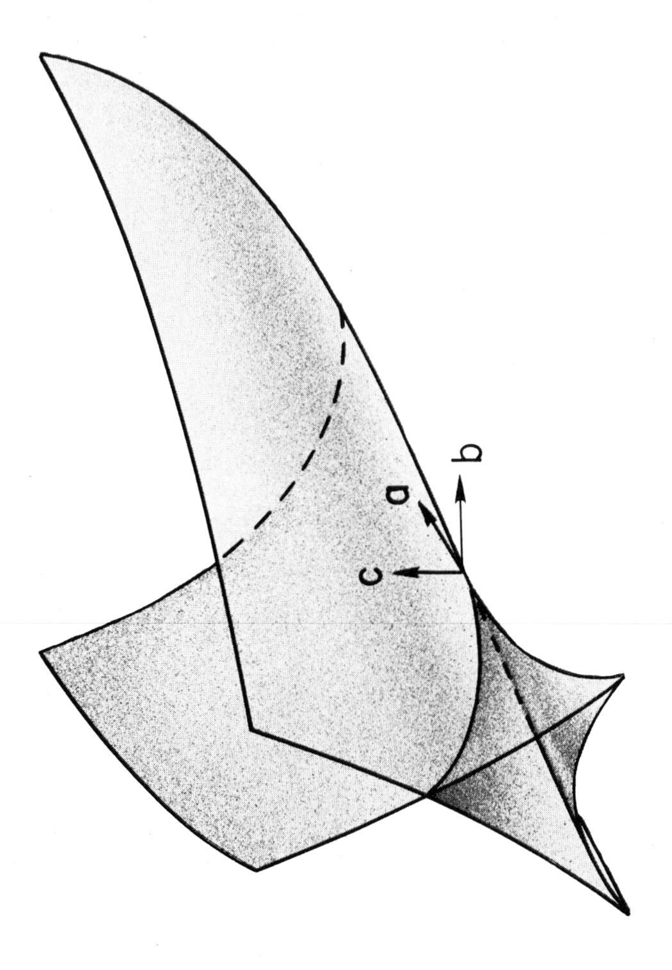

Fig. 4

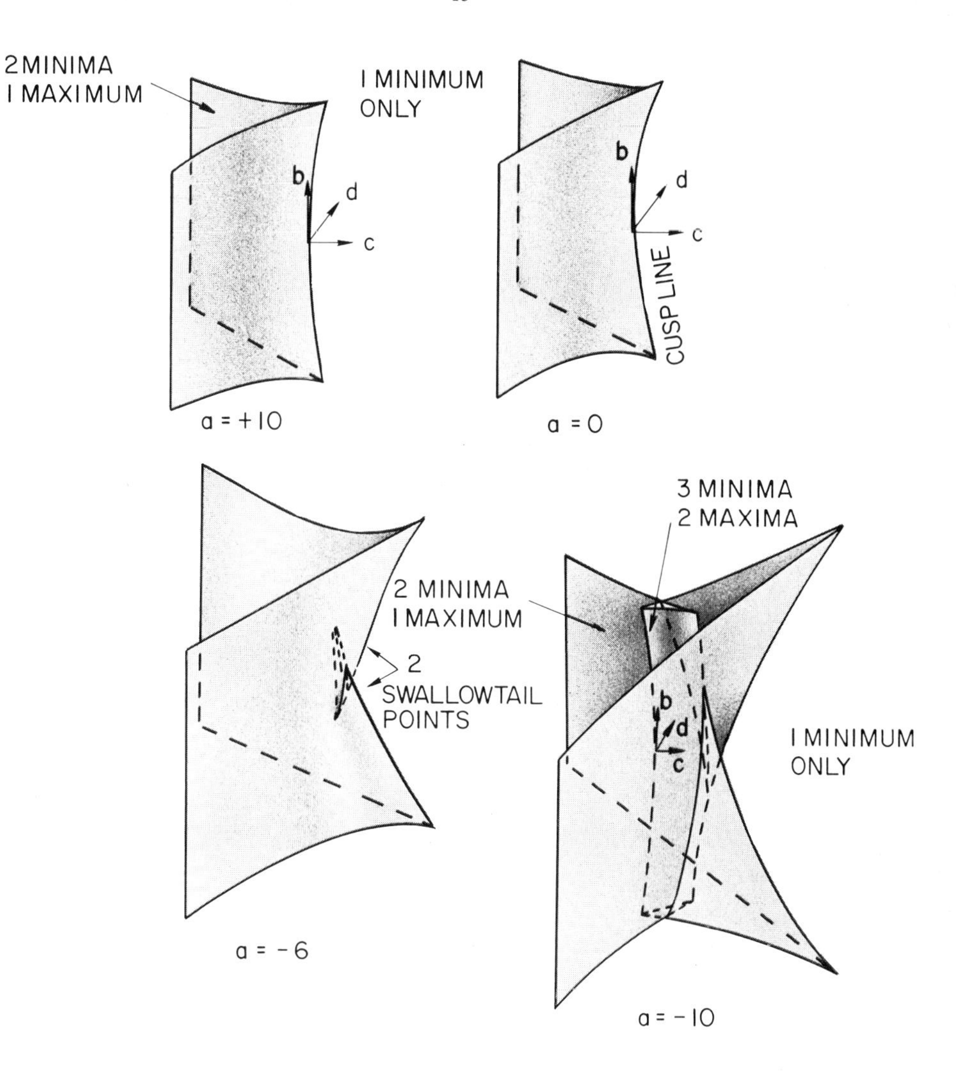

2MINIMA
1 MAXIMUM
1 MINIMUM
ONLY
b
d
c
CUSP LINE
a = +10
a = 0
3 MINIMA
2 MAXIMA
2 MINIMA
1 MAXIMUM
2
SWALLOWTAIL
POINTS
1 MINIMUM
ONLY
a = -6
a = -10

Fig. 5

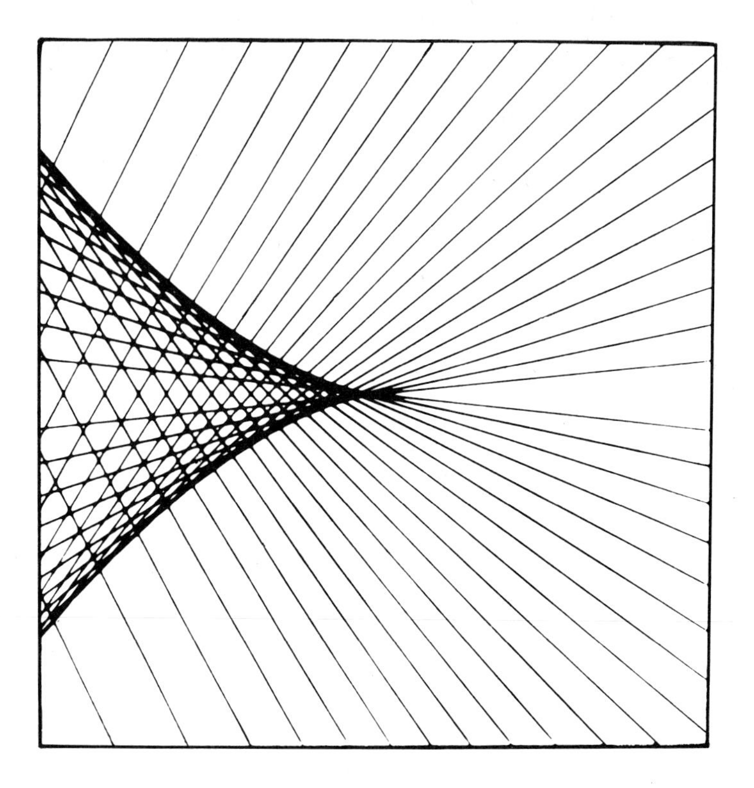

Fig. 6 Ruled Surface Projections of the Simple Cusp Catastrophe

($V = \frac{x^4}{4} + \frac{Ax^2}{2} + Bx$) onto the plane (A,B)

The two control parameters (A,B) define a unique control plane.

Fig. 7 Ruled Surface Projections of the Swallowtail Catastrophe $(V = \frac{x^5}{5} + \frac{Ax^3}{2} + \frac{Bx^2}{2} + Cx)$ onto the plane (B,C)

With A positive the projection of the ruled surface is without singularities of types higher than folds. This reflects the fact that for $x^4 + Ax^2 + Bx + C = 0$ and A positive, the equation only has one real root. For negative values of A, a three-cornered or "Swallowtail-like" region develops centered on B = C = 0 in the (B,C) plane. These overlapping surfaces indicate that, for the relevant values of A, B and C, the equation has more than one real root. This behavior is also demonstrated in the three-dimensional reconstruction in Fig. 5. Furthermore, in the picture with A = -10.0, for example, it can be clearly seen that the Swallowtail sections are constructed from two Simple Cusps with a common limb. (See Fig. 9(B)).

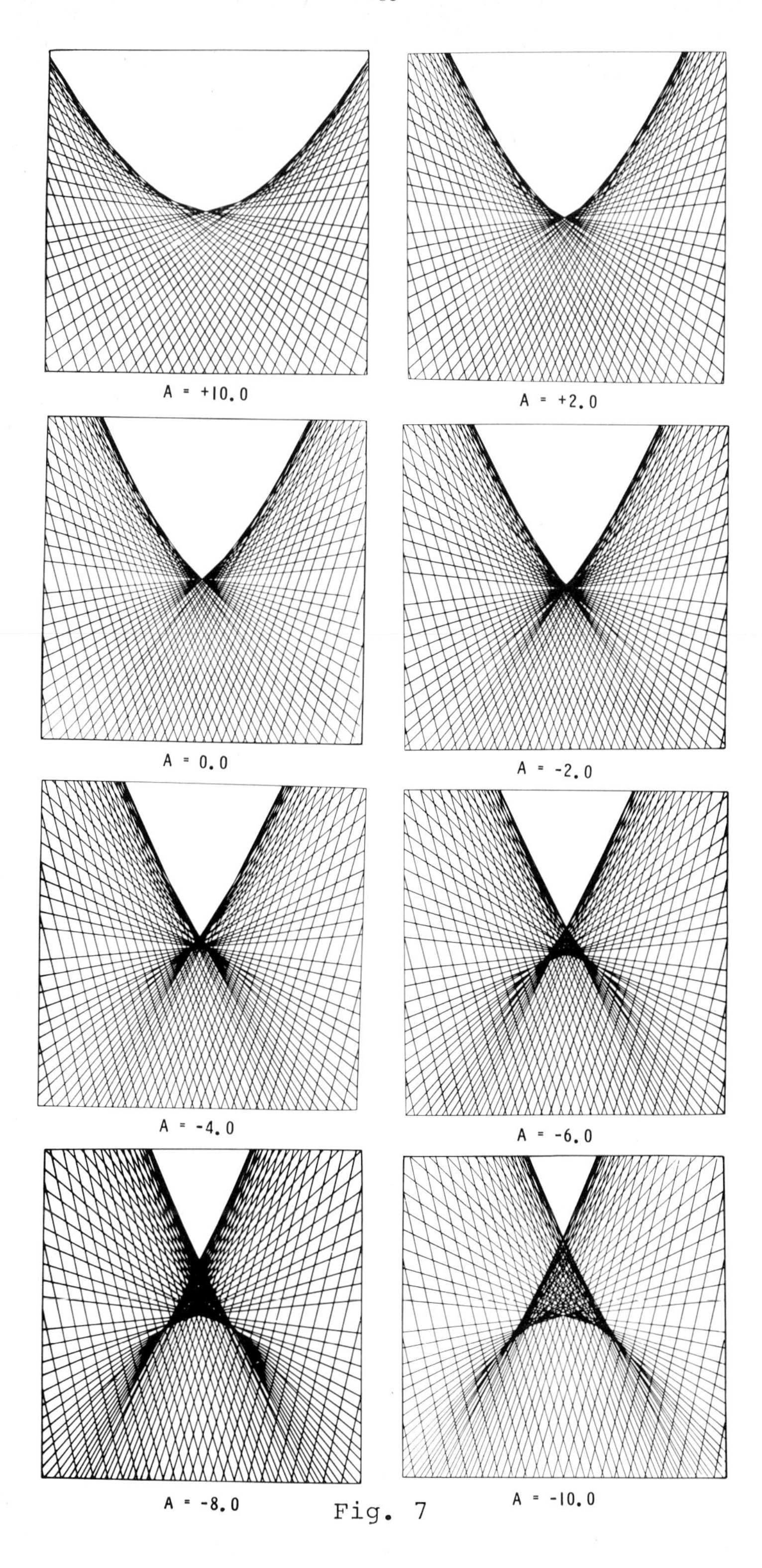
A = +10.0
A = +2.0
A = 0.0
A = -2.0
A = -4.0
A = -6.0
A = -8.0
A = -10.0

Fig. 7

Fig. 8 (A and B) Ruled Surface Projections of the Butterfly Catastrophe $V = \frac{x^6}{6} + \frac{Ax^4}{4} + \frac{Bx^3}{3} + \frac{Cx^2}{2} + Dx$ onto the plane (C,D)

For A positive and B zero, these sections are typical of the Simple Cusp shown in Figs. 3 and 6. As B runs from negative to positive values (A positive) to outer limb of the cusp swings through an arc, the magnitude of which is dependent upon the magnitude of A. This angle increases for a given change of B as A is reduced. For A negative, the typical "Butterfly" wings develop in the region C = D = 0, and become very pronounced (for example for A = -10.0, B = 0.0). This picture (and Fig. 19) show how the Butterfly characteristic shape is generated from two Swallowtail sections with a common limb. As B is varied from negative to positive (A constant negative) the small Swallowtail on the lowest limb of the surface grows from zero until it completely dominates the picture and the originally dominant Swallowtail disappears. This behavior is also shown in the three-dimensional reconstructions shown in Fig. 5.

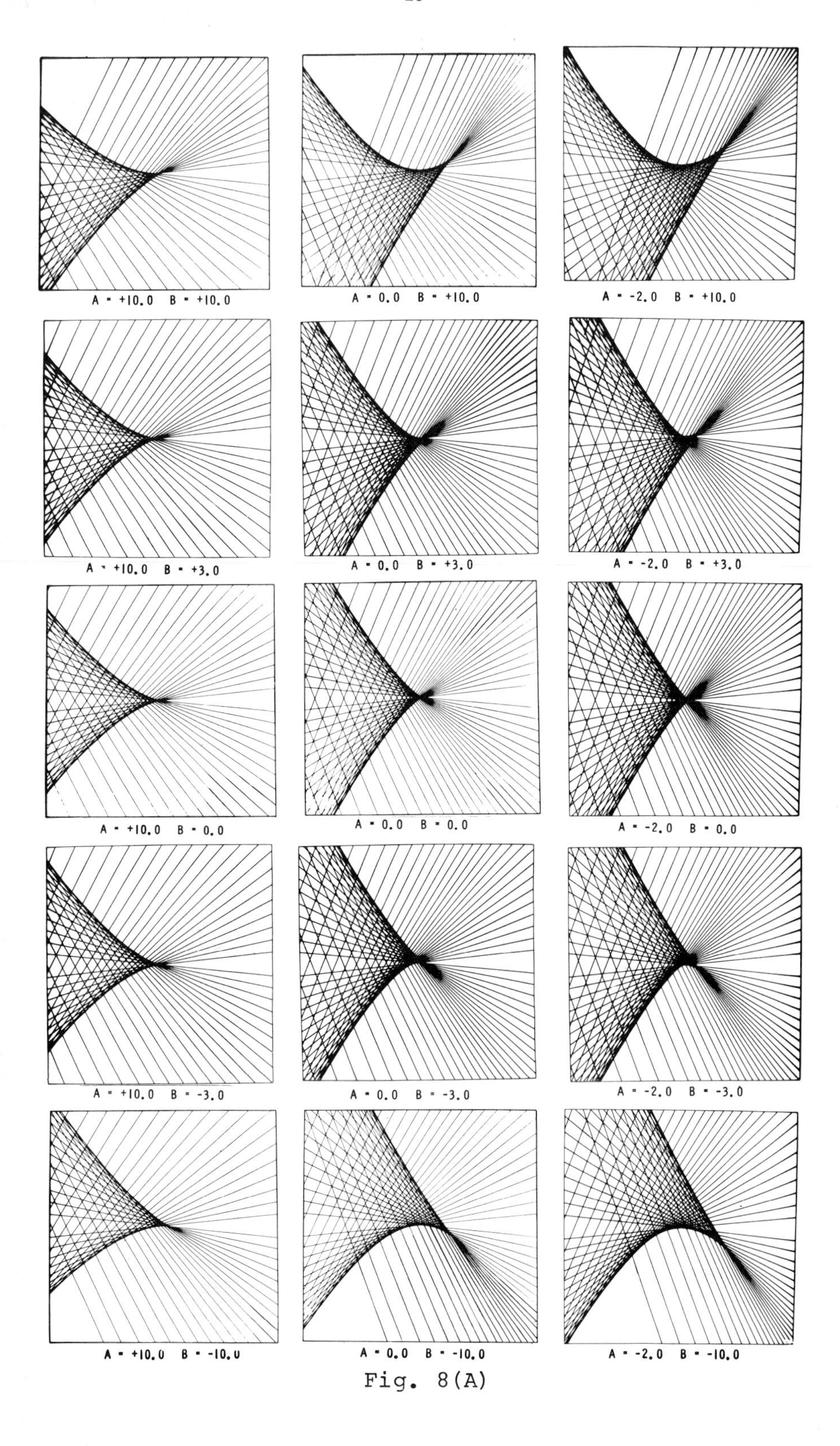
A = +10.0 B = +10.0
A = 0.0 B = +10.0
A = -2.0 B = +10.0
A = +10.0 B = +3.0
A = 0.0 B = +3.0
A = -2.0 B = +3.0
A = +10.0 B = 0.0
A = 0.0 B = 0.0
A = -2.0 B = 0.0
A = +10.0 B = -3.0
A = 0.0 B = -3.0
A = -2.0 B = -3.0
A = +10.0 B = -10.0
A = 0.0 B = -10.0
A = -2.0 B = -10.0

Fig. 8(A)

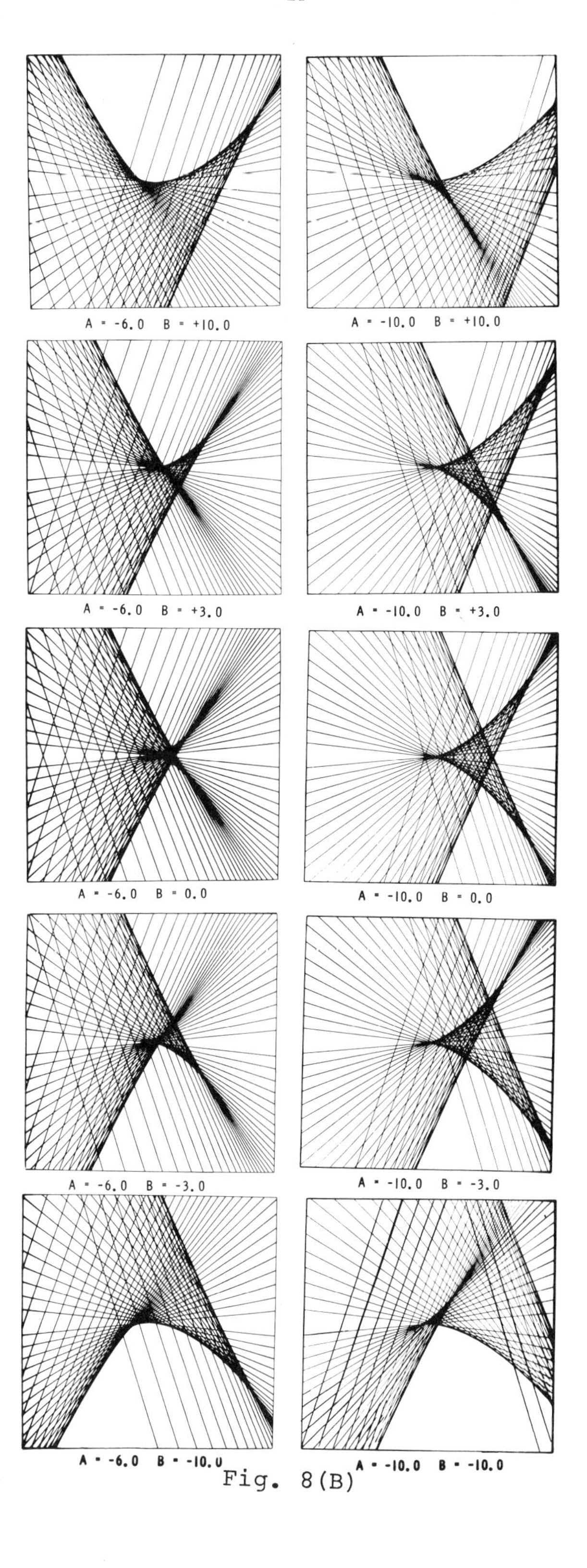
A = -6.0 B = +10.0
A = -10.0 B = +10.0
A = -6.0 B = +3.0
A = -10.0 B = +3.0
A = -6.0 B = 0.0
A = -10.0 B = 0.0
A = -6.0 B = -3.0
A = -10.0 B = -3.0
A = -6.0 B = -10.0
A = -10.0 B = -10.0

Fig. 8(B)

Fig. 9 (A and B) Ruled Surface Projections of Swallowtail Sections of the Butterfly Catastrophe ($V = \frac{x^6}{6} + \frac{Ax^4}{4} + \frac{Bx^3}{3} + \frac{Cx^2}{2} + Dx$) onto the Plane (B, C)

When D = 0, $\frac{dV}{dx} = x^5 + Ax^3 + Bx^2 + Cx = 0$ becomes $x^4 + Ax^2 + Bx + C = 0$; the typical "Swallowtail" equation. In Figs 9 (A and B) the series of pictures with D zero are therefore equivalent to those of Fig. 7. For A positive and with D running from negative to positive, it is possible to see how the two initially separate leaves of the surface fuse at D zero to give a rounded-valley like configuration. With A negative and D varying as above, the surgery which results in the production of a true Swallowtail section at D zero is evident.

The preceding pictures are examples of the general statement that each of the higher dimensional catastrophes has as sections those catastrophes of lower dimensional order. Thus the Swallowtail contains the Simple Cusp, and the Butterfly both the Swallowtail and Simple Cusp, as particular sections.

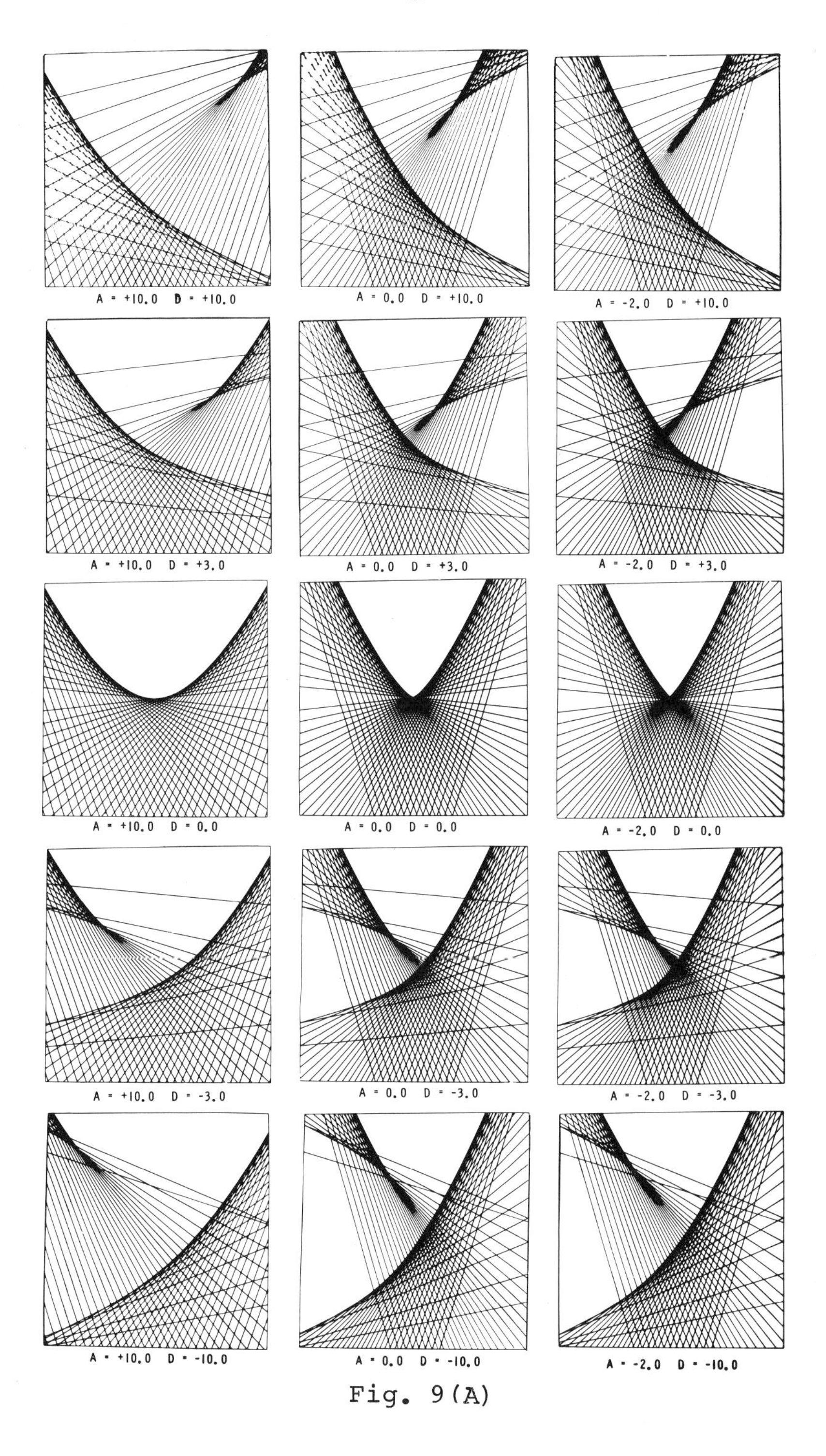
A = +10.0 D = +10.0
A = 0.0 D = +10.0
A = -2.0 D = +10.0
A = +10.0 D = +3.0
A = 0.0 D = +3.0
A = -2.0 D = +3.0
A = +10.0 D = 0.0
A = 0.0 D = 0.0
A = -2.0 D = 0.0
A = +10.0 D = -3.0
A = 0.0 D = -3.0
A = -2.0 D = -3.0
A = +10.0 D = -10.0
A = 0.0 D = -10.0
A = -2.0 D = -10.0

Fig. 9(A)

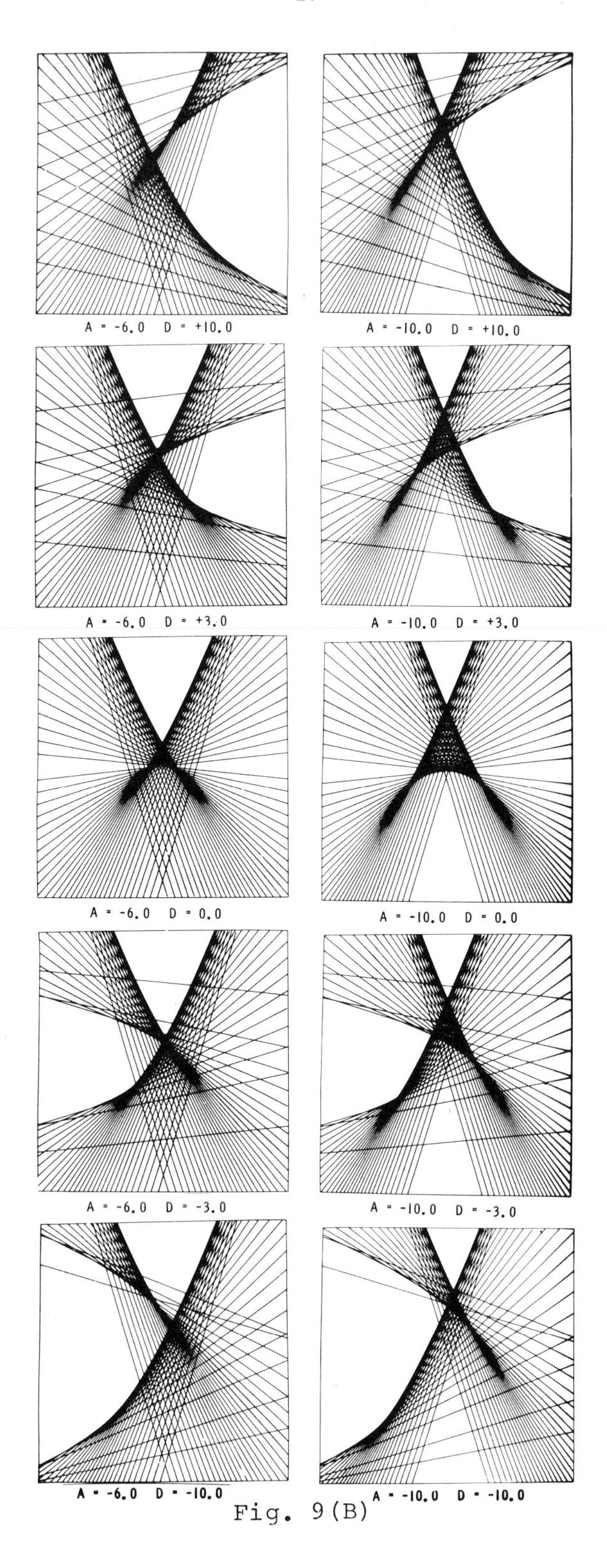
A = -6.0 D = +10.0
A = -10.0 D = +10.0
A = -6.0 D = +3.0
A = -10.0 D = +3.0
A = -6.0 D = 0.0
A = -10.0 D = 0.0
A = -6.0 D = -3.0
A = -10.0 D = -3.0
A = -6.0 D = -10.0
A = -10.0 D = -10.0

Fig. 9(B)

Ruled Surface Projections of the Wigwam Catastrophe

$V = \frac{x^7}{7} + \frac{Ax^5}{5} + \frac{Bx^4}{4} + \frac{Cx^3}{3} + \frac{Dx^2}{2} + EX$ onto the plane (D,E)

Figures 10 and 11 in which A is either positive or zero, show that the projected surface exhibits the same general behavior, as C is varied from positive to negative, as does the usual Swallowtail when A is varied in a similar manner (see Fig 7). For A negative, and small compared to C positive, two initially small Swallowtails begin to develop at the edge of the rounded valley (See Figs. 12, 13 and 14). As C is reduced these Swallowtails grow in relative importance and, for example, in Fig. 14, with C = +20, they dominate the picture. In reducing the value of C, the curve joining the two Swallowtails is reduced and eventually, merging the two into one, disappears, when C reaches zero. For C negative, all that exists is a Swallowtail-like region with no "internal" structure. Figures 10 through 14 were drawn with B zero. With B nonzero, however, the symmetry between the swallowtails is destroyed. This is shown in Figures 15 and 16 where sections were taken either side of the B zero axis and this shows clearly the, distorted, Swallowtails. (The name "Wigwam" is due to the appearance of Fig. 13, A = -7.0, B=0.0, C = +15.0, for example, when inverted.)

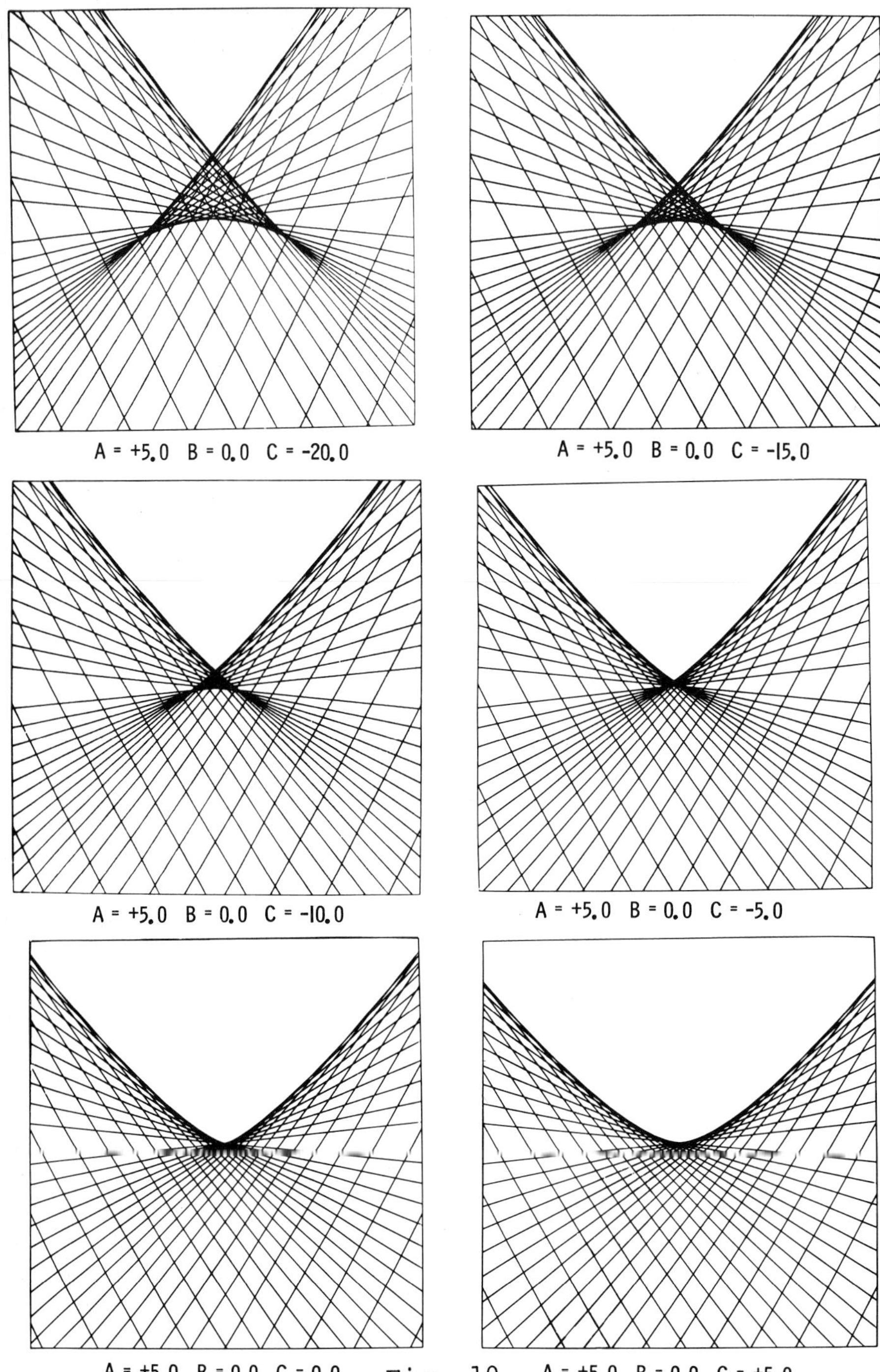
A = +5.0 B = 0.0 C = -20.0
A = +5.0 B = 0.0 C = -15.0
A = +5.0 B = 0.0 C = -10.0
A = +5.0 B = 0.0 C = -5.0
A = +5.0 B = 0.0 C = 0.0
A = +5.0 B = 0.0 C = +5.0

Fig. 10

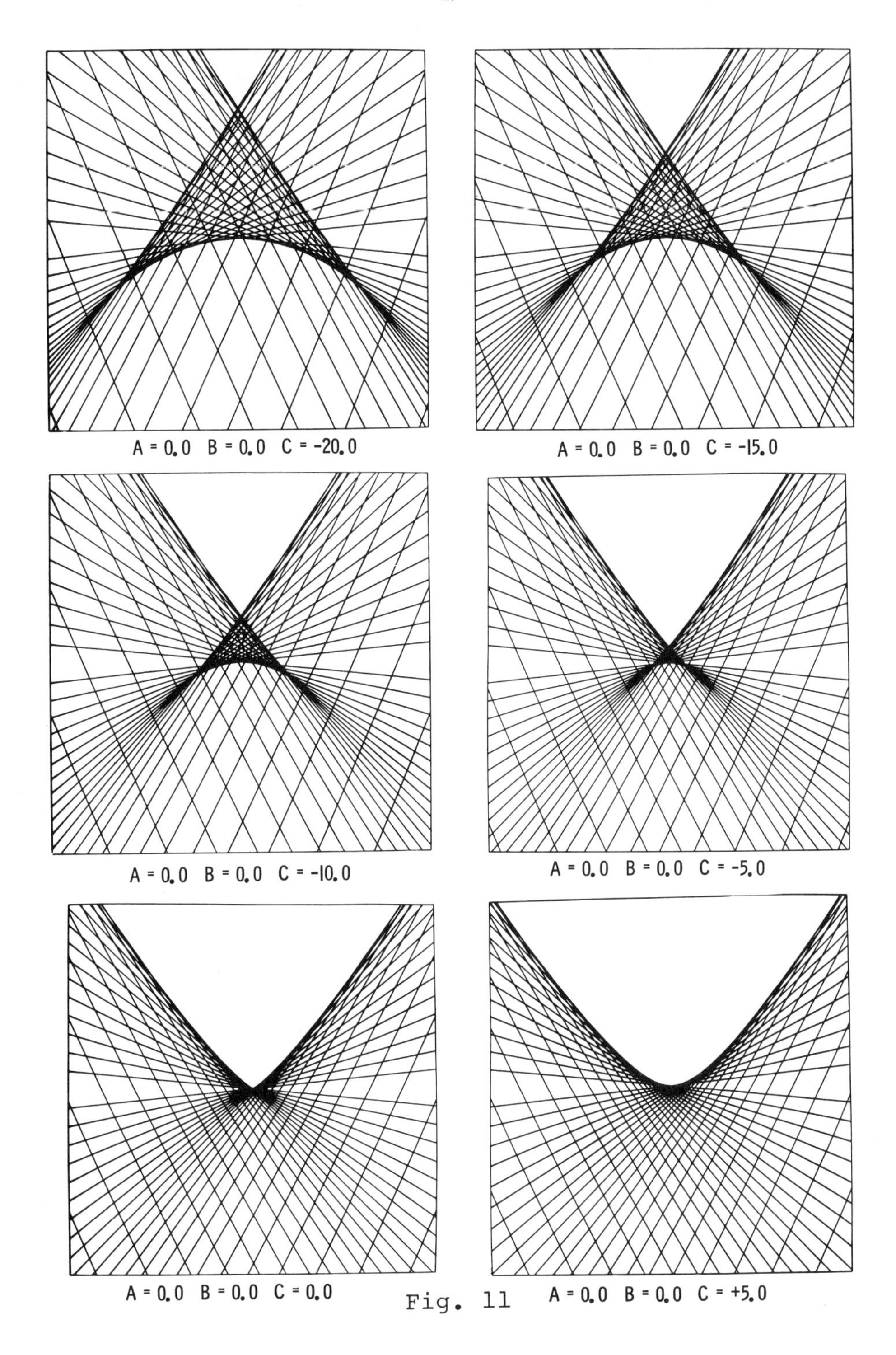
A = 0.0 B = 0.0 C = -20.0
A = 0.0 B = 0.0 C = -15.0
A = 0.0 B = 0.0 C = -10.0
A = 0.0 B = 0.0 C = -5.0
A = 0.0 B = 0.0 C = 0.0
A = 0.0 B = 0.0 C = +5.0

Fig. 11

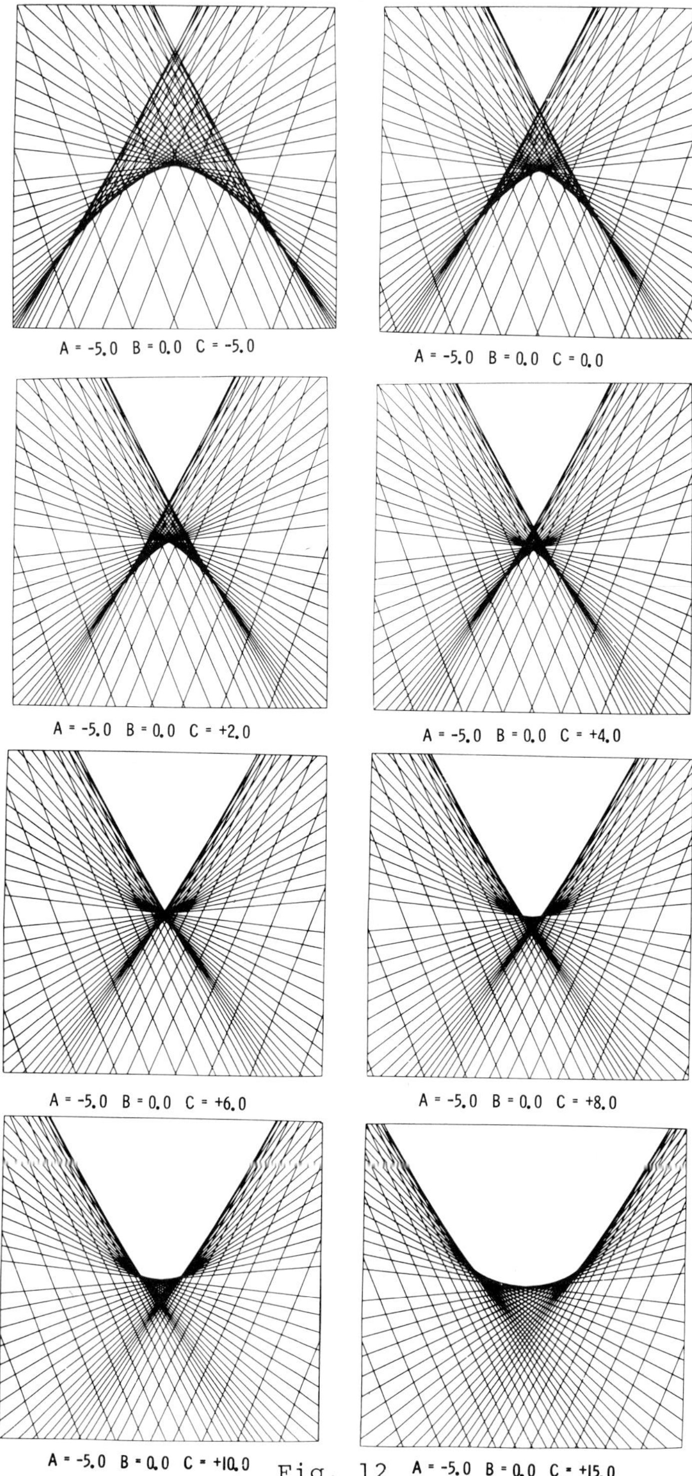
A = -5.0 B = 0.0 C = -5.0
A = -5.0 B = 0.0 C = 0.0
A = -5.0 B = 0.0 C = +2.0
A = -5.0 B = 0.0 C = +4.0
A = -5.0 B = 0.0 C = +6.0
A = -5.0 B = 0.0 C = +8.0
A = -5.0 B = 0.0 C = +10.0
A = -5.0 B = 0.0 C = +15.0

Fig. 12

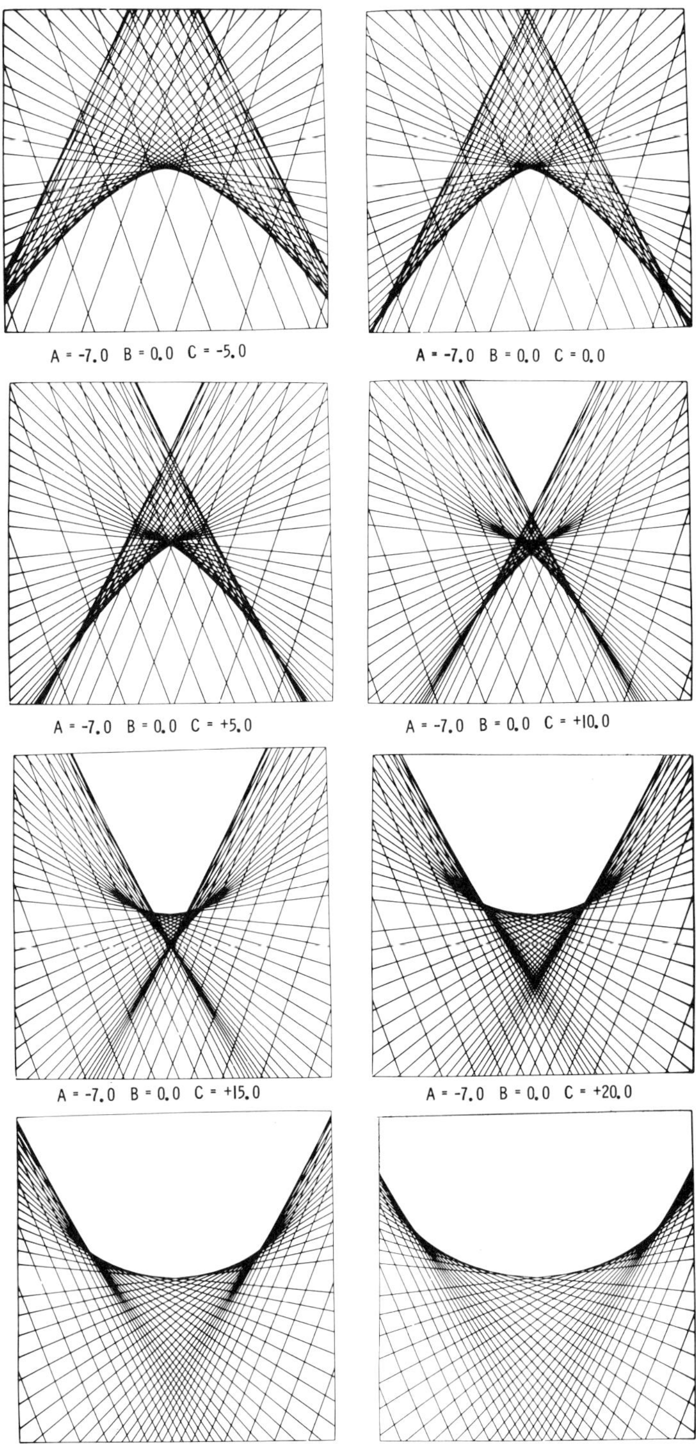
A = -7.0 B = 0.0 C = -5.0
A = -7.0 B = 0.0 C = 0.0
A = -7.0 B = 0.0 C = +5.0
A = -7.0 B = 0.0 C = +10.0
A = -7.0 B = 0.0 C = +15.0
A = -7.0 B = 0.0 C = +20.0
A = -7.0 B = 0.0 C = +25.0
A = -7.0 B = 0.0 C = +30.0

Fig. 13

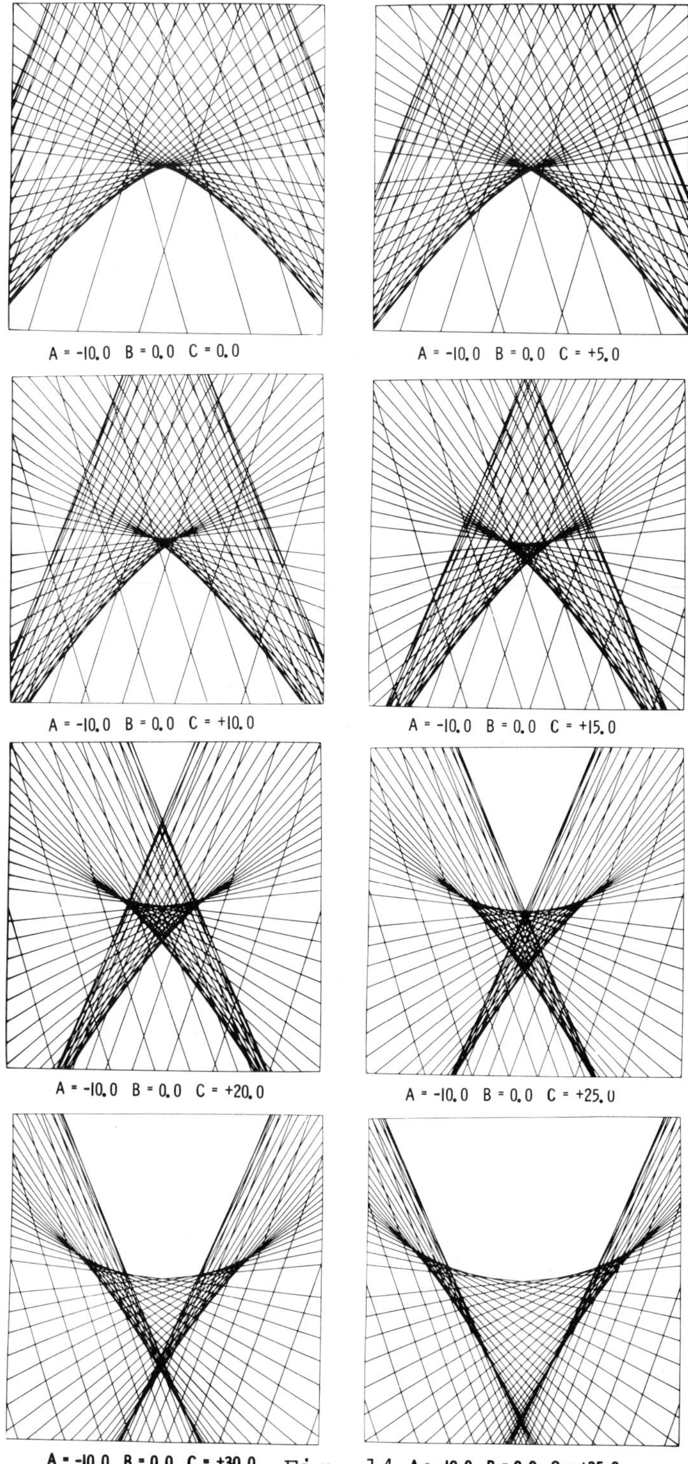
A = -10.0 B = 0.0 C = 0.0
A = -10.0 B = 0.0 C = +5.0
A = -10.0 B = 0.0 C = +10.0
A = -10.0 B = 0.0 C = +15.0
A = -10.0 B = 0.0 C = +20.0
A = -10.0 B = 0.0 C = +25.0
A = -10.0 B = 0.0 C = +30.0
A = -10.0 B = 0.0 C = +35.0

Fig. 14

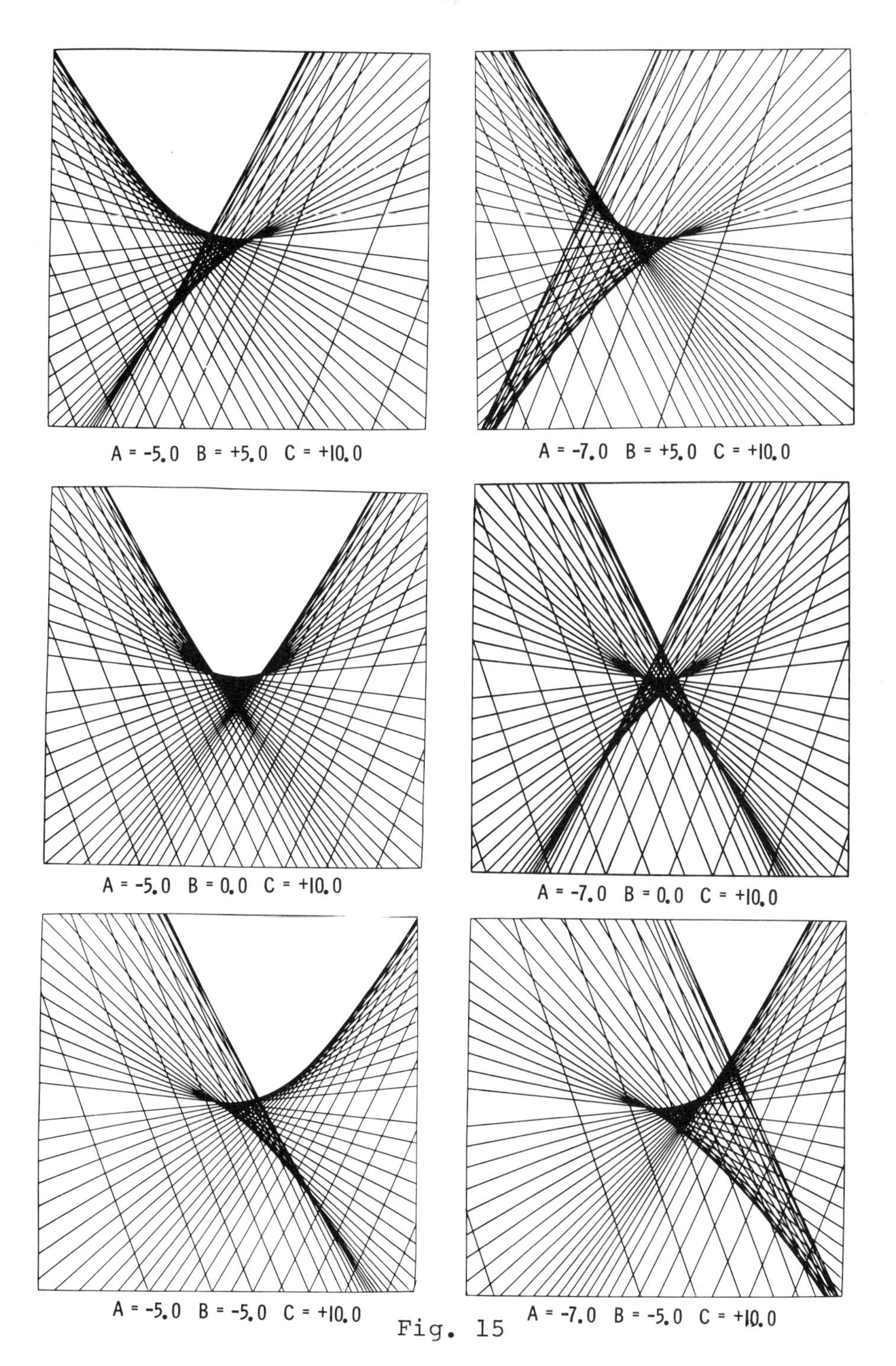
A = -5.0 B = +5.0 C = +10.0
A = -7.0 B = +5.0 C = +10.0
A = -5.0 B = 0.0 C = +10.0
A = -7.0 B = 0.0 C = +10.0
A = -5.0 B = -5.0 C = +10.0
A = -7.0 B = -5.0 C = +10.0

Fig. 15

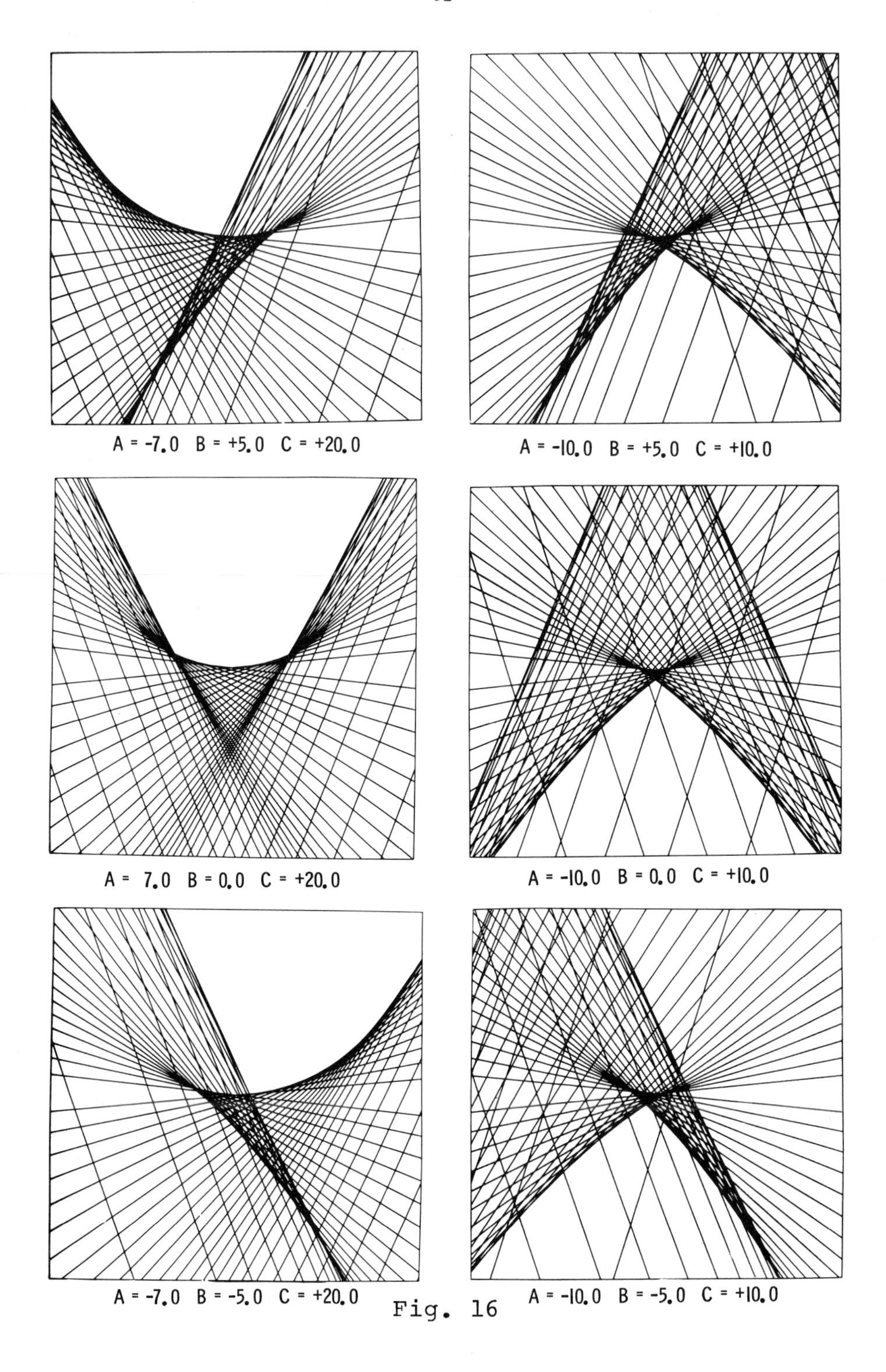
A = -7.0 B = +5.0 C = +20.0
A = -10.0 B = +5.0 C = +10.0
A = 7.0 B = 0.0 C = +20.0
A = -10.0 B = 0.0 C = +10.0
A = -7.0 B = -5.0 C = +20.0
A = -10.0 B = -5.0 C = +10.0

Fig. 16

Ruled Surface Projections of Butterfly Sections of the Wigwam Catastrophe $V = \frac{x^7}{7} + \frac{Ax^5}{5} + \frac{Bx^4}{4} + \frac{Cx^3}{c} + \frac{Dx^2}{2} + Ex$ on to the plane (C,D)

As in Figure 8 (A and B), the surfaces characteristic of the Butterfly Catastrophe exist when A is negative. Fig. 17 shows a typical Butterfly section when A is - 8 and B and E are zero. With B zero and E negative the picture consists of two overlapping cusps. At E zero, the abutting edges fuse to give the Butterfly and at E positive the Butterfly degenerates into two overlying surfaces each containing a Swallowtail configuration. Sections with A and E negative and B either zero or $\pm$ 4 shows movement of one cusp relative to the other. (Fig. 18). Figure 19 shows similar sections with E positive. Sections in which A is positive or zero and B and E are also zero are typically cuspoid (Figs. 20 and 21). However, with E negative the single Cusp breaks up into two component Cusps; with E positive the surface becomes two nonconnecting sheets (Figs. 20 and 21). Figs. 22 and 23 show B non-zero (off-center) sections for A positive and zero.

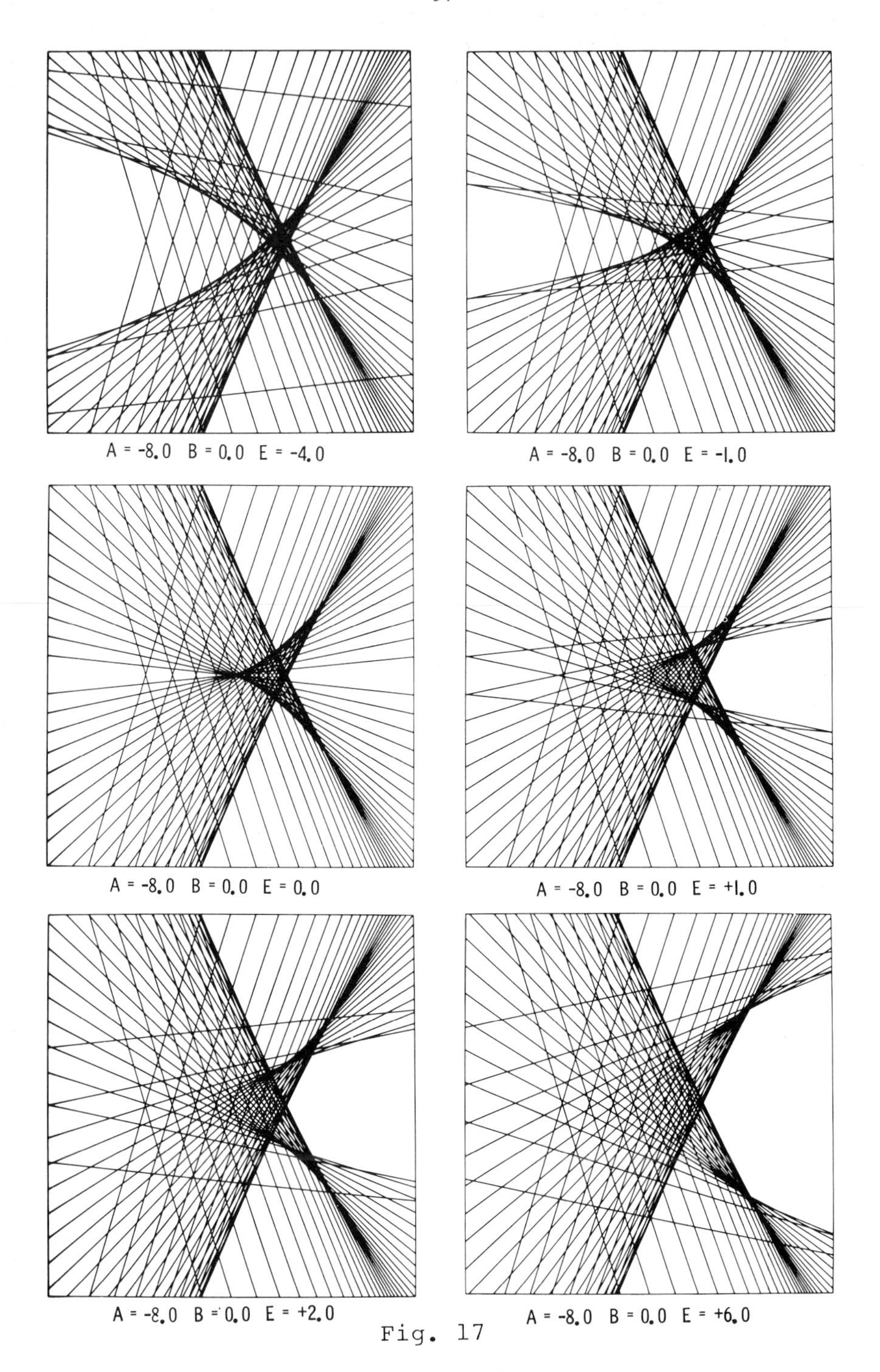
A = -8.0 B = 0.0 E = -4.0
A = -8.0 B = 0.0 E = -1.0
A = -8.0 B = 0.0 E = 0.0
A = -8.0 B = 0.0 E = +1.0
A = -8.0 B = 0.0 E = +2.0
A = -8.0 B = 0.0 E = +6.0

Fig. 17

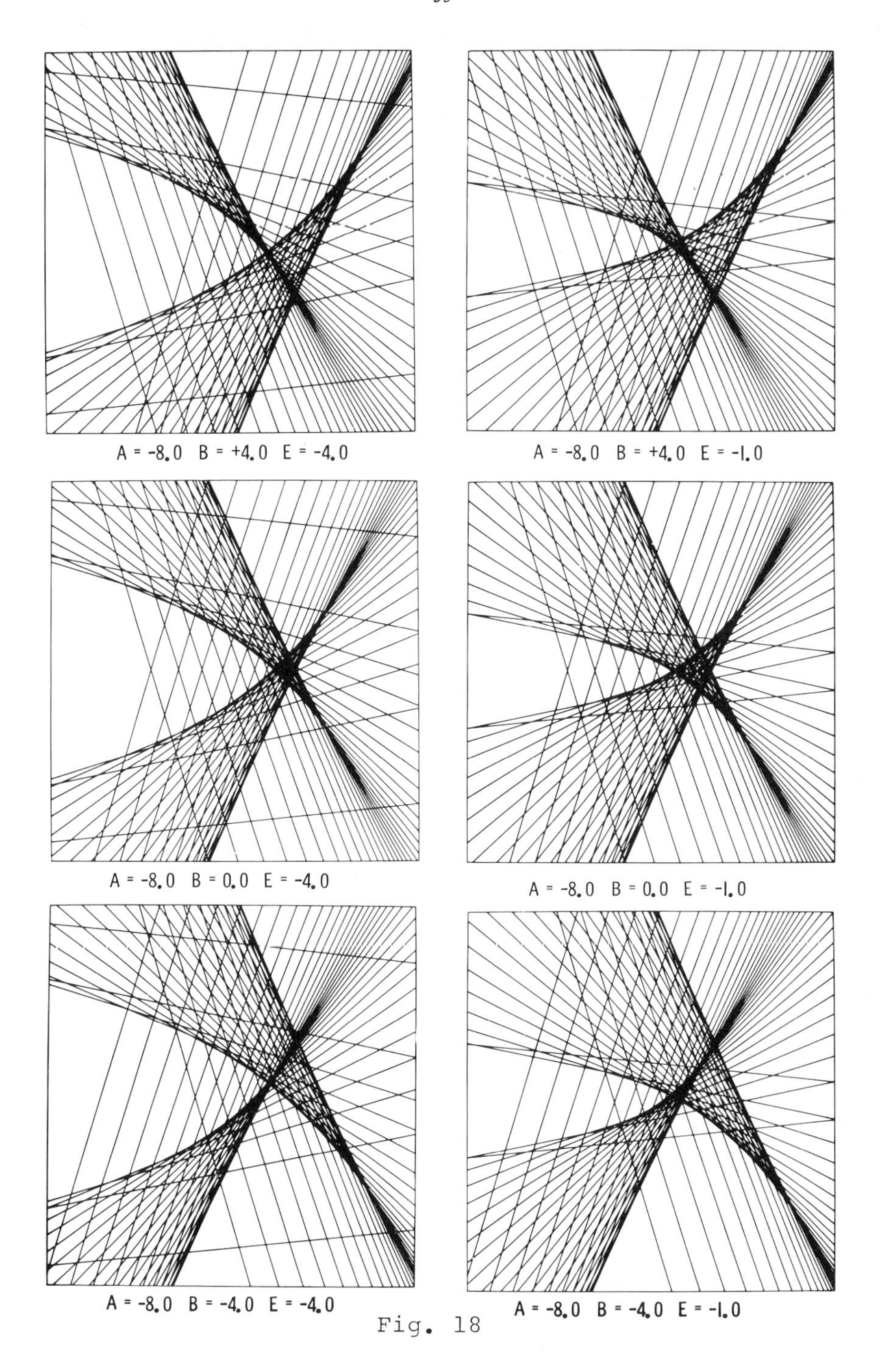
A = -8.0 B = +4.0 E = -4.0
A = -8.0 B = +4.0 E = -1.0
A = -8.0 B = 0.0 E = -4.0
A = -8.0 B = 0.0 E = -1.0
A = -8.0 B = -4.0 E = -4.0
A = -8.0 B = -4.0 E = -1.0

Fig. 18

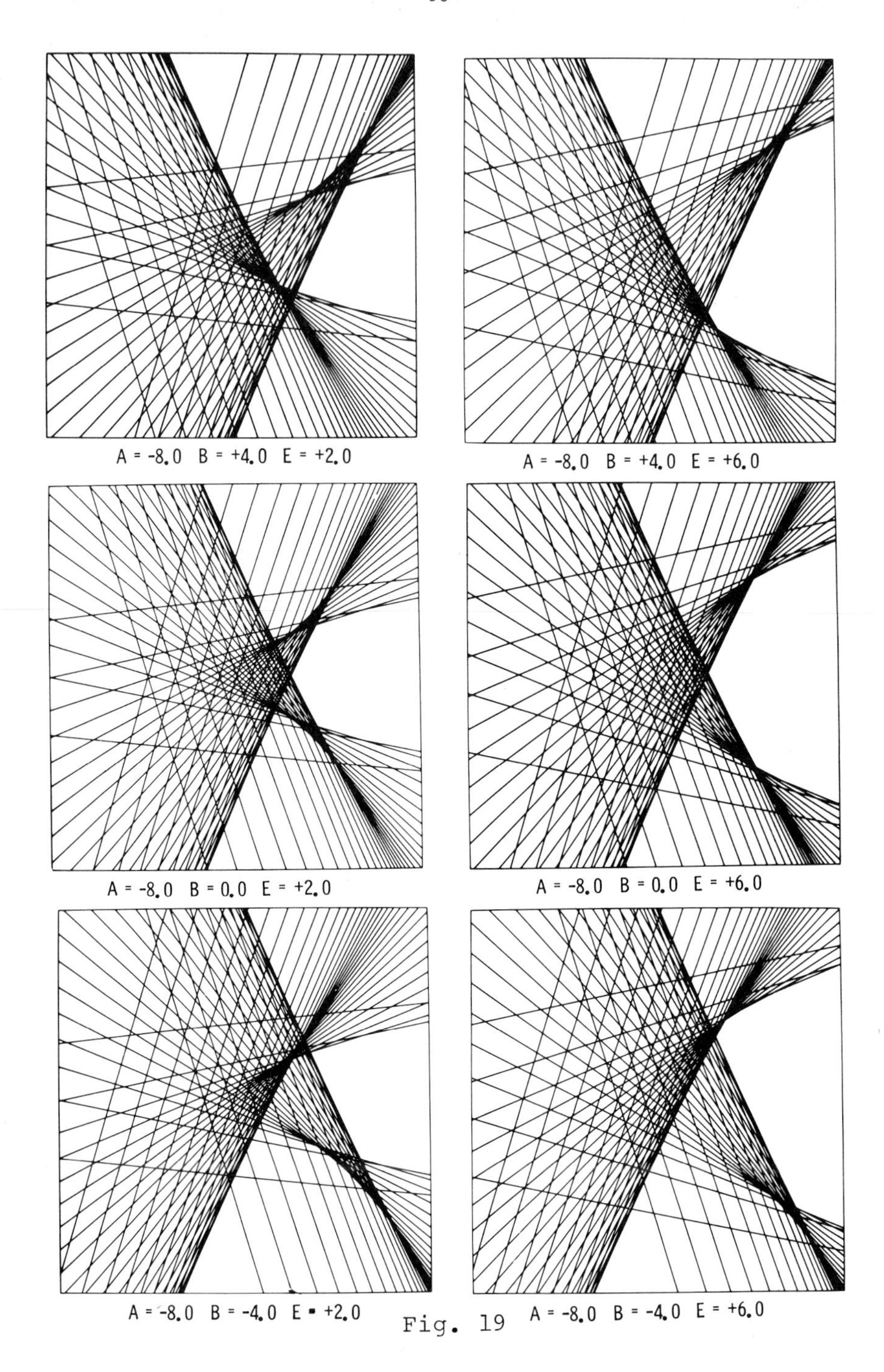
A = -8.0 B = +4.0 E = +2.0
A = -8.0 B = +4.0 E = +6.0
A = -8.0 B = 0.0 E = +2.0
A = -8.0 B = 0.0 E = +6.0
A = -8.0 B = -4.0 E = +2.0
A = -8.0 B = -4.0 E = +6.0

Fig. 19

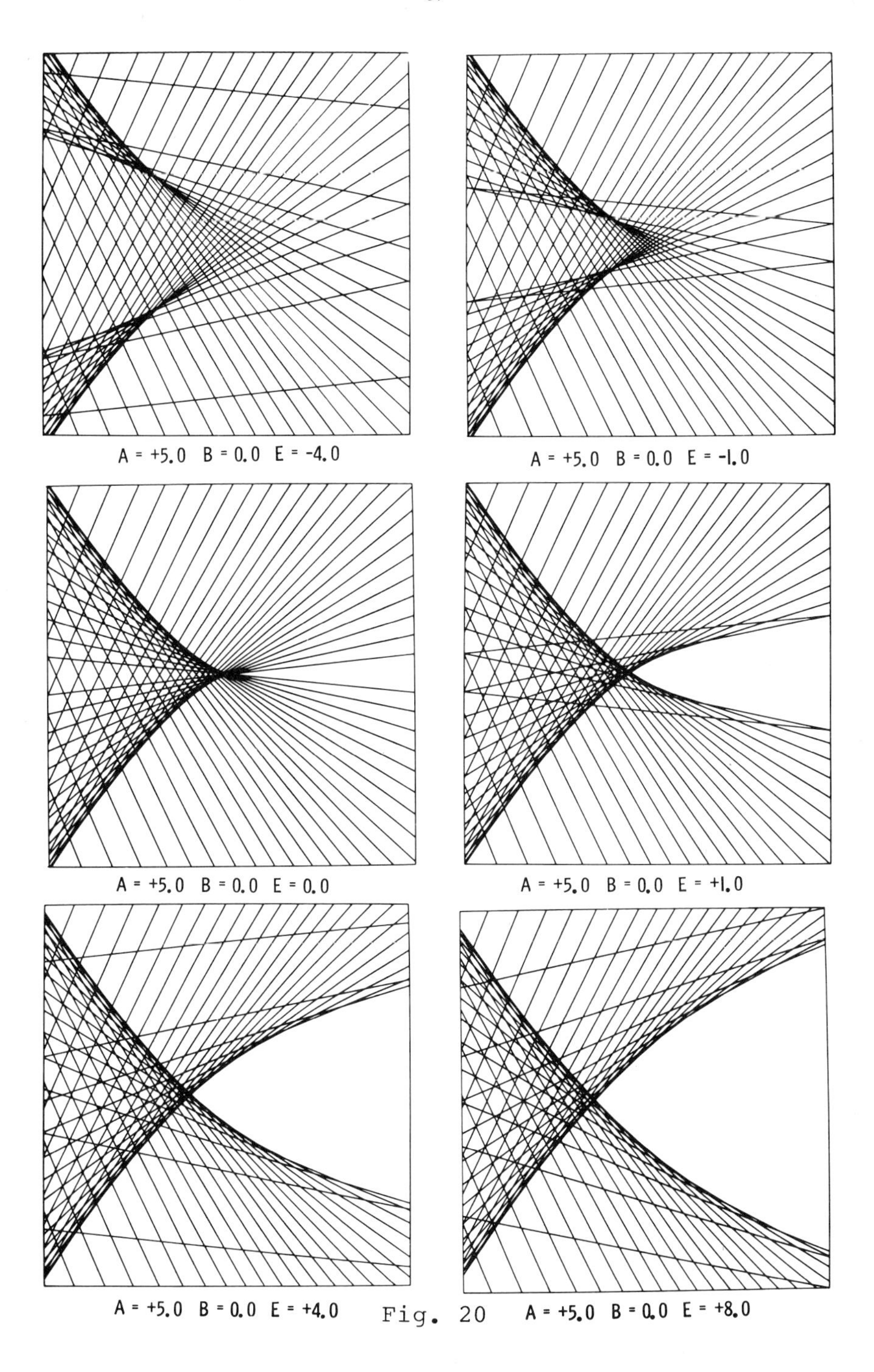
A = +5.0 B = 0.0 E = -4.0
A = +5.0 B = 0.0 E = -1.0
A = +5.0 B = 0.0 E = 0.0
A = +5.0 B = 0.0 E = +1.0
A = +5.0 B = 0.0 E = +4.0
A = +5.0 B = 0.0 E = +8.0

Fig. 20

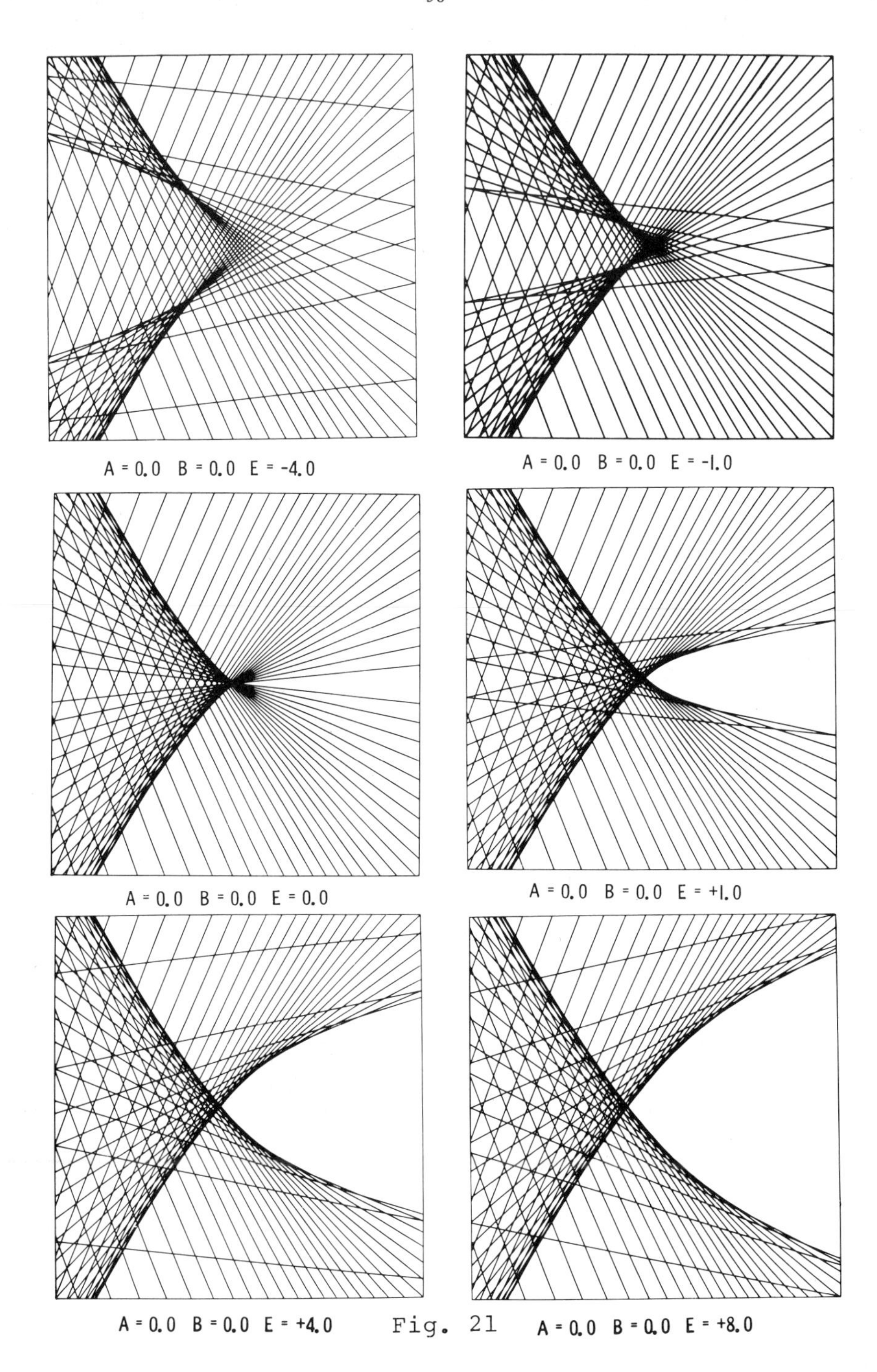
A = 0.0 B = 0.0 E = -4.0
A = 0.0 B = 0.0 E = -1.0
A = 0.0 B = 0.0 E = 0.0
A = 0.0 B = 0.0 E = +1.0
A = 0.0 B = 0.0 E = +4.0
A = 0.0 B = 0.0 E = +8.0

Fig. 21

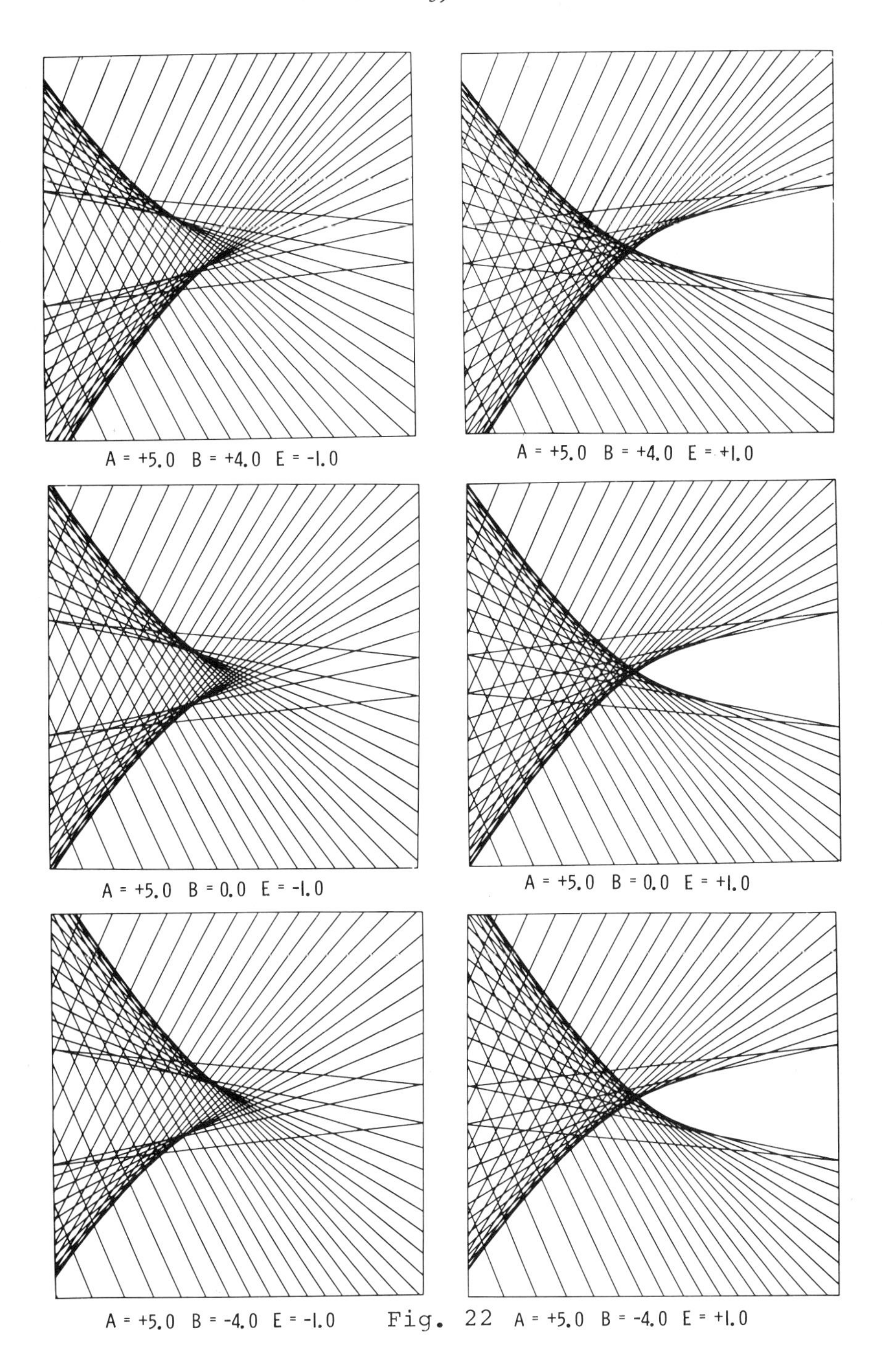
A = +5.0 B = +4.0 E = -1.0
A = +5.0 B = +4.0 E = +1.0
A = +5.0 B = 0.0 E = -1.0
A = +5.0 B = 0.0 E = +1.0
A = +5.0 B = -4.0 E = -1.0
A = +5.0 B = -4.0 E = +1.0

Fig. 22

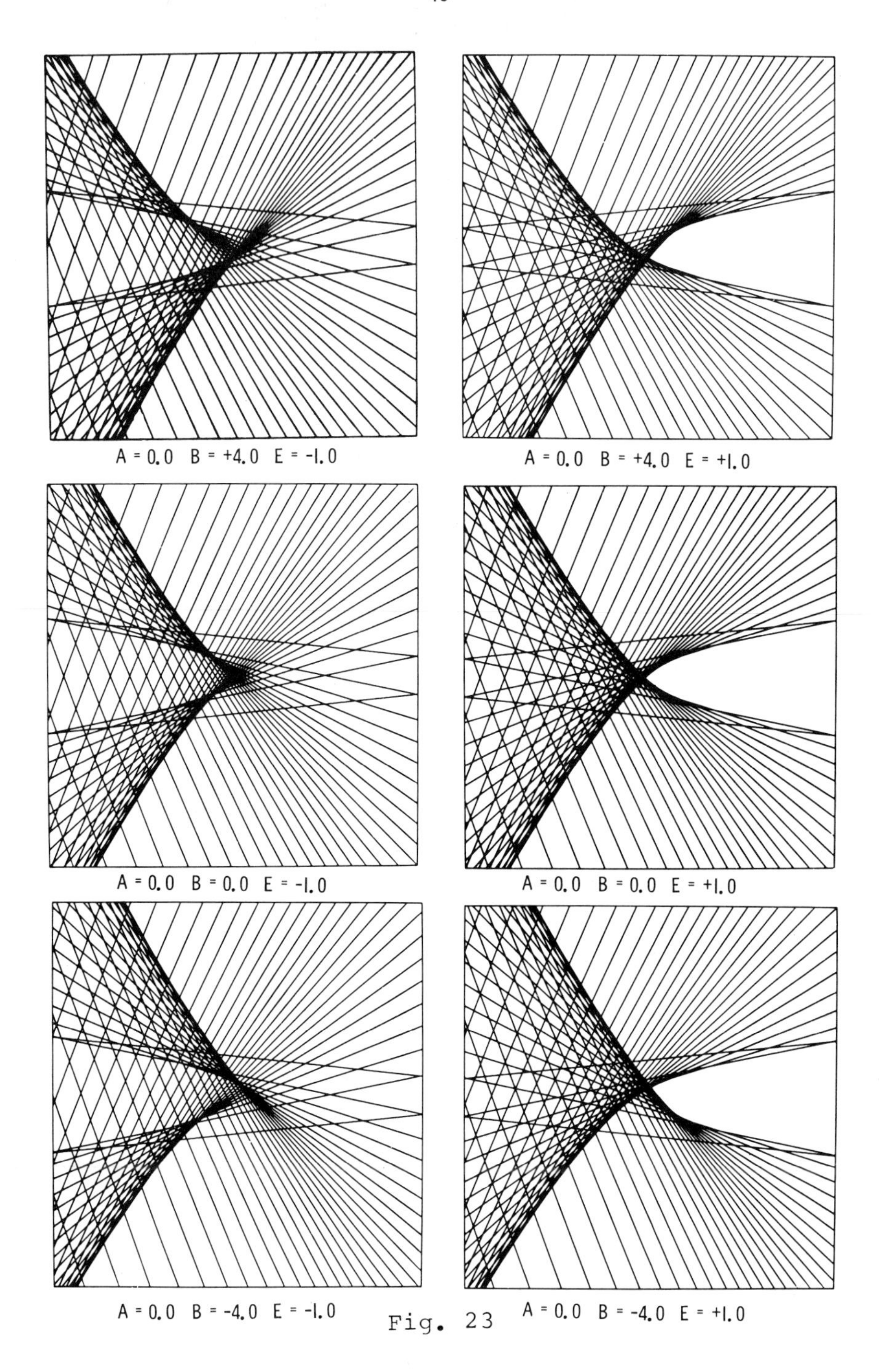
A = 0.0 B = +4.0 E = -1.0
A = 0.0 B = +4.0 E = +1.0
A = 0.0 B = 0.0 E = -1.0
A = 0.0 B = 0.0 E = +1.0
A = 0.0 B = -4.0 E = -1.0
A = 0.0 B = -4.0 E = +1.0

Fig. 23

Ruled Surface Projections of the Star Catastrophe

$$V = \frac{x^8}{8} + \frac{Ax^6}{6} + \frac{Bx^5}{5} + \frac{Cx^4}{4} + \frac{Dx^3}{3} + \frac{Ex^2}{2} + Fx.$$ onto the plane (E,F)

When A is negative, C large and positive and B and D zero, the singularity is simply cuspoid (Fig. 24, 25 and 26). However, as C is reduced, two Swallowtails begin to develop one on either edge of the cusp (see, for example, Fig. 24, C = +35 and +30); as C is further reduced than these Swallowtails grow to dominate the picture with the initially larger cusp region decaying to insignificance. The general picture is of a Butterfly surface with a second small Butterfly replacing the central cusp in the area in which E and F are small (Fig. 24 C = 20, for example, also Figs. 25 and 26.) For C negative the Star Catastrophe collapses to become simply the Butterfly Catastrophe. For A zero or positive the picture is also typical of the Butterfly with the two wings appearing when C is negative (Figs. 27 and 28).

Figs 29 (A and B), 30, 31 and 32 show the complex changes that occur in the projected surface when B is varied (A negative, C positive D zero). Note, for example the Wigwam-like morphology of part of Fig. 31, B = $\pm$ 1.0. Fig. 33 shows the changes produced when D is varied.

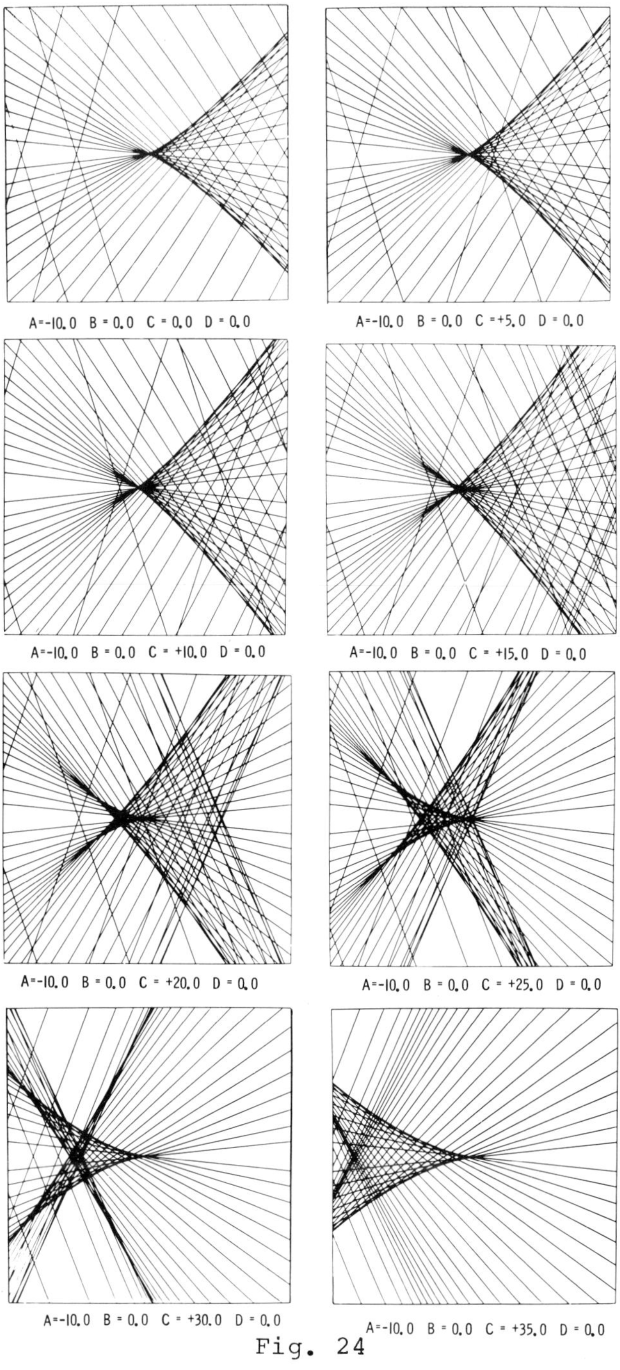
A=-10.0 B = 0.0 C = 0.0 D = 0.0
A=-10.0 B = 0.0 C =+5.0 D = 0.0
A=-10.0 B = 0.0 C = +10.0 D = 0.0
A=-10.0 B = 0.0 C = +15.0 D = 0.0
A=-10.0 B = 0.0 C = +20.0 D = 0.0
A=-10.0 B = 0.0 C = +25.0 D = 0.0
A=-10.0 B = 0.0 C = +30.0 D = 0.0
A=-10.0 B = 0.0 C = +35.0 D = 0.0

Fig. 24

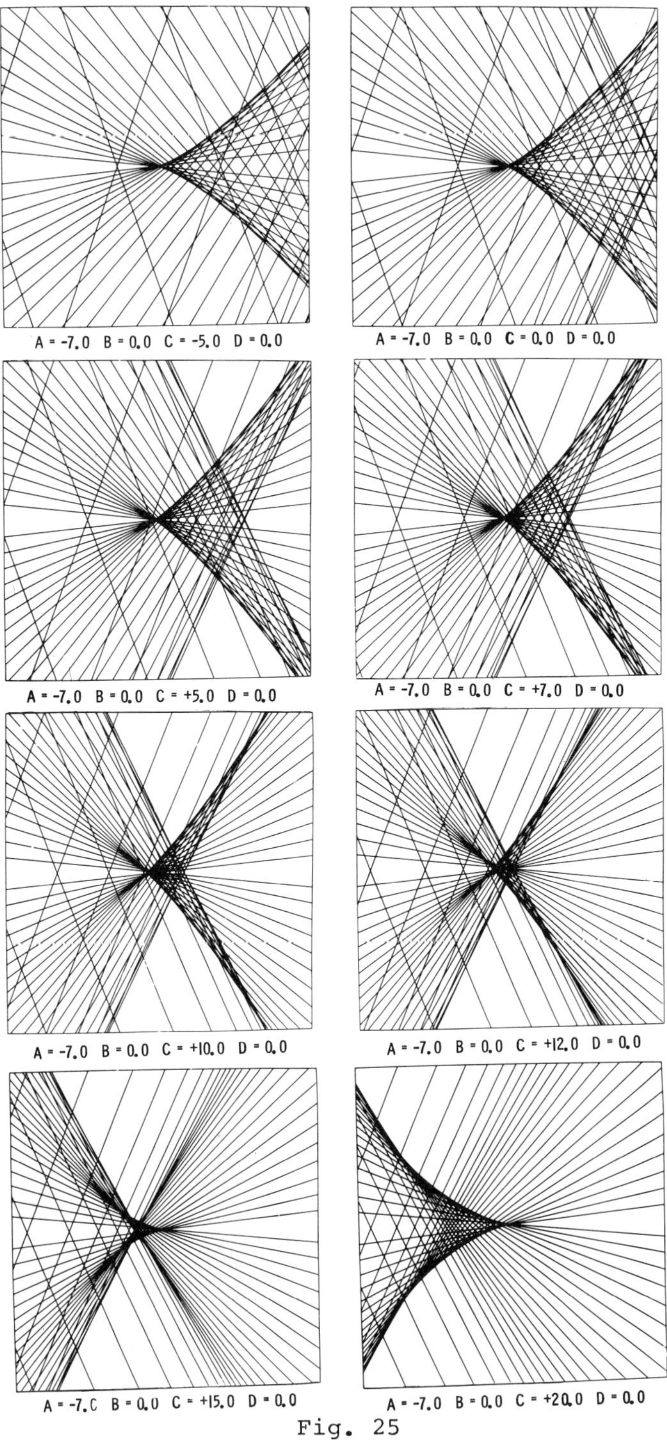
A = -7.0 B = 0.0 C = -5.0 D = 0.0
A = -7.0 B = 0.0 C = 0.0 D = 0.0
A = -7.0 B = 0.0 C = +5.0 D = 0.0
A = -7.0 B = 0.0 C = +7.0 D = 0.0
A = -7.0 B = 0.0 C = +10.0 D = 0.0
A = -7.0 B = 0.0 C = +12.0 D = 0.0
A = -7.C B = 0.0 C = +15.0 D = 0.0
A = -7.0 B = 0.0 C = +20.0 D = 0.0

Fig. 25

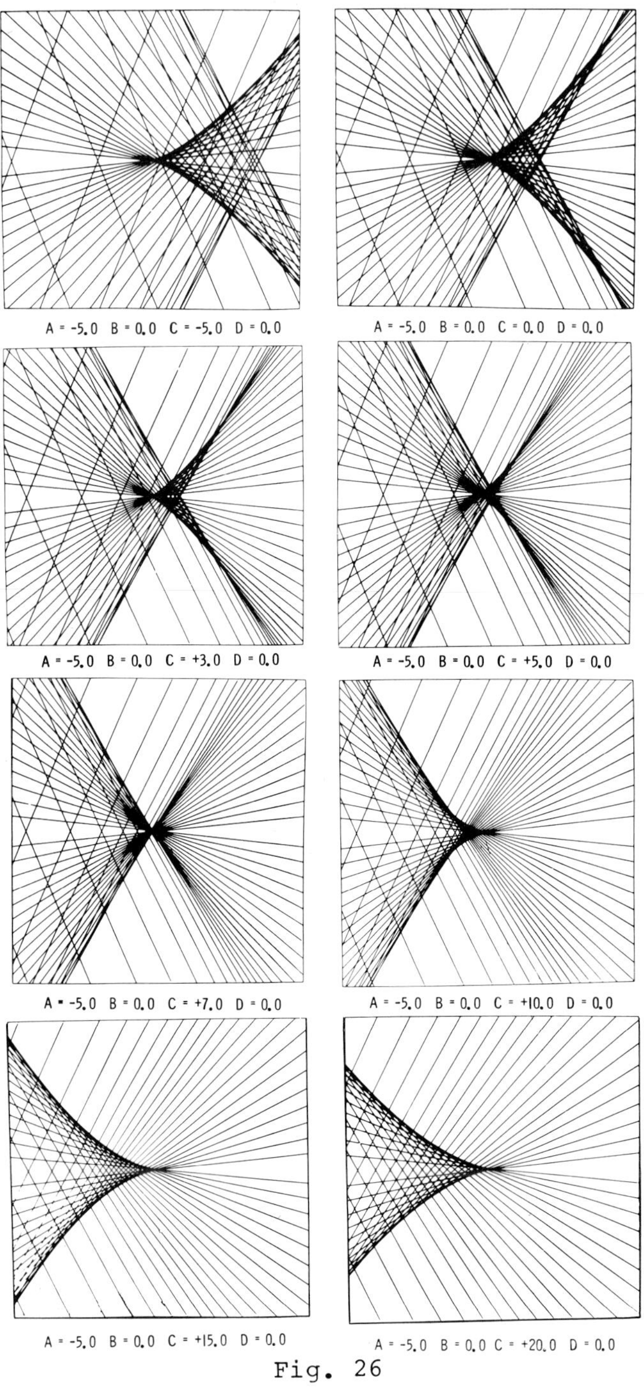
A = -5.0 B = 0.0 C = -5.0 D = 0.0
A = -5.0 B = 0.0 C = 0.0 D = 0.0
A = -5.0 B = 0.0 C = +3.0 D = 0.0
A = -5.0 B = 0.0 C = +5.0 D = 0.0
A = -5.0 B = 0.0 C = +7.0 D = 0.0
A = -5.0 B = 0.0 C = +10.0 D = 0.0
A = -5.0 B = 0.0 C = +15.0 D = 0.0
A = -5.0 B = 0.0 C = +20.0 D = 0.0

Fig. 26

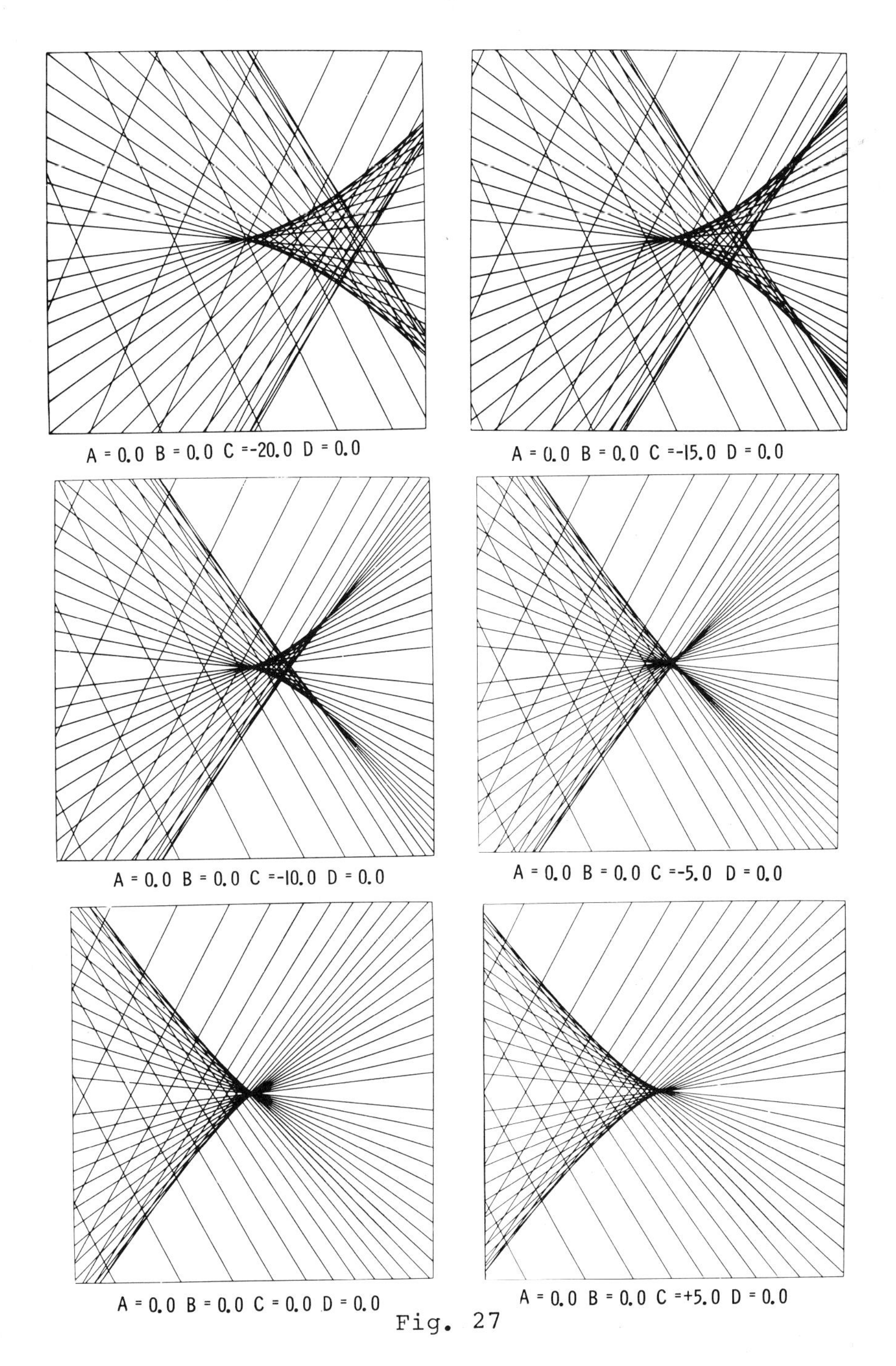
A = 0.0 B = 0.0 C =-20.0 D = 0.0
A = 0.0 B = 0.0 C =-15.0 D = 0.0
A = 0.0 B = 0.0 C =-10.0 D = 0.0
A = 0.0 B = 0.0 C =-5.0 D = 0.0
A = 0.0 B = 0.0 C = 0.0 D = 0.0
A = 0.0 B = 0.0 C =+5.0 D = 0.0

Fig. 27

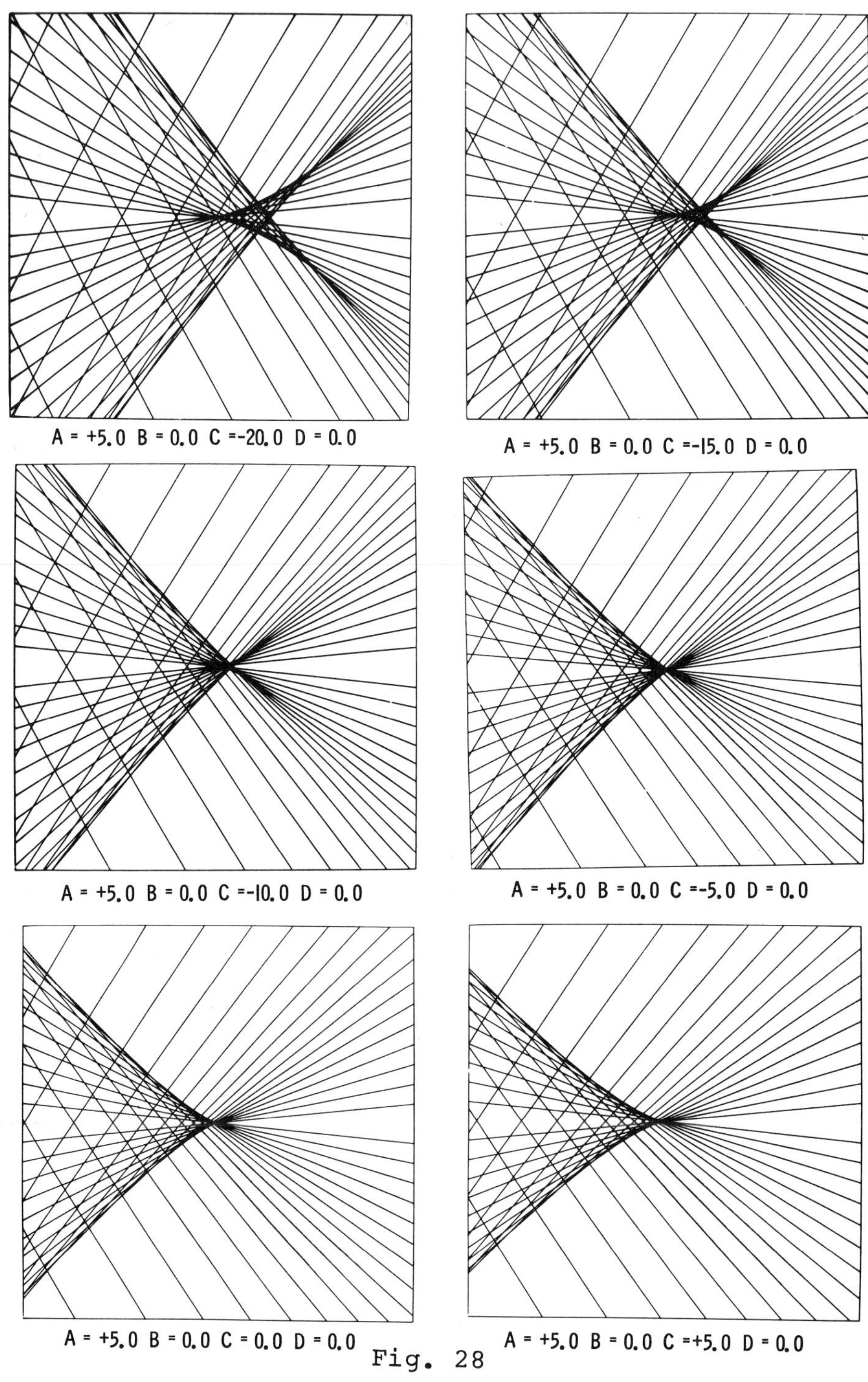
A = +5.0 B = 0.0 C =-20.0 D = 0.0
A = +5.0 B = 0.0 C =-15.0 D = 0.0
A = +5.0 B = 0.0 C =-10.0 D = 0.0
A = +5.0 B = 0.0 C =-5.0 D = 0.0
A = +5.0 B = 0.0 C = 0.0 D = 0.0
A = +5.0 B = 0.0 C =+5.0 D = 0.0

Fig. 28

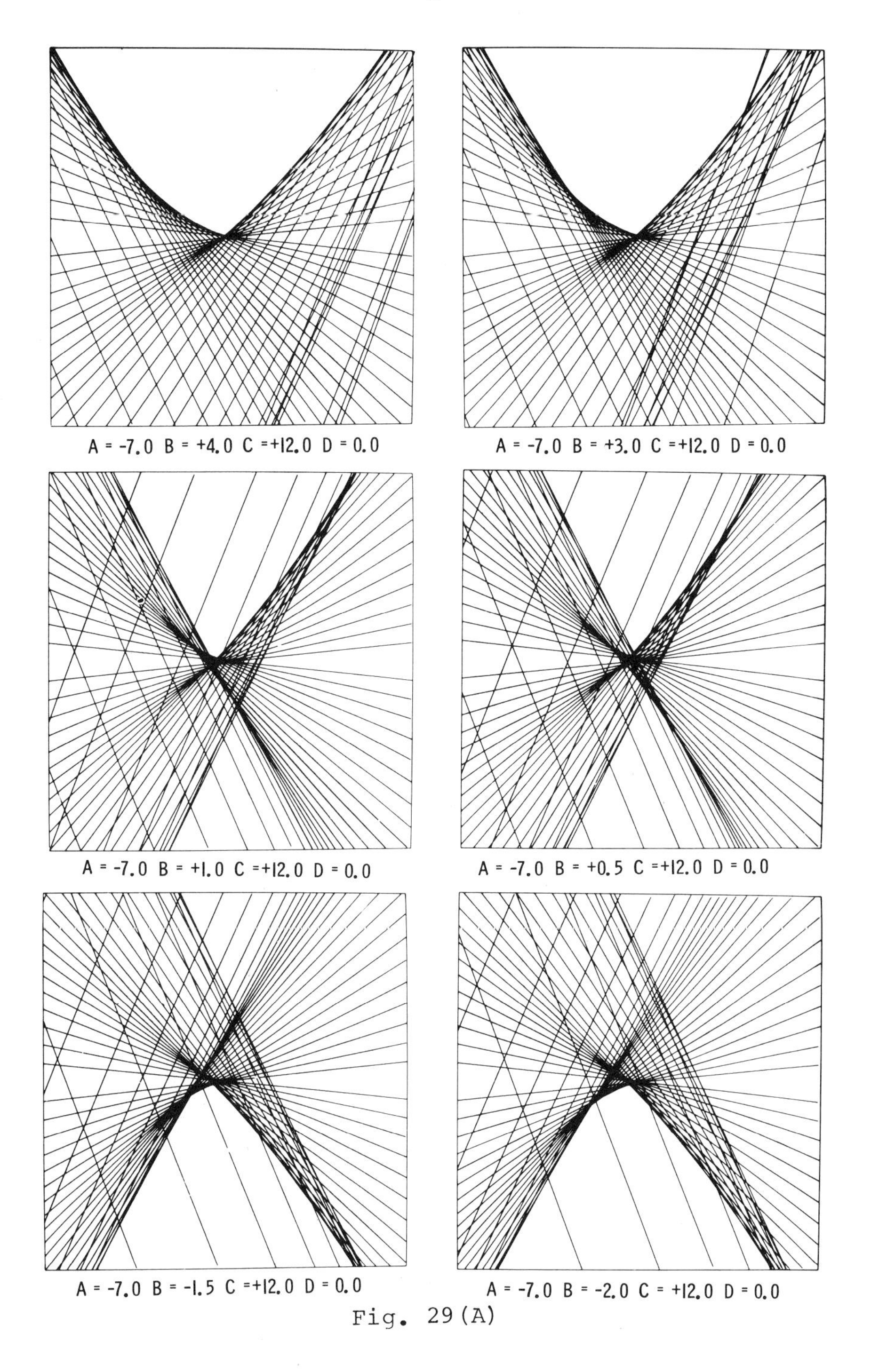
A = -7.0 B = +4.0 C =+12.0 D = 0.0
A = -7.0 B = +3.0 C =+12.0 D = 0.0
A = -7.0 B = +1.0 C =+12.0 D = 0.0
A = -7.0 B = +0.5 C =+12.0 D = 0.0
A = -7.0 B = -1.5 C =+12.0 D = 0.0
A = -7.0 B = -2.0 C = +12.0 D = 0.0

Fig. 29(A)

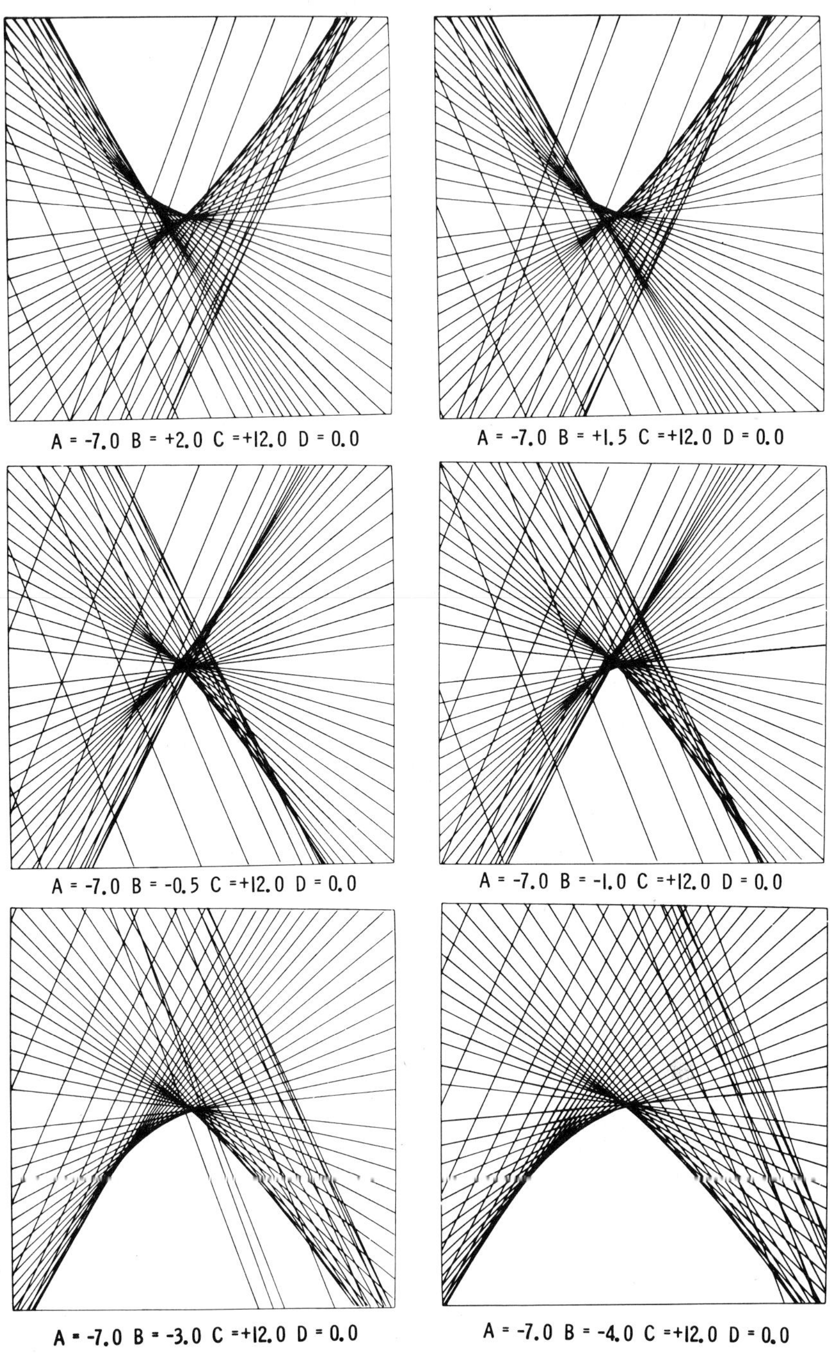
A = -7.0 B = +2.0 C =+12.0 D = 0.0
A = -7.0 B = +1.5 C =+12.0 D = 0.0
A = -7.0 B = -0.5 C =+12.0 D = 0.0
A = -7.0 B = -1.0 C =+12.0 D = 0.0
A = -7.0 B = -3.0 C =+12.0 D = 0.0
A = -7.0 B = -4.0 C =+12.0 D = 0.0

Fig. 29(B)

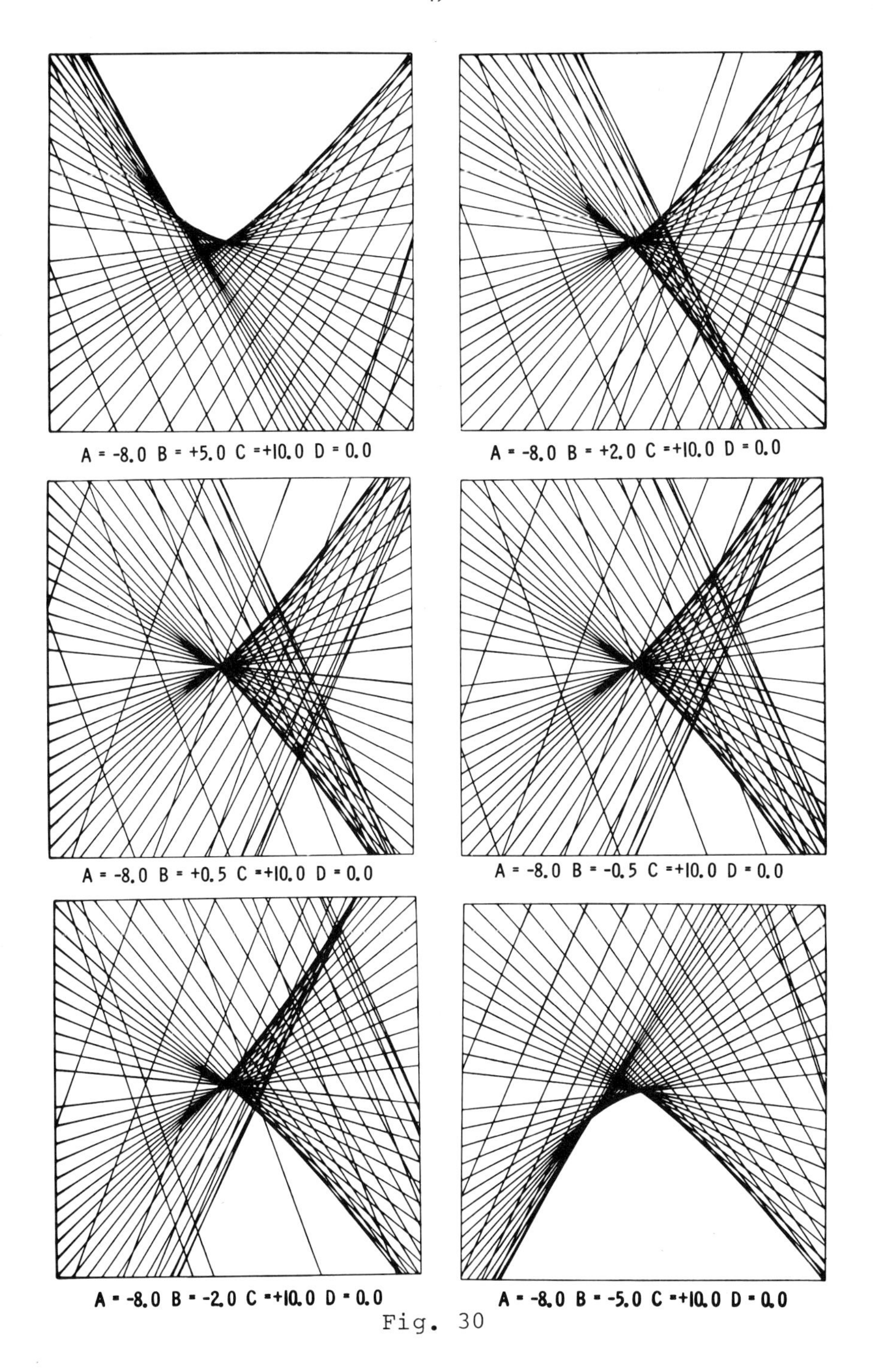
A = -8.0 B = +5.0 C =+10.0 D = 0.0
A = -8.0 B = +2.0 C =+10.0 D = 0.0
A = -8.0 B = +0.5 C =+10.0 D = 0.0
A = -8.0 B = -0.5 C =+10.0 D = 0.0
A = -8.0 B = -2.0 C =+10.0 D = 0.0
A = -8.0 B = -5.0 C =+10.0 D = 0.0

Fig. 30

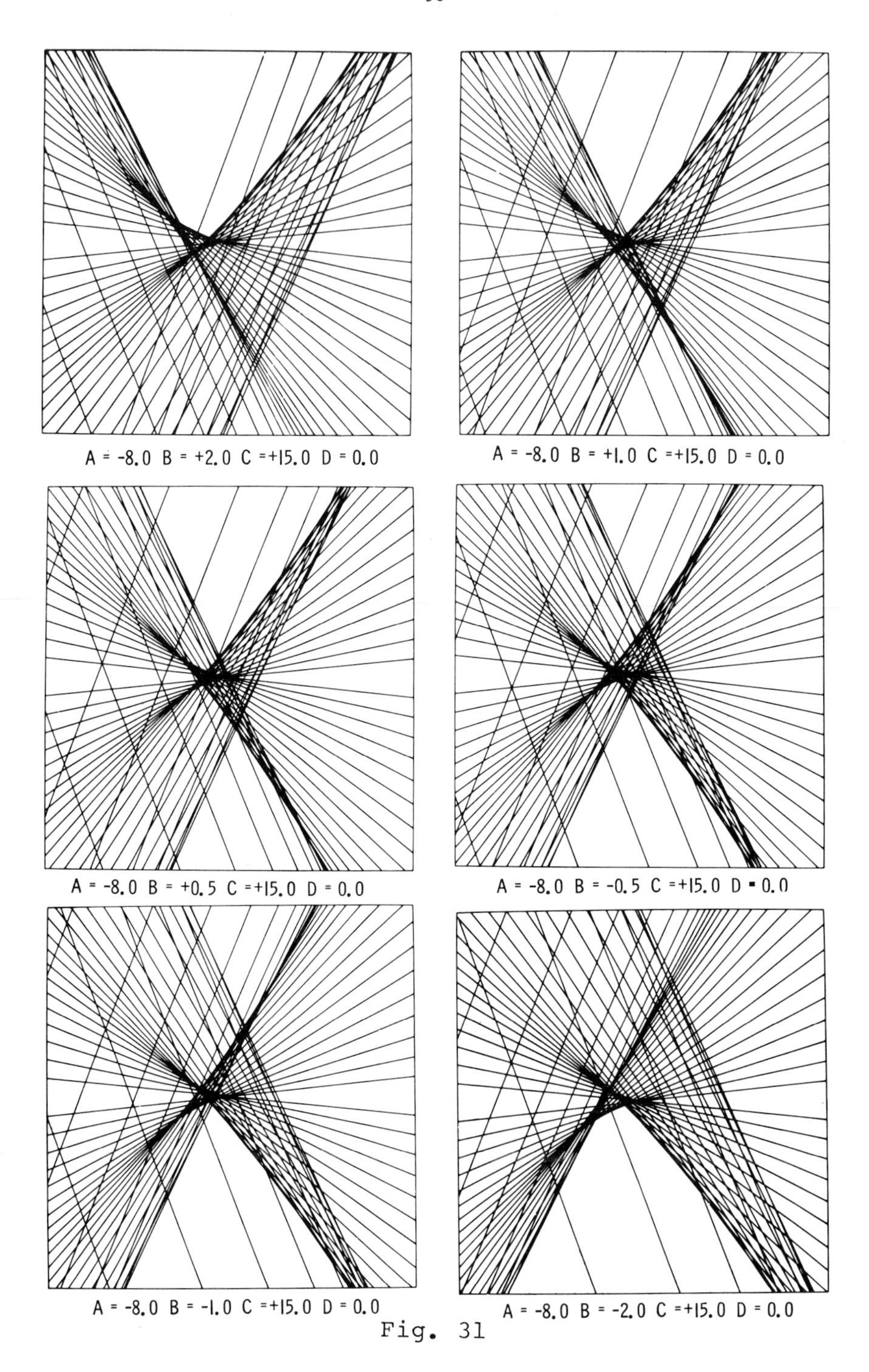
A = -8.0 B = +2.0 C =+15.0 D = 0.0
A = -8.0 B = +1.0 C =+15.0 D = 0.0
A = -8.0 B = +0.5 C =+15.0 D = 0.0
A = -8.0 B = -0.5 C =+15.0 D = 0.0
A = -8.0 B = -1.0 C =+15.0 D = 0.0
A = -8.0 B = -2.0 C =+15.0 D = 0.0

Fig. 31

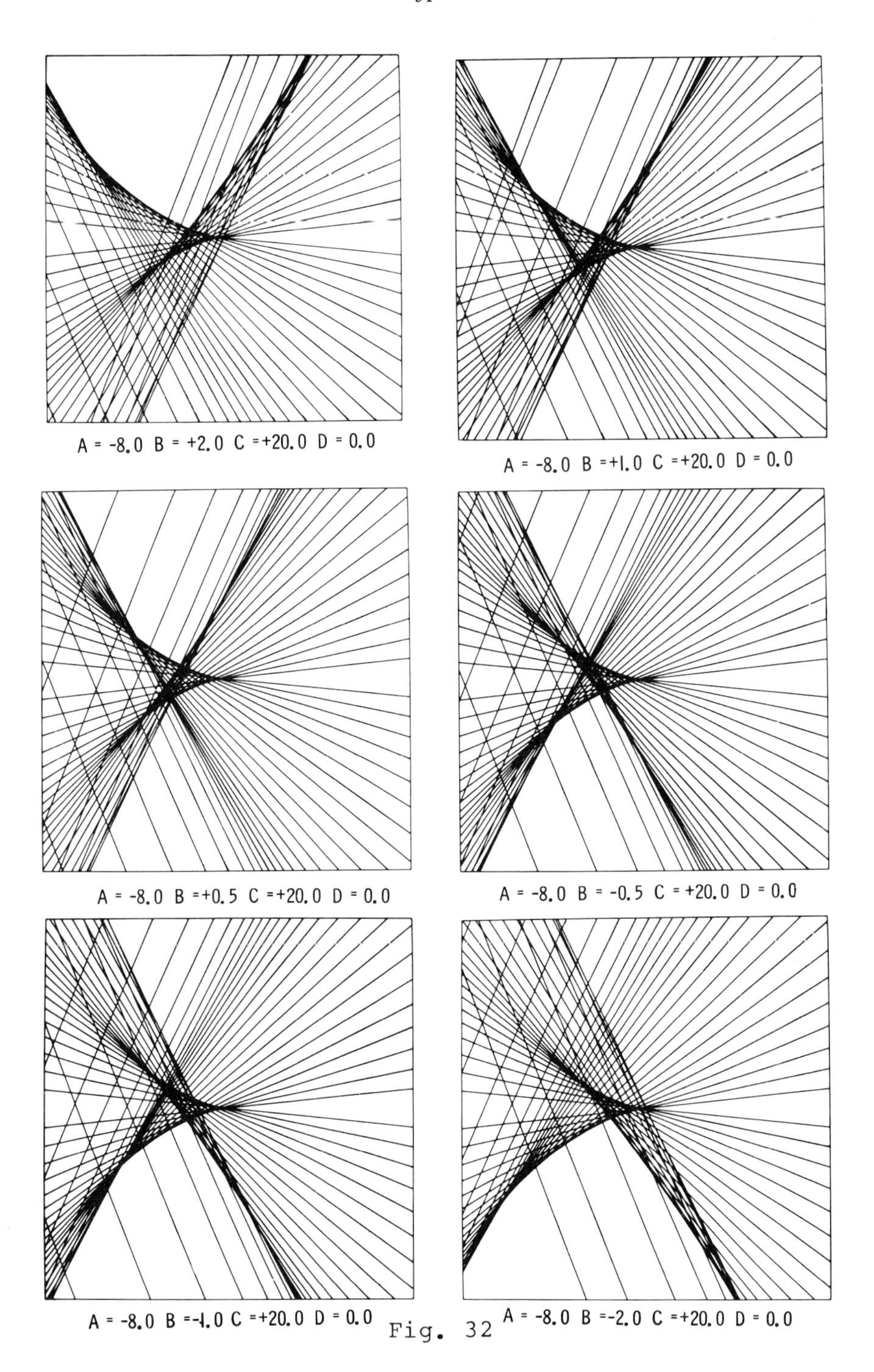
A = -8.0 B = +2.0 C =+20.0 D = 0.0
A = -8.0 B =+1.0 C =+20.0 D = 0.0
A = -8.0 B =+0.5 C =+20.0 D = 0.0
A = -8.0 B = -0.5 C = +20.0 D = 0.0
A = -8.0 B =-1.0 C =+20.0 D = 0.0
A = -8.0 B =-2.0 C =+20.0 D = 0.0

Fig. 32

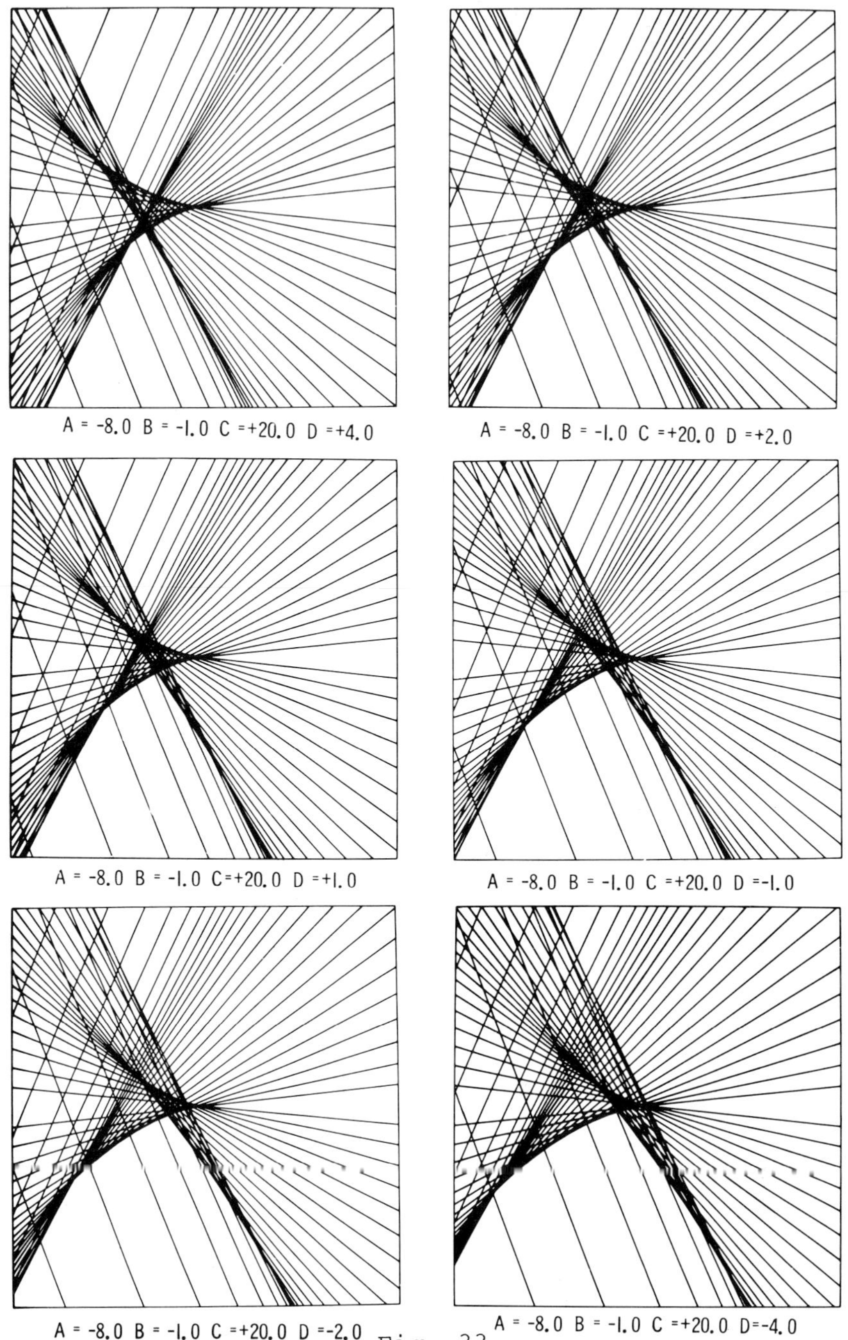
A = -8.0 B = -1.0 C =+20.0 D =+4.0
A = -8.0 B = -1.0 C =+20.0 D =+2.0
A = -8.0 B = -1.0 C=+20.0 D =+1.0
A = -8.0 B = -1.0 C =+20.0 D =-1.0
A = -8.0 B = -1.0 C =+20.0 D =-2.0
A = -8.0 B = -1.0 C =+20.0 D=-4.0

Fig. 33

Ruled Surface Projections of Wigwam Sections of the Star Catastrophe $V = \frac{x^8}{8} + \frac{Ax^6}{6} + \frac{Bx^5}{5} + \frac{Cx^4}{4} + \frac{Dx^3}{3} + \frac{Ex^2}{2} + Fx.$ on the planc (D,E)

A negative, B and F zero and C running from +20 to zero (Fig. 34) produces a series of sections typical of the Wigwam Catastrophe (See, for example Fig. 13). As before, it is possible to elucidate the structure of the cuspoids by unfolding along a moving slice of the higher order catastrophes. Figs. 35, 36, 37, 38 and 39 show this unfolding along the F axis with A negative, B zero and C positive. These pictures show how the Wigwam sections can be constructed by surgery of two previously separate surfaces.

Figs. 40 (A and B) and 41 (A and B) with A either zero or positive simply show the superficial production of Swallowtail-like sections (as shown also in Figs. 9 (A and B)).

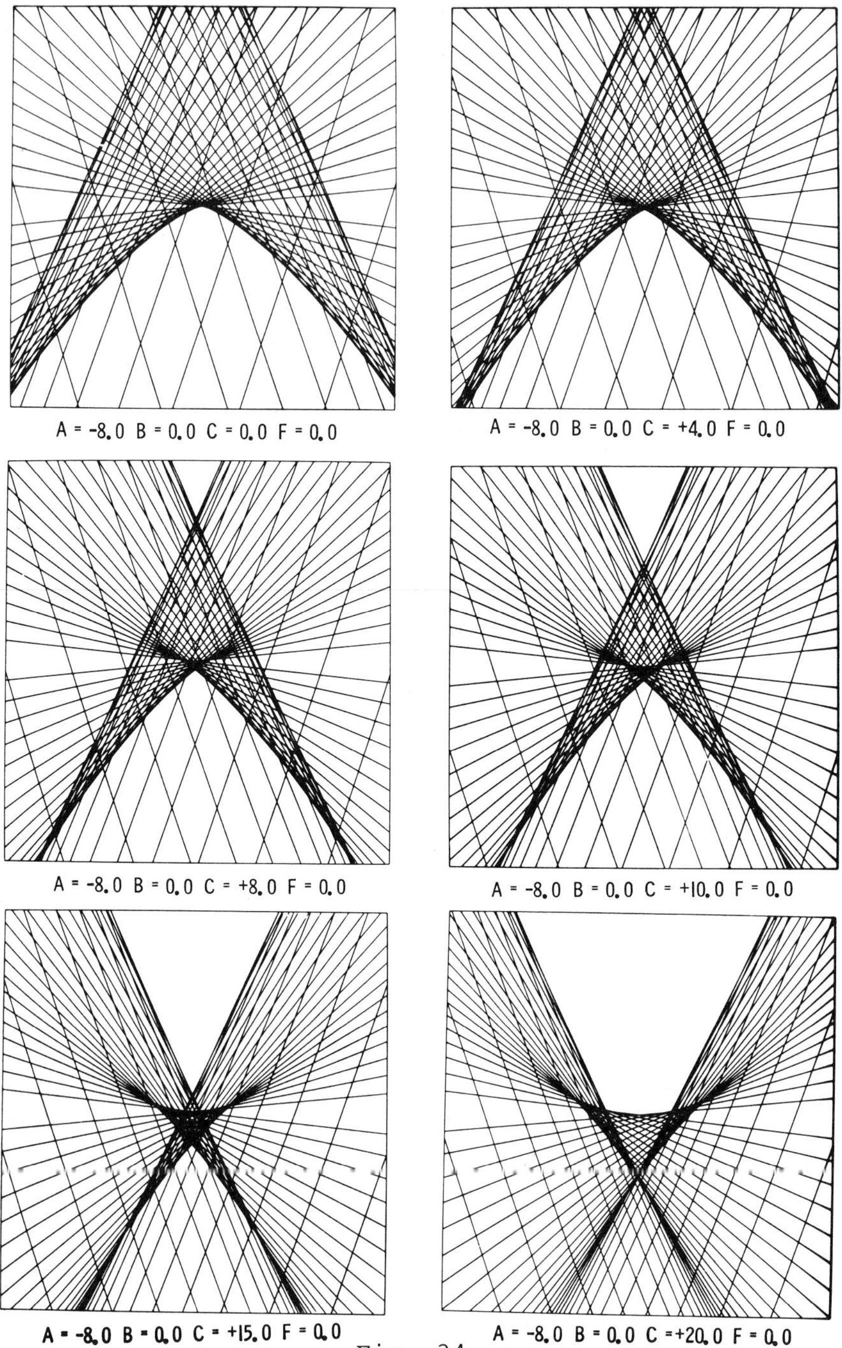
A = -8.0 B = 0.0 C = 0.0 F = 0.0
A = -8.0 B = 0.0 C = +4.0 F = 0.0
A = -8.0 B = 0.0 C = +8.0 F = 0.0
A = -8.0 B = 0.0 C = +10.0 F = 0.0
A = -8.0 B = 0.0 C = +15.0 F = 0.0
A = -8.0 B = 0.0 C = +20.0 F = 0.0

Fig. 34

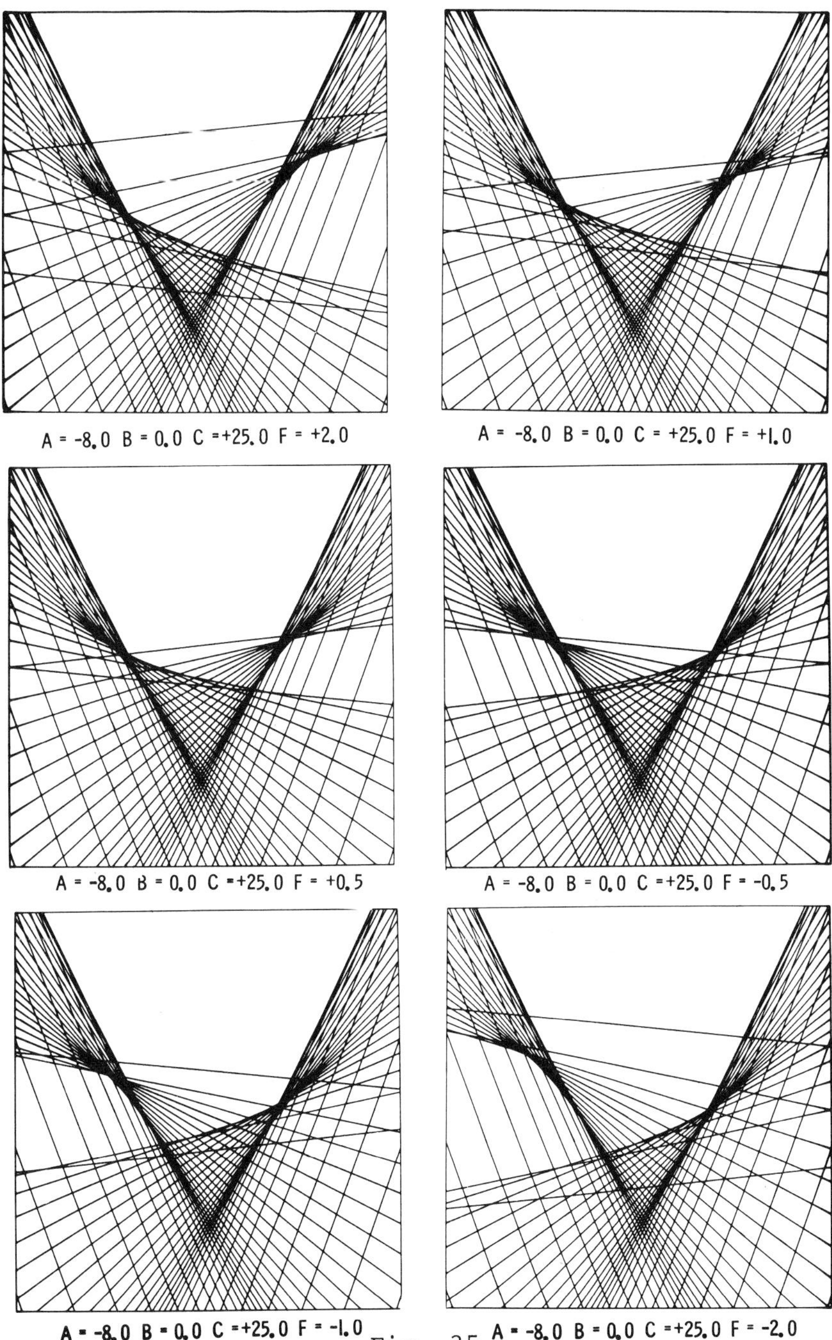
A = -8.0 B = 0.0 C =+25.0 F = +2.0
A = -8.0 B = 0.0 C = +25.0 F = +1.0
A = -8.0 B = 0.0 C =+25.0 F = +0.5
A = -8.0 B = 0.0 C =+25.0 F = -0.5
A = -8.0 B = 0.0 C =+25.0 F = -1.0
A = -8.0 B = 0.0 C =+25.0 F = -2.0

Fig. 35

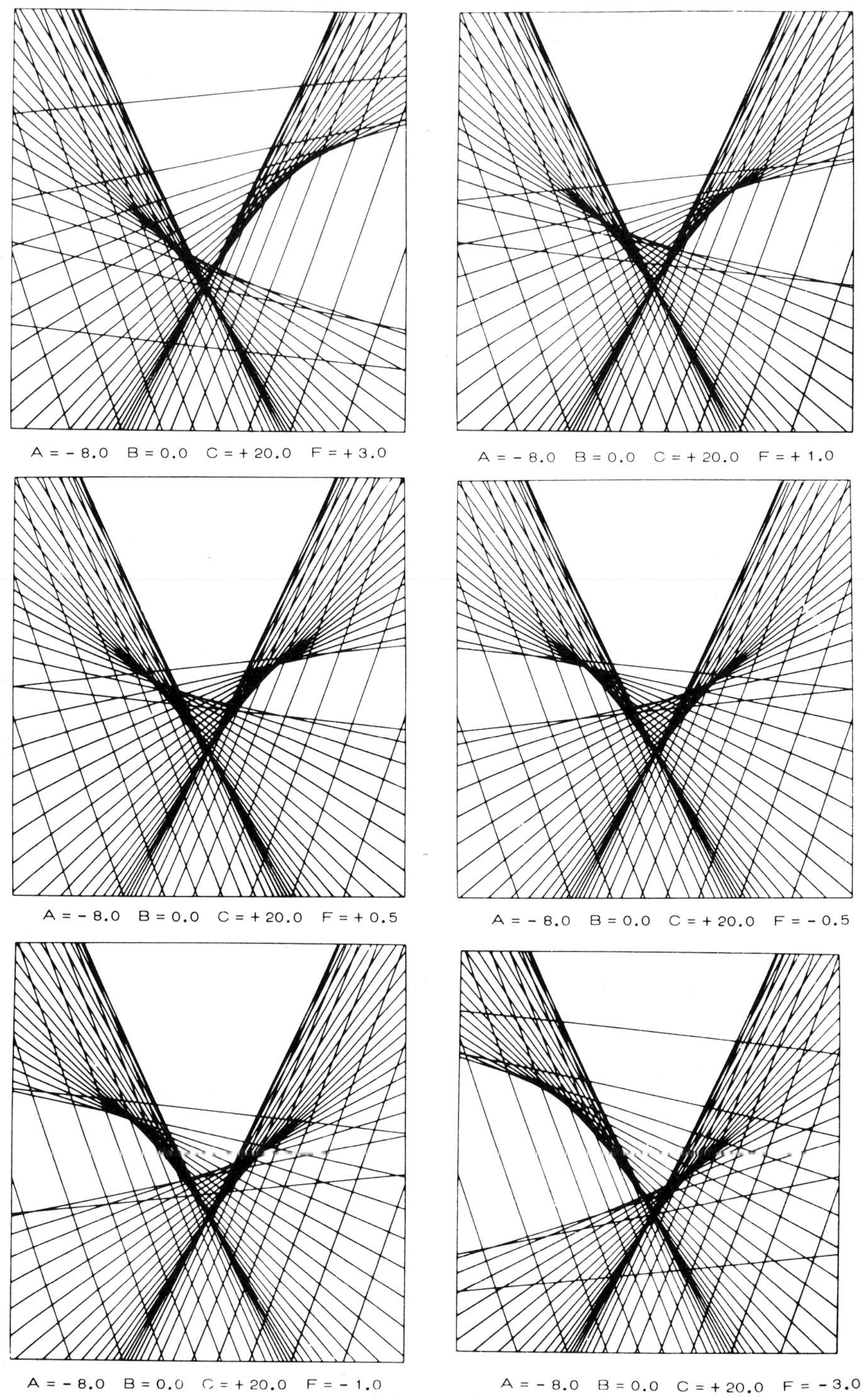
A = - 8.0 B = 0.0 C = + 20.0 F = + 3.0
A = - 8.0 B = 0.0 C = + 20.0 F = + 1.0
A = - 8.0 B = 0.0 C = + 20.0 F = + 0.5
A = - 8.0 B = 0.0 C = + 20.0 F = - 0.5
A = - 8.0 B = 0.0 C = + 20.0 F = - 1.0
A = - 8.0 B = 0.0 C = + 20.0 F = - 3.0

Fig. 36

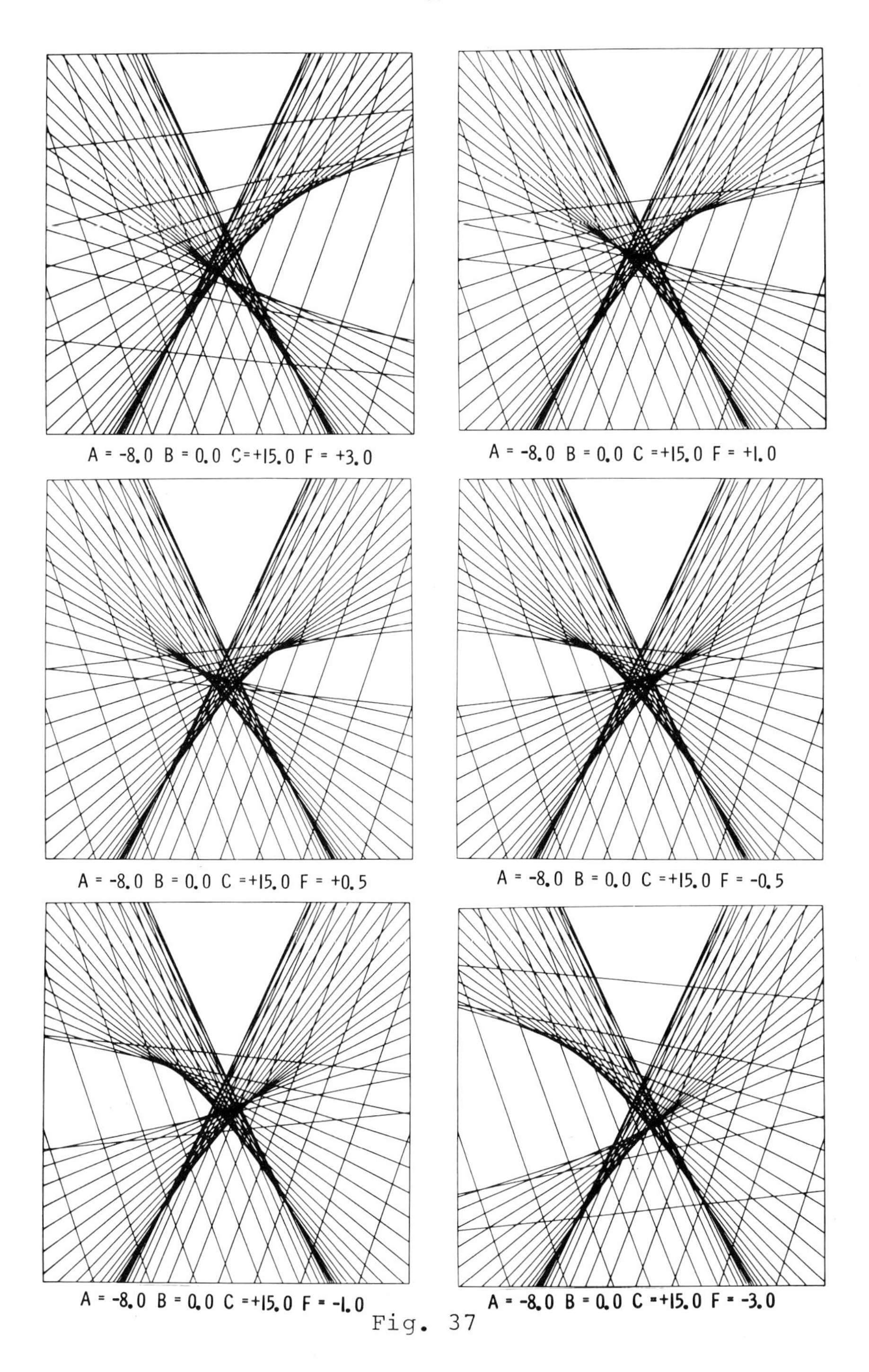
A = -8.0 B = 0.0 C=+15.0 F = +3.0
A = -8.0 B = 0.0 C =+15.0 F = +1.0
A = -8.0 B = 0.0 C =+15.0 F = +0.5
A = -8.0 B = 0.0 C =+15.0 F = -0.5
A = -8.0 B = 0.0 C =+15.0 F = -1.0
A = -8.0 B = 0.0 C =+15.0 F = -3.0

Fig. 37

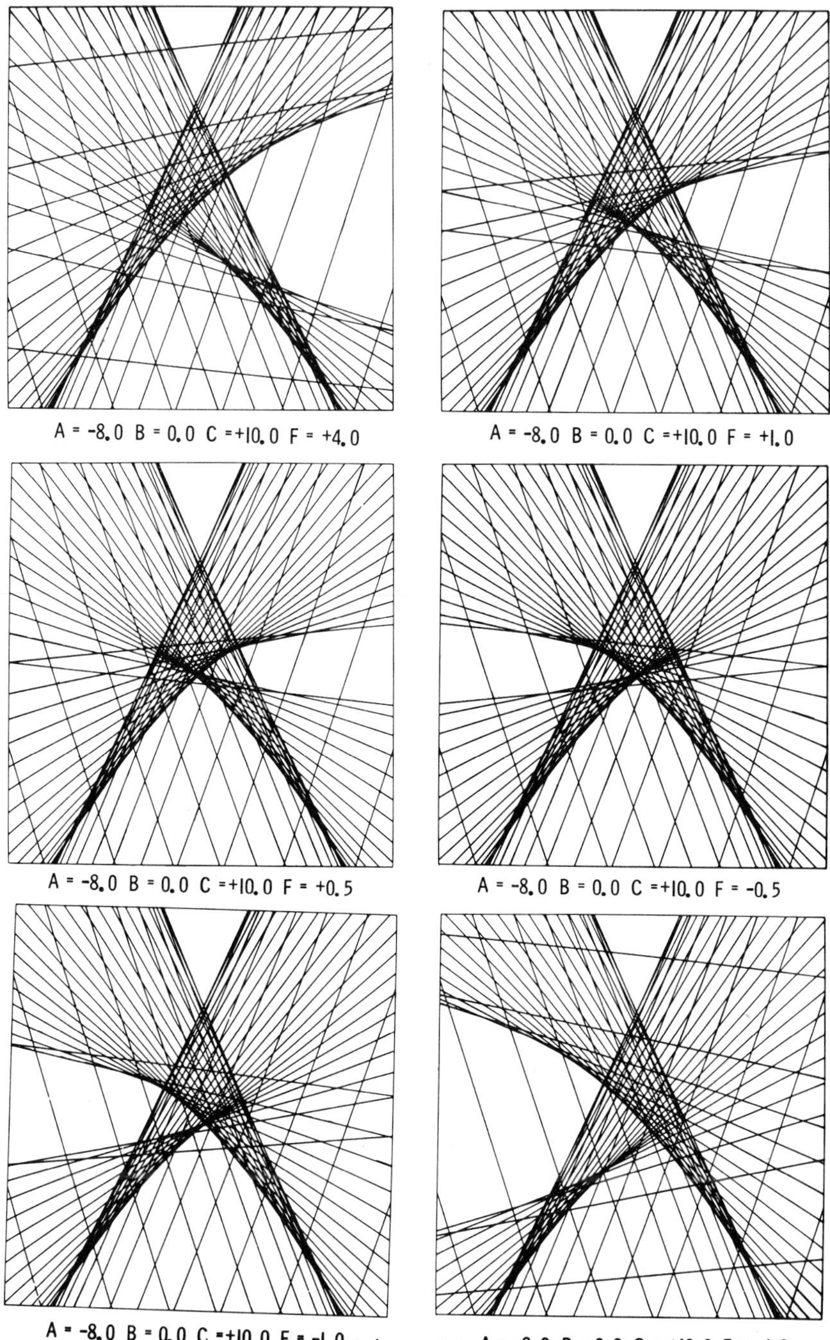
A = -8.0 B = 0.0 C =+10.0 F = +4.0
A = -8.0 B = 0.0 C =+10.0 F = +1.0
A = -8.0 B = 0.0 C =+10.0 F = +0.5
A = -8.0 B = 0.0 C =+10.0 F = -0.5
A = -8.0 B = 0.0 C =+10.0 F = -1.0
A = -8.0 B = 0.0 C = +10.0 F = -4.0

Fig. 38

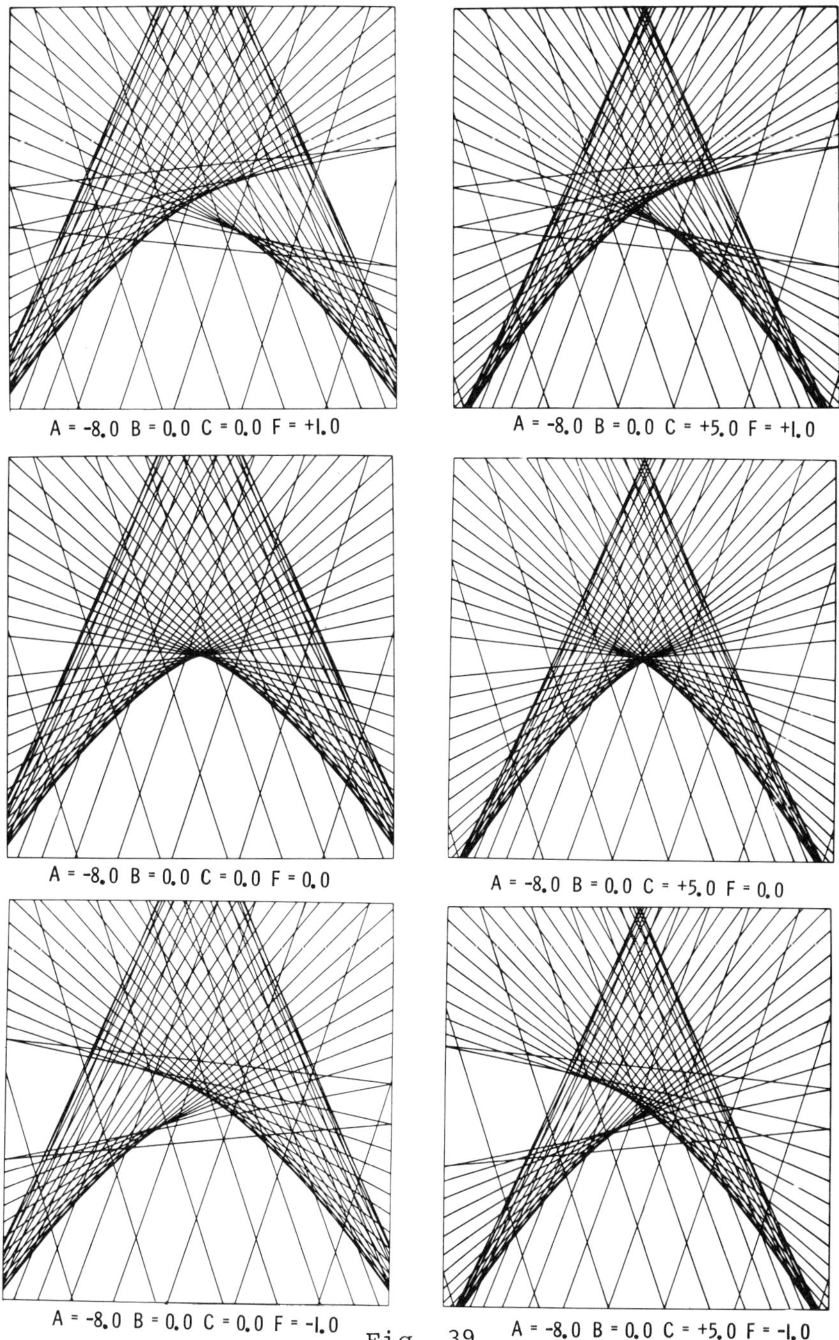
A = -8.0 B = 0.0 C = 0.0 F = +1.0
A = -8.0 B = 0.0 C = +5.0 F = +1.0
A = -8.0 B = 0.0 C = 0.0 F = 0.0
A = -8.0 B = 0.0 C = +5.0 F = 0.0
A = -8.0 B = 0.0 C = 0.0 F = -1.0
A = -8.0 B = 0.0 C = +5.0 F = -1.0

Fig. 39

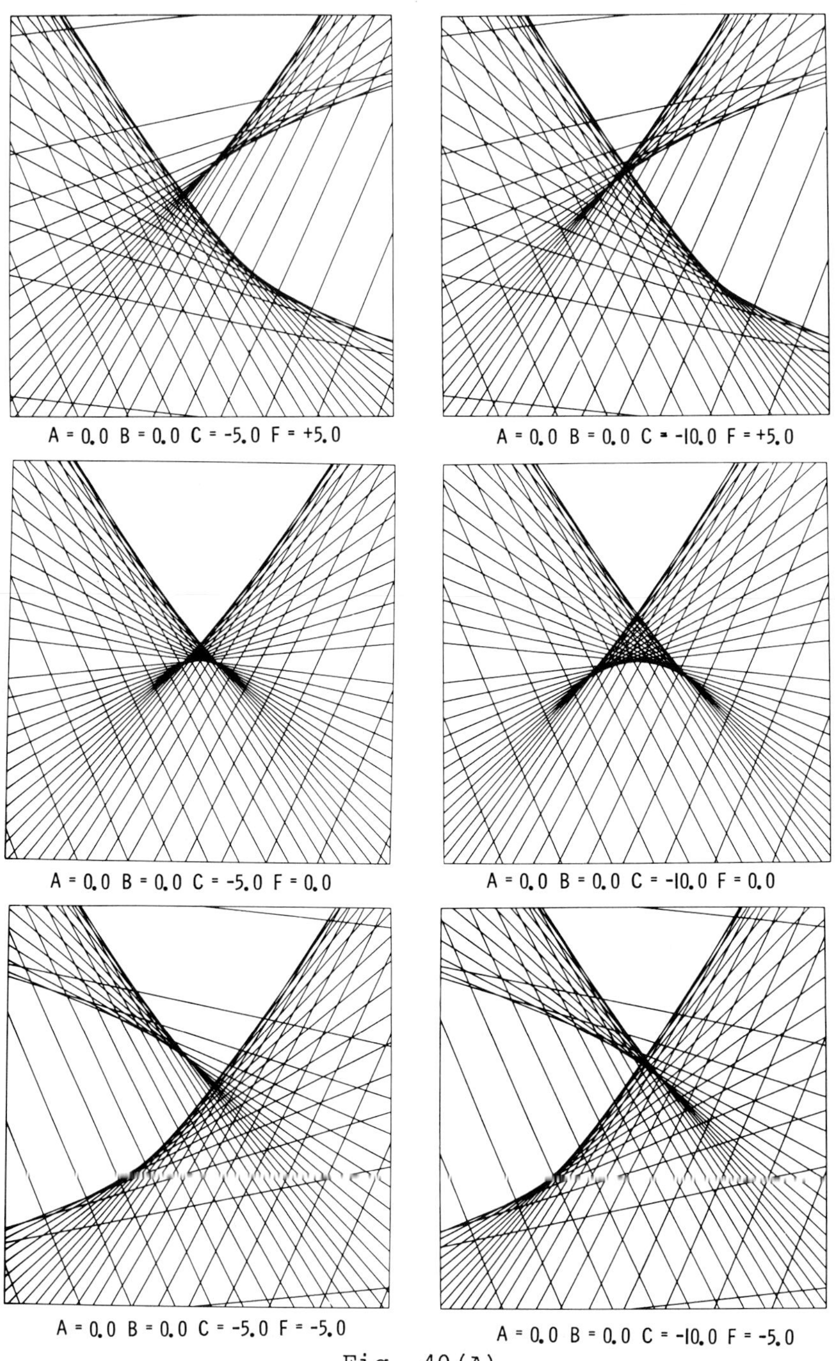
A = 0.0 B = 0.0 C = -5.0 F = +5.0
A = 0.0 B = 0.0 C = -10.0 F = +5.0
A = 0.0 B = 0.0 C = -5.0 F = 0.0
A = 0.0 B = 0.0 C = -10.0 F = 0.0
A = 0.0 B = 0.0 C = -5.0 F = -5.0
A = 0.0 B = 0.0 C = -10.0 F = -5.0

Fig. 40(A)

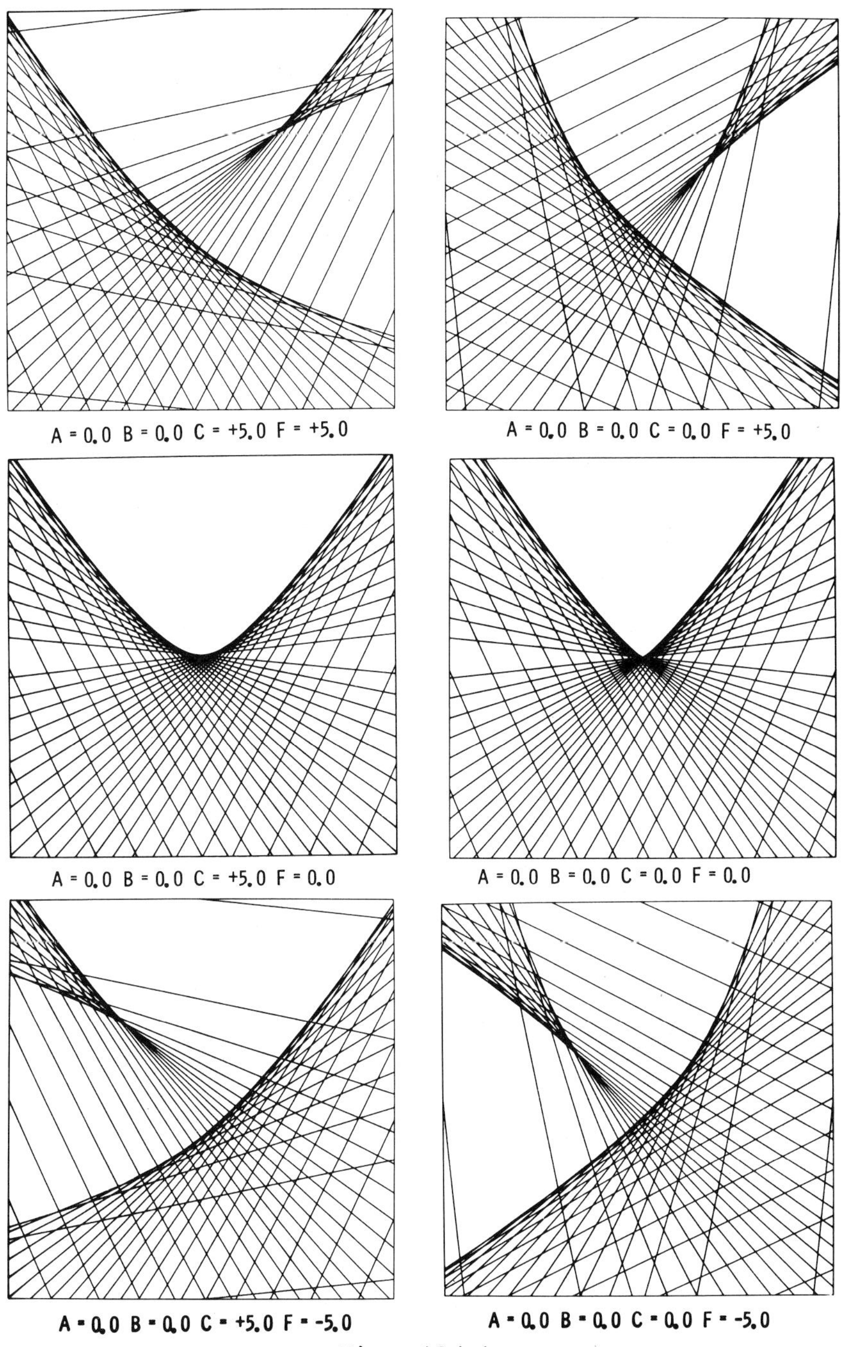
A = 0.0 B = 0.0 C = +5.0 F = +5.0
A = 0.0 B = 0.0 C = 0.0 F = +5.0
A = 0.0 B = 0.0 C = +5.0 F = 0.0
A = 0.0 B = 0.0 C = 0.0 F = 0.0
A = 0.0 B = 0.0 C = +5.0 F = -5.0
A = 0.0 B = 0.0 C = 0.0 F = -5.0

Fig. 40(B)

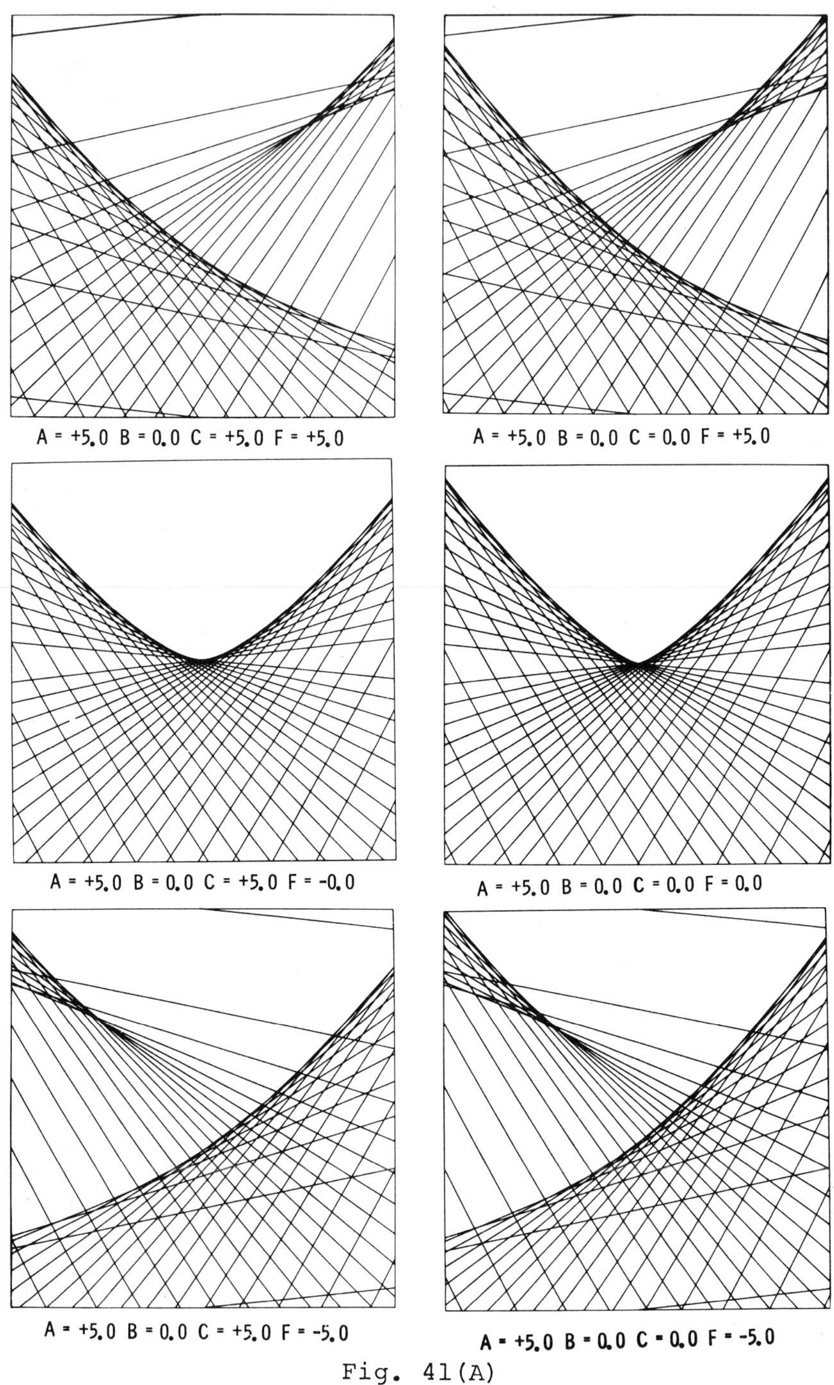
A = +5.0 B = 0.0 C = +5.0 F = +5.0
A = +5.0 B = 0.0 C = 0.0 F = +5.0
A = +5.0 B = 0.0 C = +5.0 F = -0.0
A = +5.0 B = 0.0 C = 0.0 F = 0.0
A = +5.0 B = 0.0 C = +5.0 F = -5.0
A = +5.0 B = 0.0 C = 0.0 F = -5.0

Fig. 41(A)

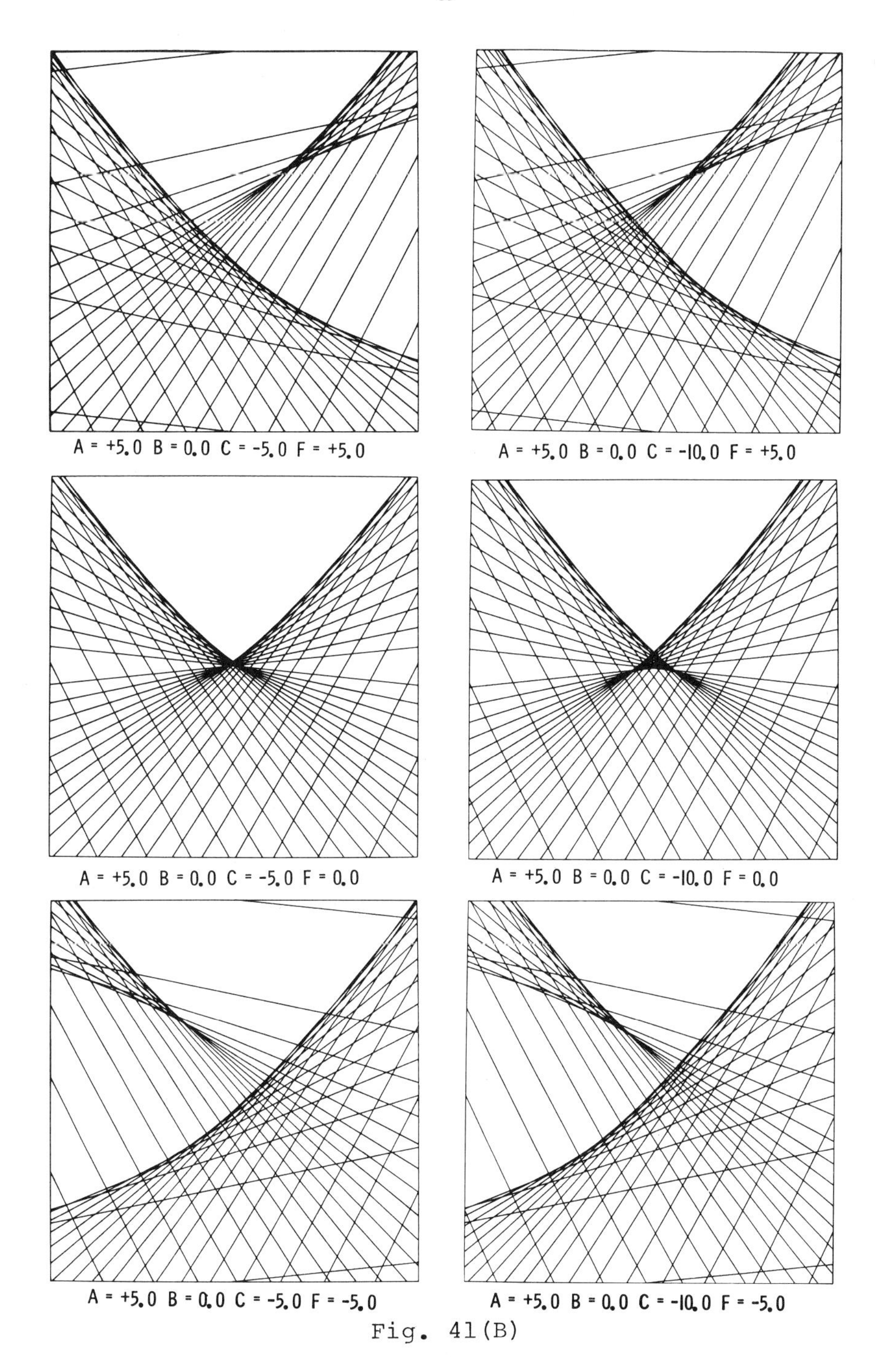
A = +5.0 B = 0.0 C = -5.0 F = +5.0
A = +5.0 B = 0.0 C = -10.0 F = +5.0
A = +5.0 B = 0.0 C = -5.0 F = 0.0
A = +5.0 B = 0.0 C = -10.0 F = 0.0
A = +5.0 B = 0.0 C = -5.0 F = -5.0
A = +5.0 B = 0.0 C = -10.0 F = -5.0

Fig. 41(B)

THE GEOMETRY OF THE ELEMENTARY CATASTROPHES:
(2). The Hyperbolic and Elliptic Umbilics.

by

A. E. R. Woodcock
IBM Thomas J. Watson Research Center
Yorktown Heights, N.Y. 10598, U.S.A.

and

T. Poston
Instituto de Mathemática Pura e Aplicada
Rio-de-Janeiro, Brazil

ABSTRACT: This paper presents a description of the geometry of the two three-dimensional singularities which are stable with respect to variation of the families of potentials giving rise to them, though not with respect to perturbation directly of the map. They arise from three dimensional families of potentials with domain or "behavior space" at least two dimensional hence their full geometry is essentially five-dimensional.

This work was begun when both authors were at the Institute of Mathematics, University of Warwick, Coventry CV4 7AL Warwickshire, England.

INTRODUCTION:

The only structurally stable singularities of maps from surfaces to surfaces are of Fold-or Cusp-type, and these are both map-stable and occur as catastrophes (1). The divergence of Catastrophe theory from stable singularity theory begins when there are three dimensions of control, and at least two of behavior: we have then as possible catastrophes the three cuspoids up to this dimension (Fold, Cusp and Swallow-tail) and the Hyperbolic and Elliptic Umbilics. Since the full geometry of these latter is essentially five-dimensional, it becomes necessary to make use of techniques for systematically incorporating multi-dimensional information into pictures restricted to the plane. A new and more general drawing technique becomes necessary here: instead of the families of potentials on R^1 of (1), we are concerned with two families of potentials on $X = R^2$, each family parameterized by $C = R^3$.

We have:

The Elliptic Umbilic:

$$V(u,v,w)\ (x,y) = x^3 - 3xy^2 + w(x^2 + y^2) + ux + vy.$$

The Hyperbolic Umbilic:

$$V(u,v,w)\ (x,y) = x^3 + y^3 + wxy + ux + vy$$

Here: $\nabla_X V = \left(\frac{\partial V}{\partial x}, \frac{\partial V}{\partial y}\right)$

so that M becomes:

$$\{(u,v,w,x,y) \mid \frac{\partial V}{\partial x}(u,v,w,x,y) = 0 = \frac{\partial V}{\partial y}(u,v,w,x,y)\}$$

(a critical point of a potential in R^2 must be critical in both x and y directions). M is now a 3-submanifold of R^5; the two equations:

$$\frac{\partial V}{\partial x}(u,v,w,x,y) = 0, \quad \frac{\partial V}{\partial y}(u,v,w,x,y) = 0.$$

are linear in a,b and c, and together with $x = x_o$, $y = y_o$ for any choice of x_o, y_o define a line in R^5. M is thus ruled in codimension two; instead of being the union of a one-parameter family of planes, as is the three-dimensional catastrophe manifold of the Butterfly, it consists of a two-parameter family of lines. The bifurcation set remains the envelope of the projections of these lines in C, but since a computer cannot easily draw lines on three-dimensional space, the simple approach of (1) is no longer adequate. If we now take a plane slice P of C, $\pi^{-1}(P)$ is a four-dimensional hyperplane, generically meeting each of the straight lines contained in M in a single point: the projection of a selection of these points as isolated dots is again just a set of dots, with no visible envelope. Hence we must find a way of returning to the drawing of lines: but now they can no longer be straight.

For each plane P in C, $\pi^{-1}(P)$ is a 4-submanifold of R^5, and therefore generically meets the 3-manifold M in a surface. Hence $\chi|M\ \pi^{-1}(P)$ is a map between 2-manifolds. Now if we take any map $f:A \to B$ between 2-manifolds, choose a chart U on A and local coordinates (p,q) on U, and consider the images l_k under $f|U$ of the lines $q = k$, $k\varepsilon R$, it is clear that the images of any fold lines of $f|U$ must appear as the envelope Γ of the l_k (and those of any higher singularities as singularities of Γ), Fig. 1a, except in the non-generic cases where the lines $q = k$ are parallel to the fold, Fig. 1b, or perpendicular to it, Fig. 1c. Such non-generic cases can easily be avoided by changing the chosen coordinates (p,q).

Now, in the case of the umbilics, we have seen that for each plane $P \subseteq C$ the hyperplane $\pi^{-1}(P)$ meets M in generically one point (x,y) for each choice of (x,y). Where this is true for all (x,y) the dependence of p(x,y) is clearly smooth, so that x and y serve in fact as global coordinates on $M \cap \pi^{-1}(P)$, and we may apply the remarks above directly, by setting the computer to take successive constants k for y and for each one draw the curve of solutions to the linear equations on P defined by restriction of:

$$\frac{\partial V}{\partial x}(u,v,w,x,k) = 0, \frac{\partial V}{\partial y}(u,v,w,x,k) = 0. \qquad (A).$$

as x varies between prescribed limits.

If some choices of (x,y) give a line parallel to $\pi^{-1}(P)$, so that no solution exists (e.g. for the Elliptic Umbilic, if $P\{(u,v,w) | c = 1\}$ and $y = 0$), $\pi^{-1}(P) \cap M$ has more than one component, each "going to infinity" as these choices are approached, so that a complete picture over the finite region we are concerned with can be drawn simply by avoiding the offending values of (x,y).

The more drastic degeneracy of a line in M actually lying in $\pi^{-1}(P)$, giving multiple solutions to (A), corresponds to a surgery of $\chi^{-1}(P)$ as P is varied, by parallel plane, through the critical plane P_o for which degeneracy occurs. (This is exactly analogous to the surgery of the inverse image $\chi^{-1}(l_k)$ of the lines:

$$l_k = \{(a,b) | b = k\}$$

in the control space of the Cusp Catastrophe, as k varies through 0 (see Fig. 2)). Special techniques would be necessary to deal completely with this case: specifically, a combination of the methods used here with those of (1). However, notice that the development of $\chi^{-1}(l_k)$ in Fig. 2 would be clear from a sequence of values of k straddling 0 closely but not actually touching it (particularly if animated) and a similar device could be employed here. (A point of singularity, being non-generic, is of interest less for itself than for the geometry that it imposes on its surroundings: we shall return to this point in (2).). This degeneracy has not occurred in the sections that we have drawn here.

Notice that the three-dimensional appearance of the computer pictures is, as in the previous paper, highly meaningful, but in this case what is visible is a projection of a surface embedded in four dimensions down to three. Hence the apparent self-intersections, e.g. as the folds in the Hyperbolic Umbilic "pass through each other" at $w = 0$: if it is borne in mind that each curve corresponds to a distinct value of y from the rest, and each point on it to a distinct value of x, the pictures can be mentally "lifted" into four-dimensional space, and the self-intersections seen to be spurious.

Perspective drawings of the bifurcation sets in three dimensions are shown in Figs. 3a, 4a. The boundaries of the regions where minima exist (both the Elliptic and Hyperbolic Umbilics, as can be seen are self-dual) are shown in Figs. 3b, 4b, minima being usually the physically significant states in applications of catastrophe theory. (Not invariably, however; in, e.g., optics all stationary states are of importance.) Fig. 3b is identified by Thom with the hair (3,4), Fig. 4b. with the structure of the breaking wave (3,4,5); it is not yet clear whether such identification can be strengthened to constitute a fruitful tool in the theory of these phenomena.

The existence of regions of C which correspond to no minima can be an obstacle in application: this difficulty can be avoided by the addition of a "compactifying" term to the potentials, which does not alter the type of the singularity in the region extremely close to the origin but ensures that $V_c^{-1}\{x\varepsilon R | x \leq k\}$ is compact for all $c\varepsilon$ C, $k\varepsilon R$, so guaranteeing at least one minimum for each V_c. The compactified equations are:

<u>Elliptic Umbilic:</u>

either:

$$V(u,v,w)\ (x,y) = x^4 + y^4 + x^3 - 3xy^2 + w(x^2 + y^2) + ux + vy$$

or:

$$V(u,v,w)\ (x,y) = \frac{x^4}{4} + \frac{y^4}{4} + x^3 - 3xy^2 + w(x^2 - y^2) + ux + vy.$$

<u>Hyperbolic Umbilic:</u>

either:

$$V(u,v,w)\ (x,y) = x^4 + y^4 + x^3 + y^3 + wxy + ux + vy.$$

or:

$$V(u,v,w)\ (x,y) = \frac{x^4}{4} + \frac{y^4}{4} + x^3 + y^3 + wxy + ux + vy.$$

These alternatives are not importantly different from the differential point of view (close enough to the origin, they are equivalent to each other and to the uncompactified versions), but the effect on the geometry of a large or small compactifying term is of morphological interest.

Bibliography:

(1). Woodcock, A. E. R. & Poston, T., The Geometrical Properties of the Cuspoids (this volume).

(2). Woodcock, A. E. R. & Poston, T., The Geometrical Properties of the Parabolic Umbilic (this volume).

(3). Thom, R., Topology, 8 p.313f. 1969.

(4). Thom, R., Stabilité, Structurelle et Morphogénèse, published by: Benjamin-Addison Wesley, 1972.

(5). Zeeman, E. C., ln Dynamical Systems, University of Warwick Symposium Volume, published by: Springer-Verlag, 1970.

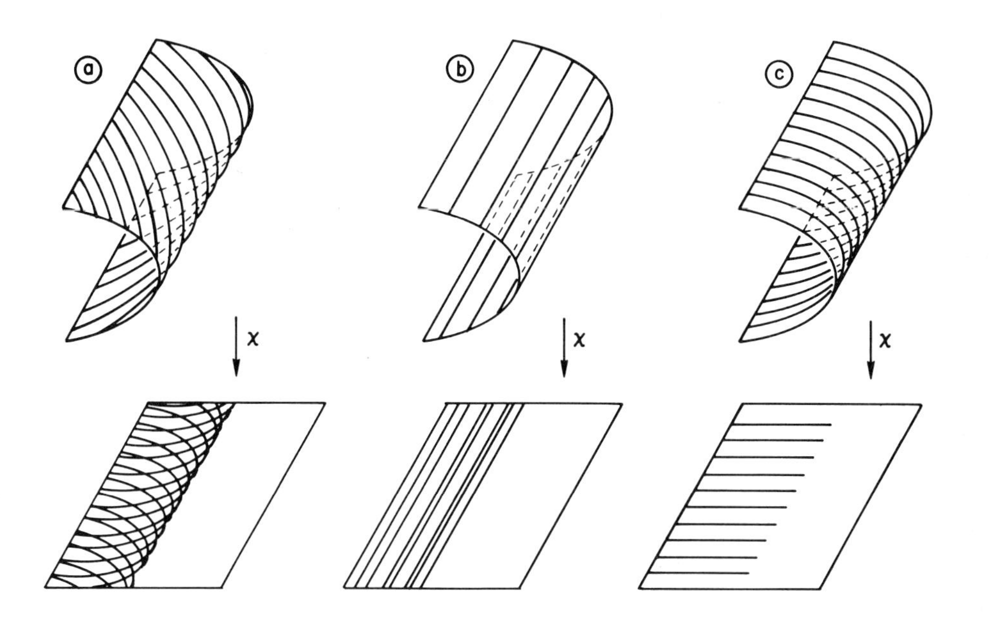

Fig. 1

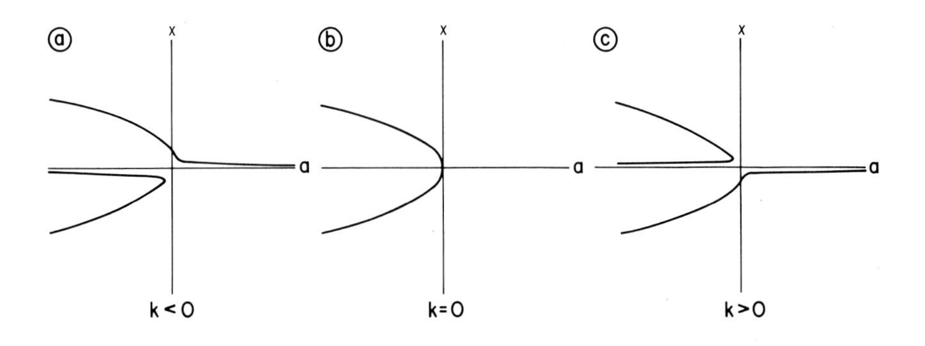

Fig. 2

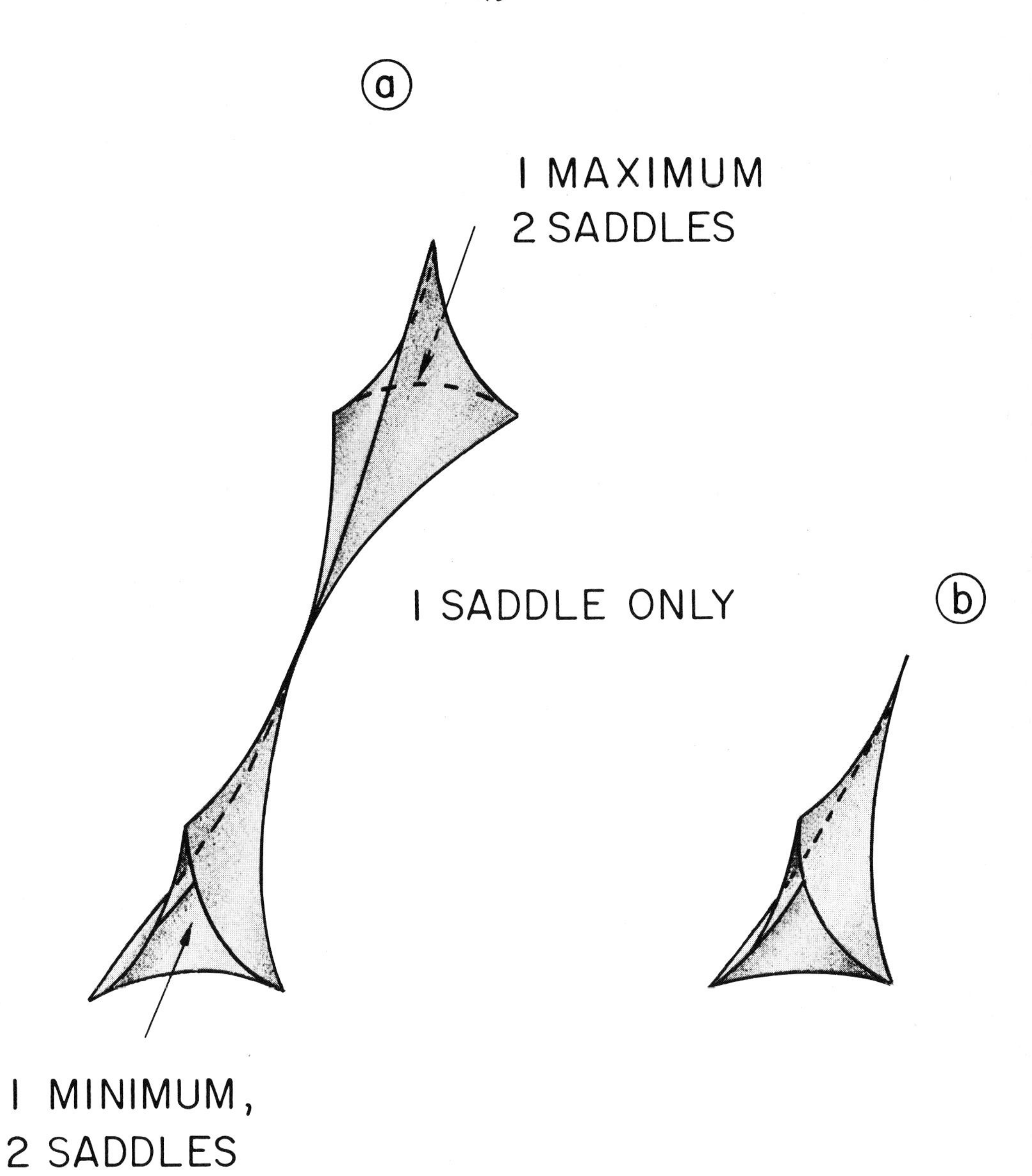

Fig. 3

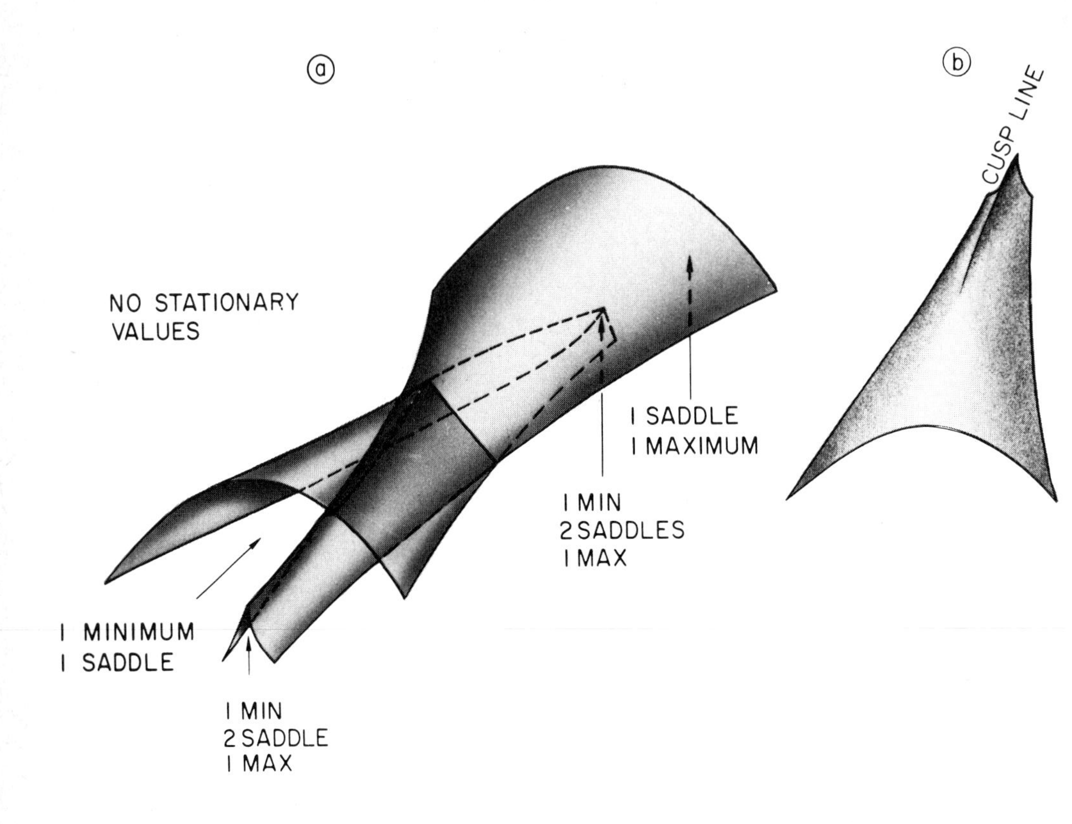
(a)
(b)
CUSP LINE
NO STATIONARY VALUES
I SADDLE
I MAXIMUM
I MIN
2 SADDLES
I MAX
I MINIMUM
I SADDLE
I MIN
2 SADDLE
I MAX

Fig. 4

<u>The Elliptic Umbilic:</u> $V=x^3 - 3xy^2 + w(x^2 + y^2) + ux + vy$.

Figs. 5A and 5B clearly show that for large magnitudes of the parameter w, the Bifurcation Set (formed as the envelope of the lines for which $\frac{\partial V}{\partial x} = \frac{\partial V}{\partial y} = 0$ and by projection onto the (u,v) plane) is typically an equilateral triangle with fluted sides. As w is decreased, the size of the triangular region diminishes until, at w=0, it dissappears, to grow again symmetrically as the sign of w is reversed.

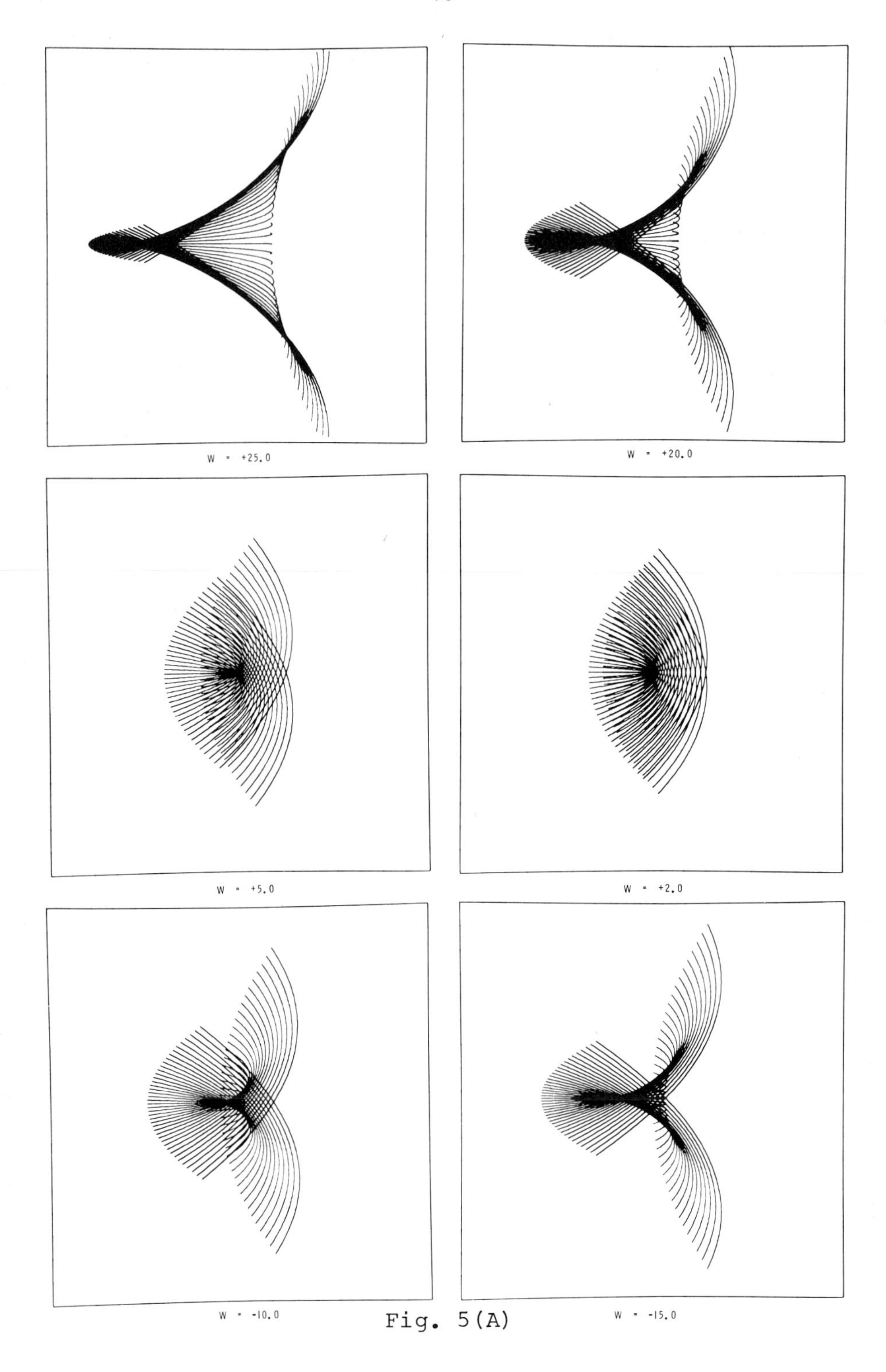
W = +25.0
W = +20.0
W = +5.0
W = +2.0
W = -10.0
W = -15.0

Fig. 5(A)

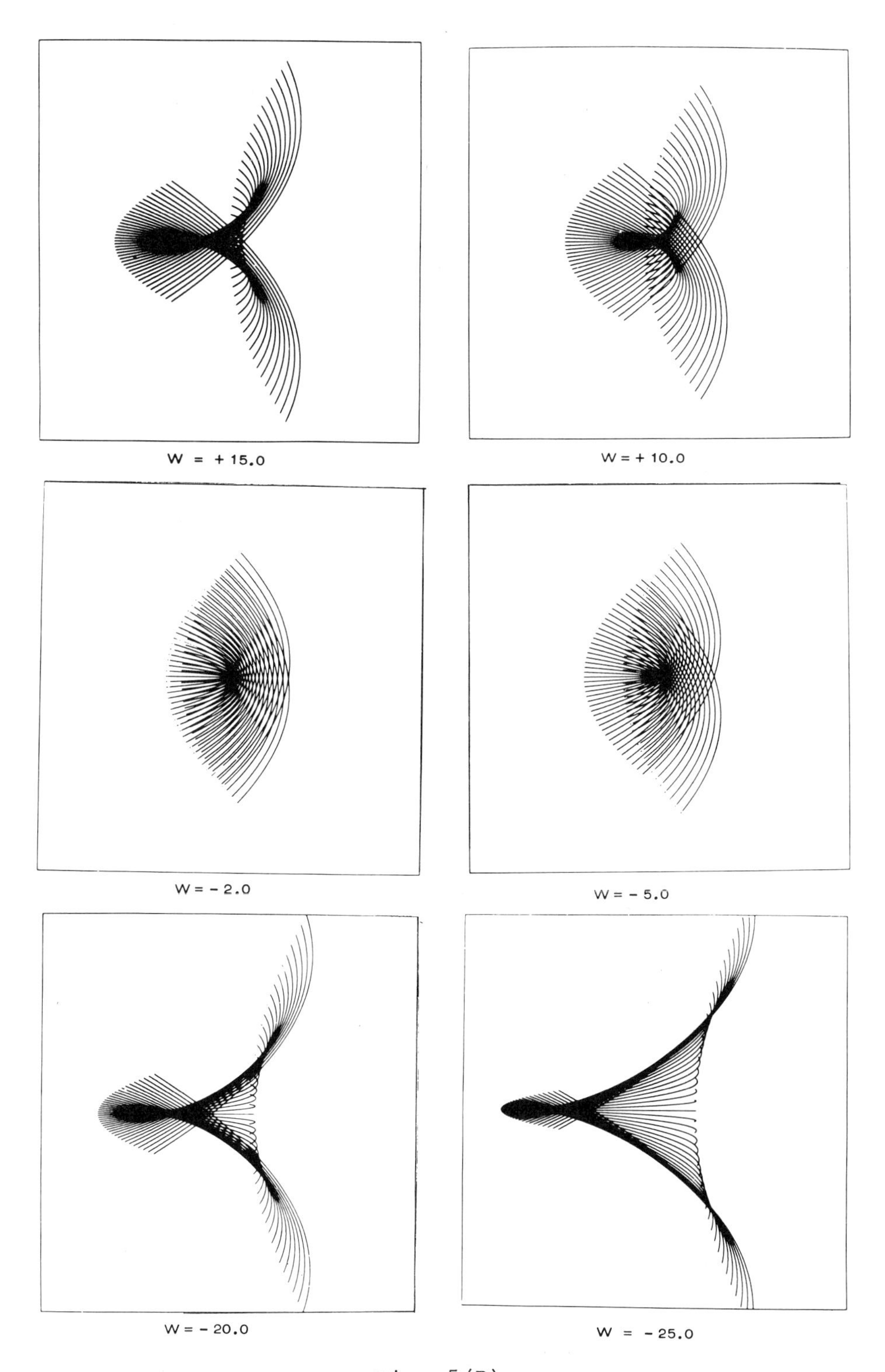
W = + 15.0
W = + 10.0
W = - 2.0
W = - 5.0
W = - 20.0
W = - 25.0

Fig. 5(B)

Compactified Versions of the Elliptic Umbilic

The Elliptic Umbilic equation may be modified so that the relevant potential function has a lower minimum value; this operation is termed a "Compactification". Examples of such a compactified Umbilic are shown in Figs. (6A and 6B) and (7A and 7B). Fig. (6A and 6B) represent projections onto the (u,v) plane of the Strongly Compactified Elliptic Umbilic, generated from a potential function of the form : $V = x^4 + y^4 + x^3 - 3xy^2 + w(x^2 + y^2) + ux + vy$. For positive values of w, the surface is flat; however, as w is reduced and made negative, the surface rolls over itself in a very complex way (see for instance w = -30.0).

This complex imagination of the surface is much more dramatic in the case of the Weekly Compactified Elliptic Umbilic, Fig. (7A and 7B) derived from a potential function of the form: $V = \frac{x^4}{4} + \frac{y^4}{4} + x^3 - 3xy^2 + w(x^2+y^2) + ux + vy$. Again, for positive values of w, the surface is relatively uncomplicated, even flat for sufficiently large positive values of w. As w becomes large and negative, the 'arms' (which first appear at w = - 5.0) become more pronounced and the central region (which at w = - 10.0, at -20.0 appears to contain an Elliptic Umbilic-like region) invaginates to form a double-doughnut-like figure (w = - 50.0 for example). (This catastrophe has been termed, (by S.H.W.), the 'Yak' due to the appearance of the figure for w = - 30.0).

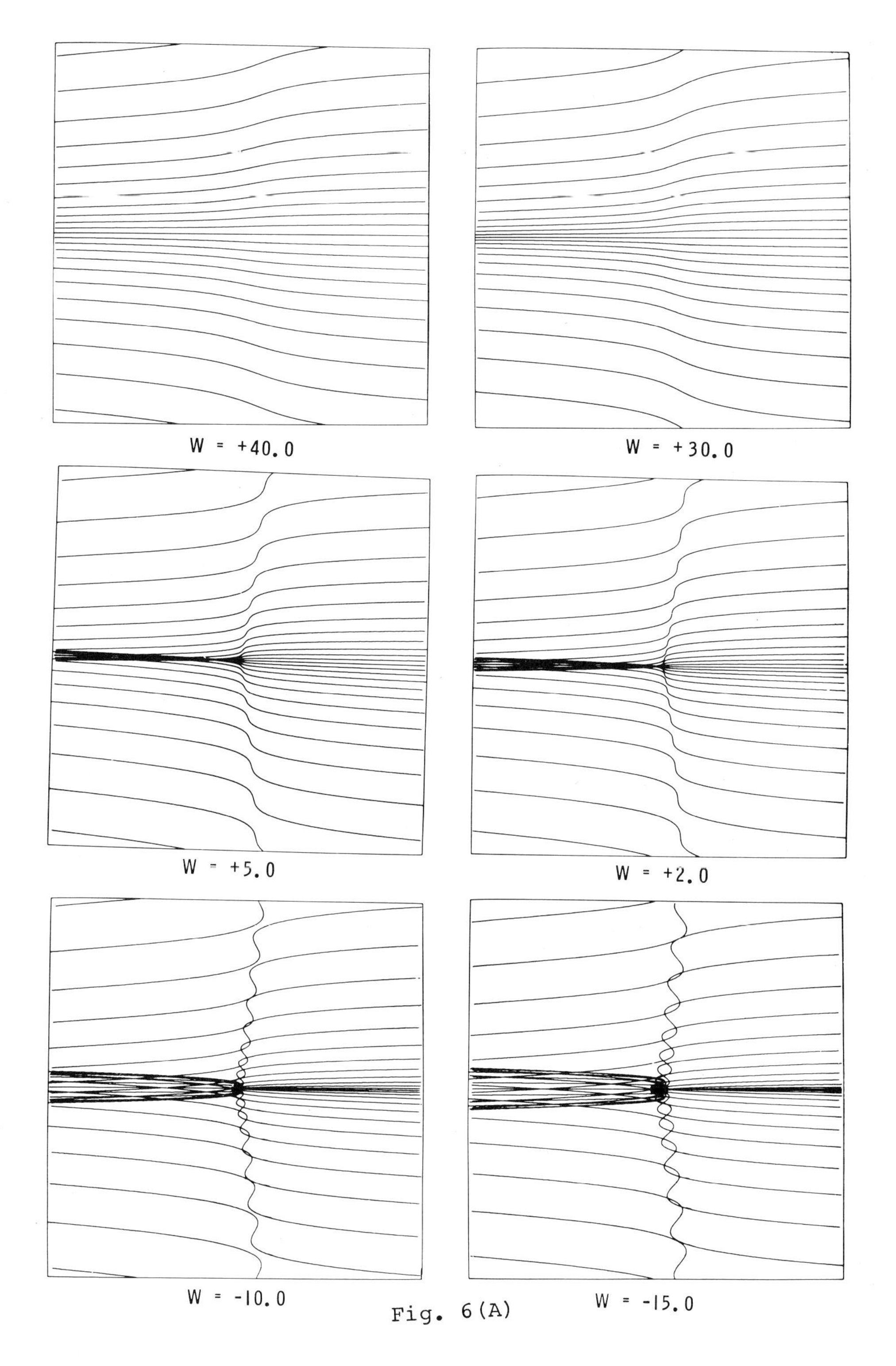

Fig. 6(A)

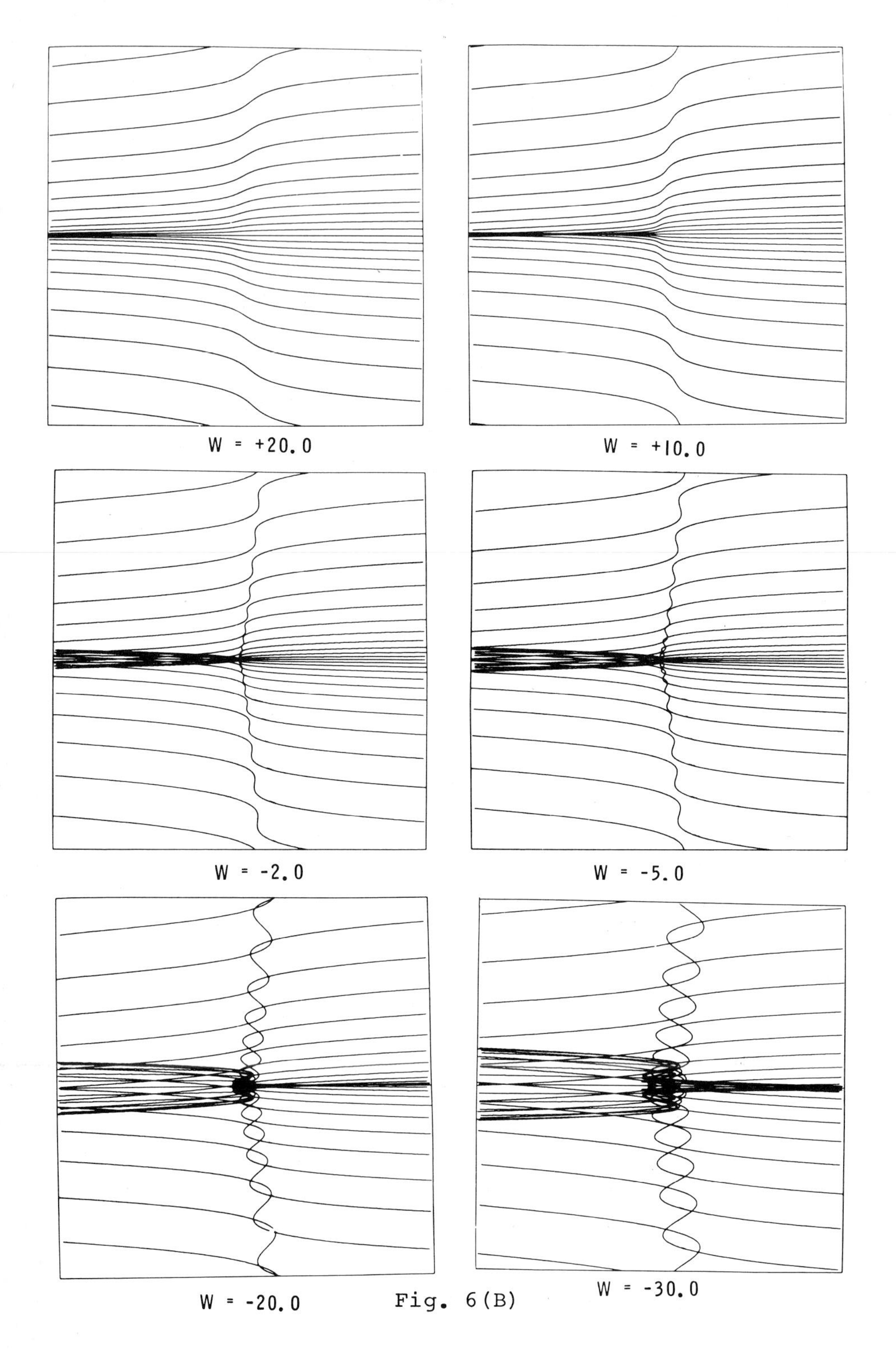
W = +20.0
W = +10.0
W = -2.0
W = -5.0
W = -20.0
W = -30.0

Fig. 6(B)

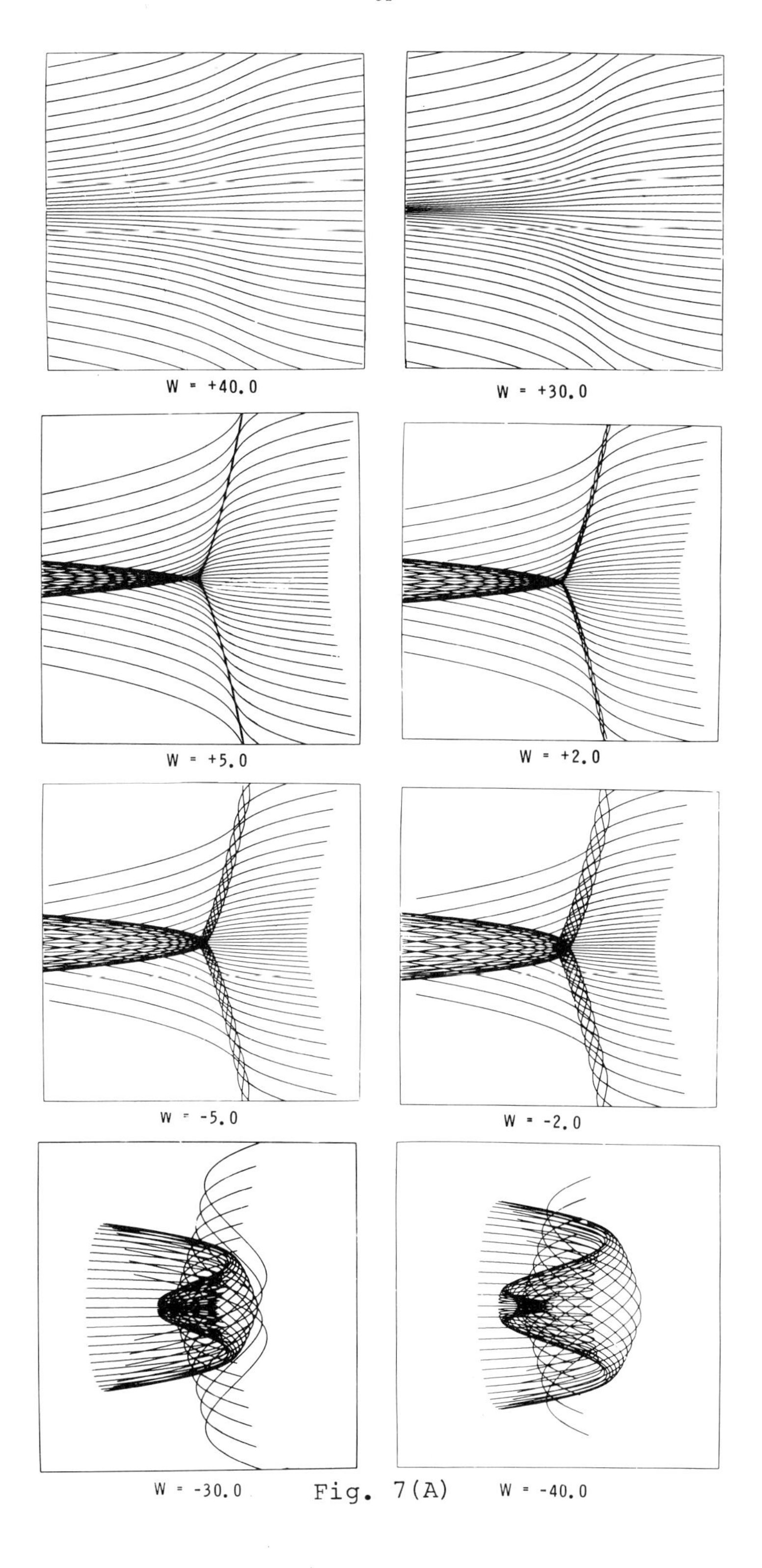
W = +40.0
W = +30.0
W = +5.0
W = +2.0
W = -5.0
W = -2.0
W = -30.0
W = -40.0

Fig. 7(A)

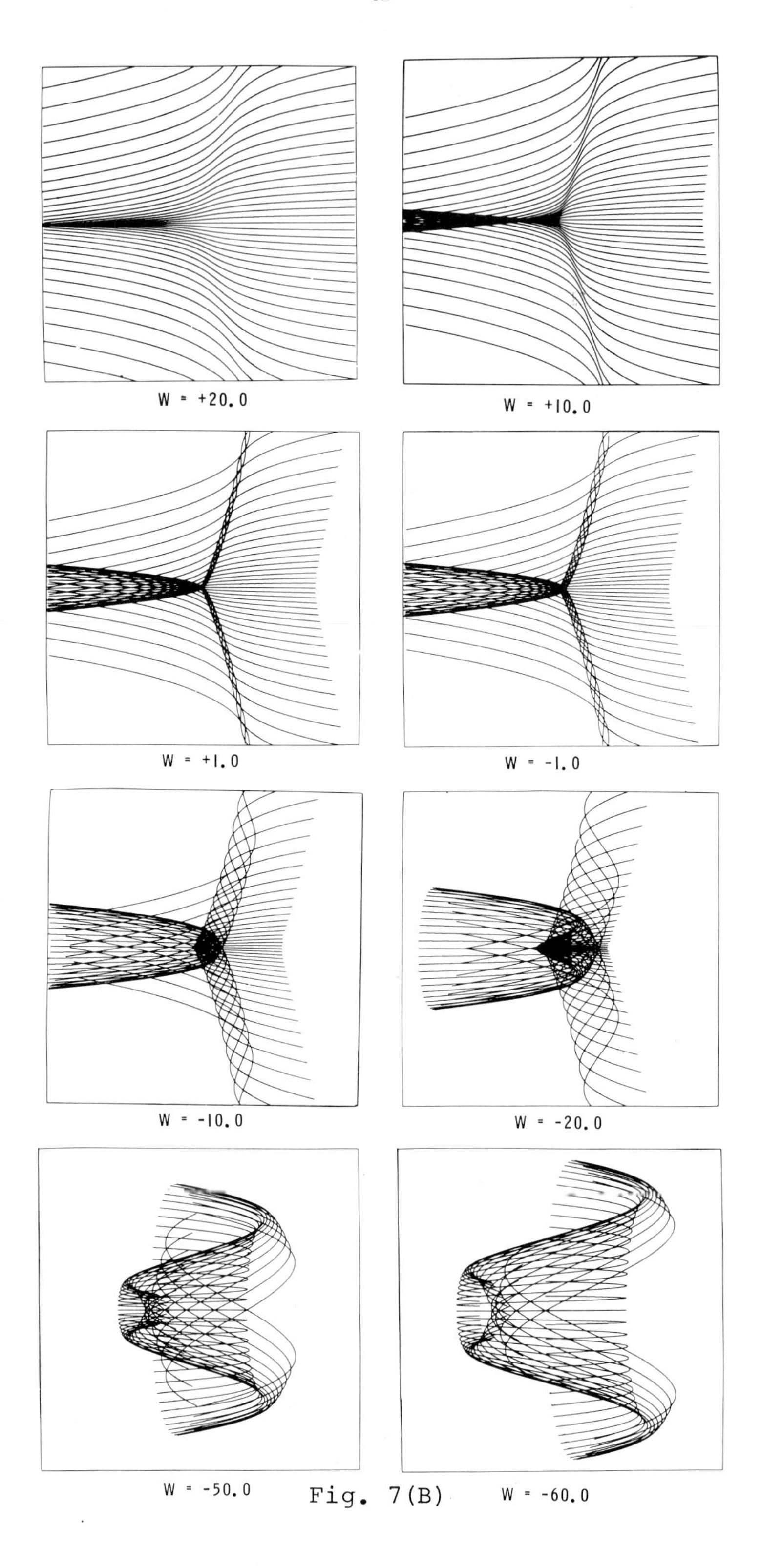
W = +20.0
W = +10.0
W = +1.0
W = -1.0
W = -10.0
W = -20.0
W = -50.0
W = -60.0

Fig. 7(B)

The Hyperbolic Umbilic: $V = x^3 + y^3 + wxy + ux + vy$.

Figs. 8A and 8B show that the projection of the Behavior Space of the Hyperbolic Umbilic onto the (u,v) control plane produces, in general, a Bifurcation Set consisting of two parts: one cusp-like and the other a more rounded curve. The Bifurcation Set has the property that as w is changed from + 40.0, for example, to -40.0 the cusp-like region (labelled A in Fig 8A) broadens and the rounded edge (B) becomes more pointed until, at w=0, they overlap to form a rectangular figure. As w is made increasingly negative, the original region A becomes further rounded while the region B becomes cusp-like; thus changing the sign of w reverses the nature of the regions A and B (Figs. 8A and 8B).

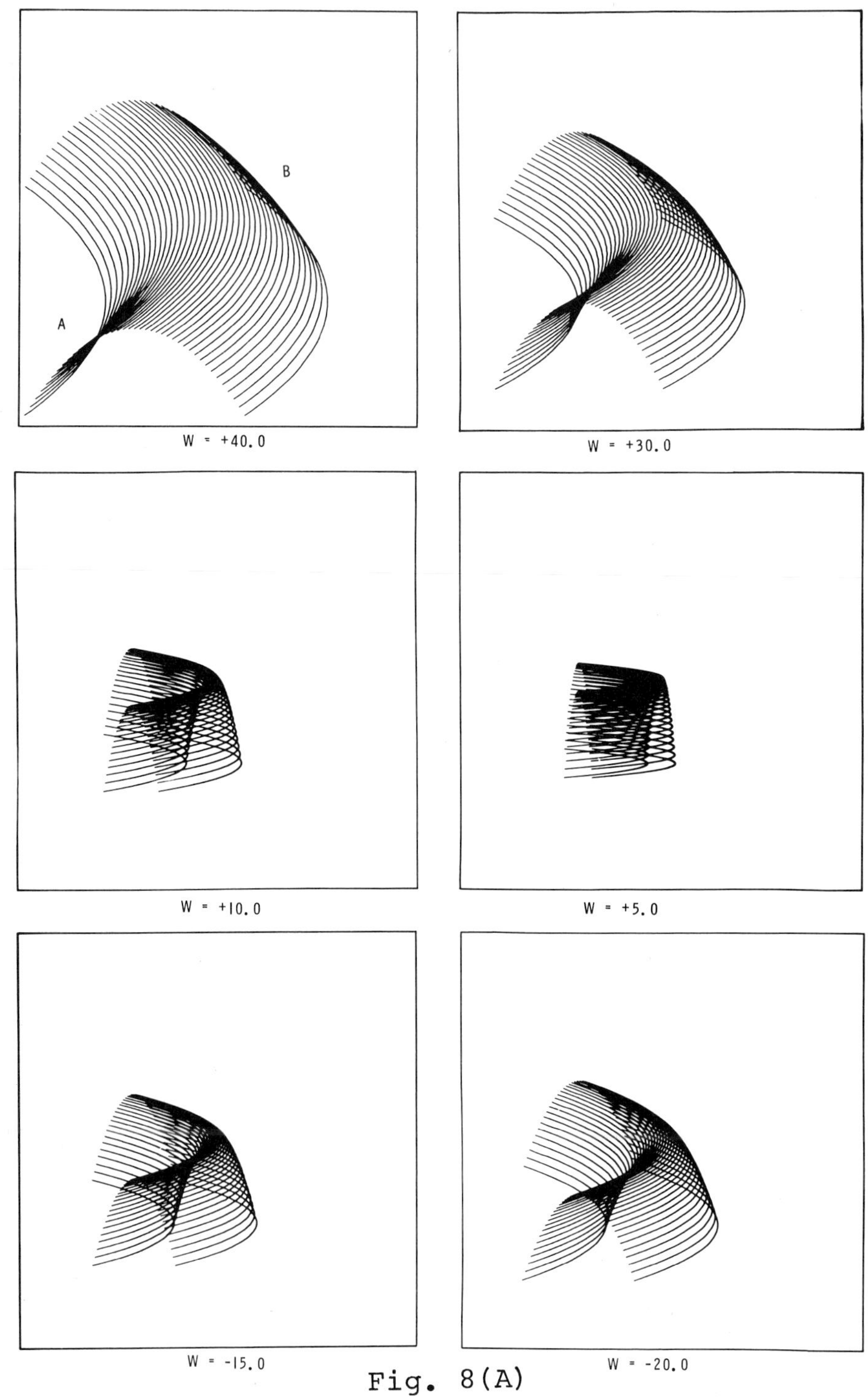

Fig. 8(A)

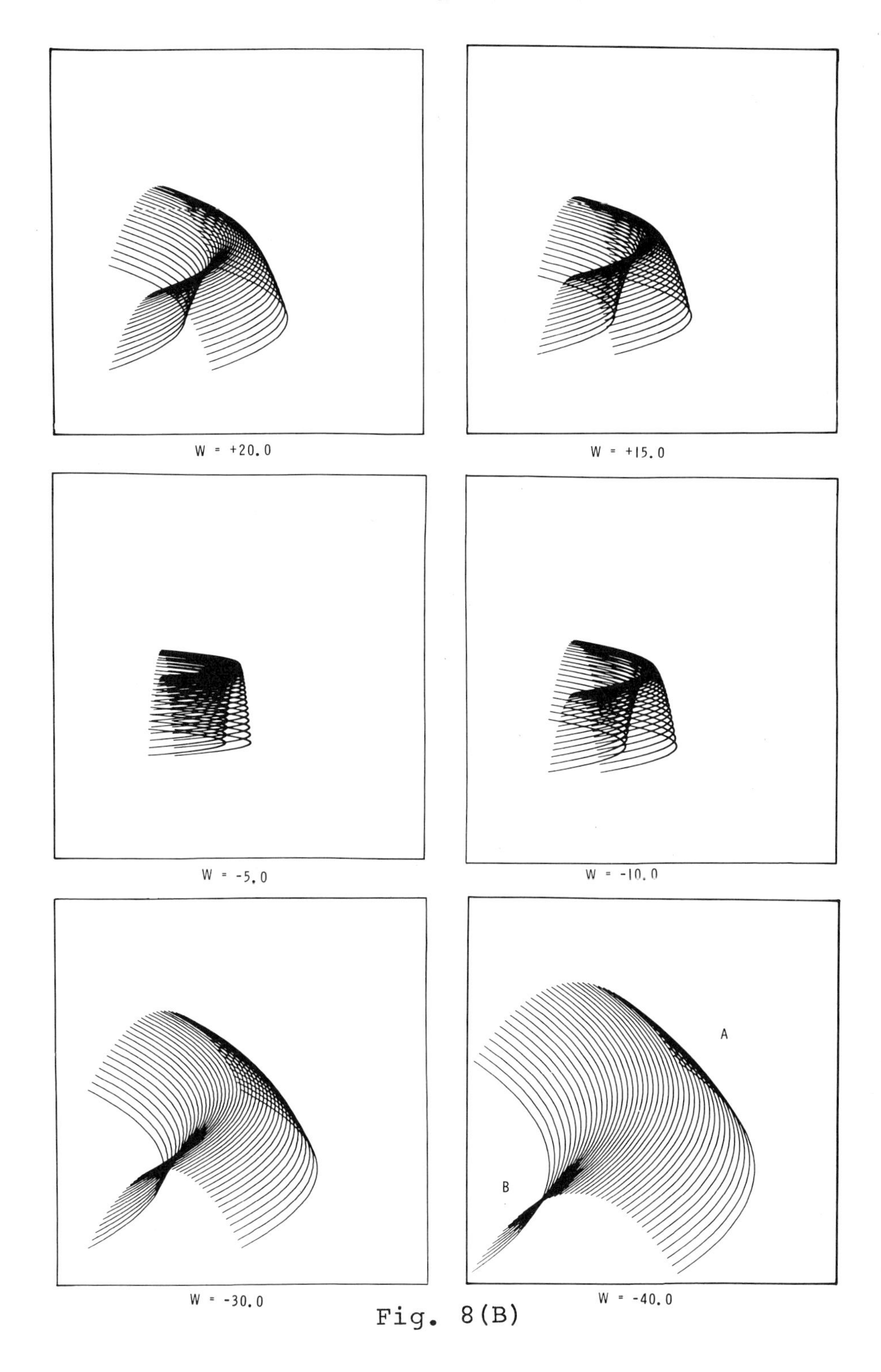

Fig. 8(B)

Compactified Versions of the Hyperbolic Umbilic:

Introduction of a lower minimum value for the Hyperbolic Umbilic potential function may be performed in a manner similar to that used for the Elliptic Umbilic. Fig. 9A and 9B show the projection of the Behavior Space onto the (u,v) Control plane for the Strongly Compactified Hyperbolic Umbilic, derived from a potential function of the form:

$$V = x^4 = y^4 + x^3 + y^3 + wxy + ux + vy.$$

For large positive values of w, the figure is simply that of two cusps pointing towards the origin of the (u,v) plane and joined by common limbs. As w is decreased to zero, the two cusps become broadened and the region between them becomes smaller as the cusps approach the origin. For small values of w, the cusps approach a rectangular outline. Reversal of the sign of w causes the axis of the two-cusp combination to turn through 90 degrees; the morphology of the cusp regions for positive and negative values of w are similar except for this rotation.

Figs. 10A and 10B show the projection for the Weakly Compactified Hyperbolic Umbilic:

$$V = \frac{x^4}{4} + \frac{y^4}{4} + x^3 + y^3 + wxy + ux + vy.$$

These figures are essentially similar to those of the Strongly Compactified case except that the edge of the (x,y) plane becomes observable and the central region is relatively larger in the present case.

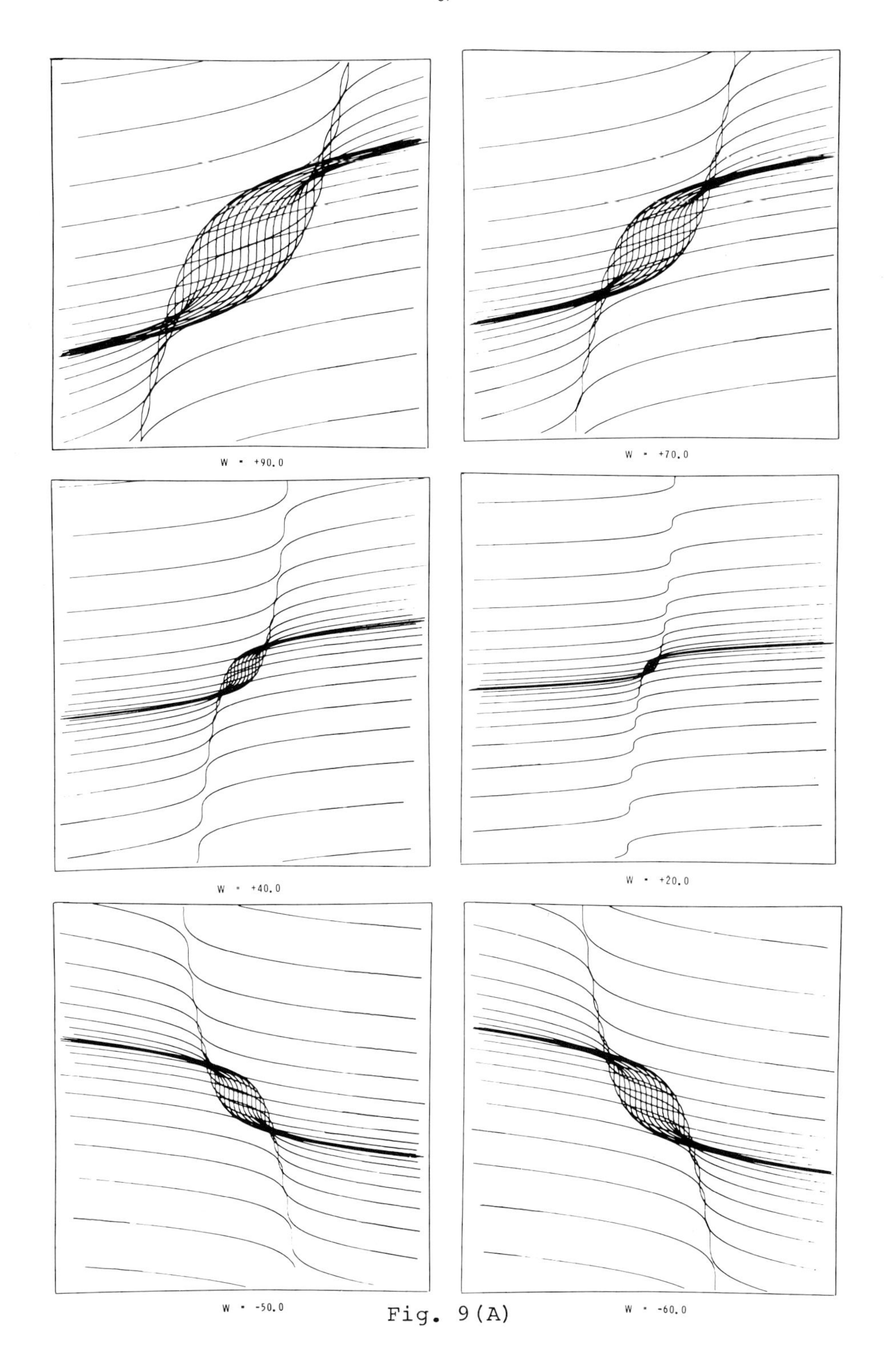

Fig. 9(A)

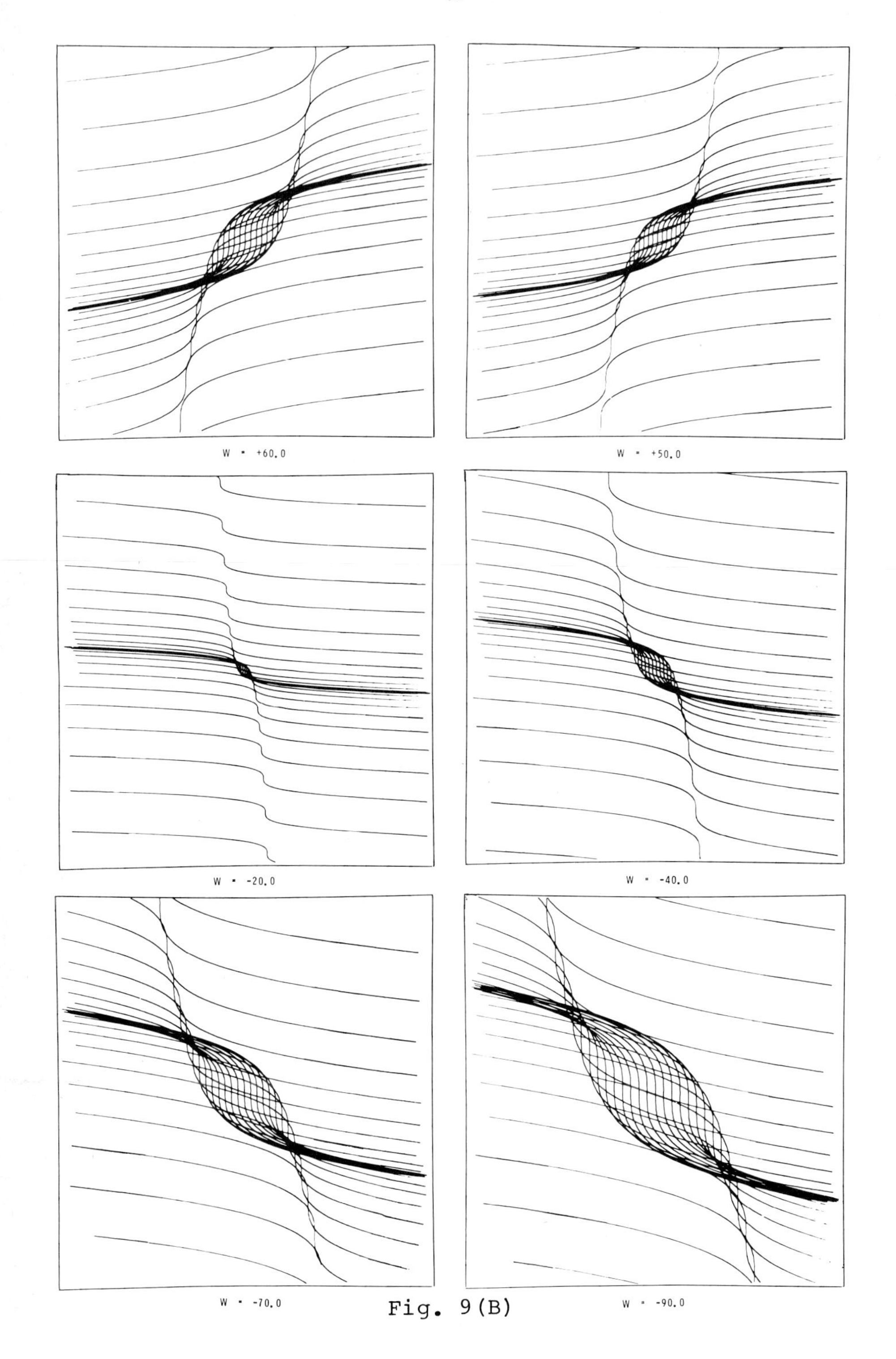

Fig. 9(B)

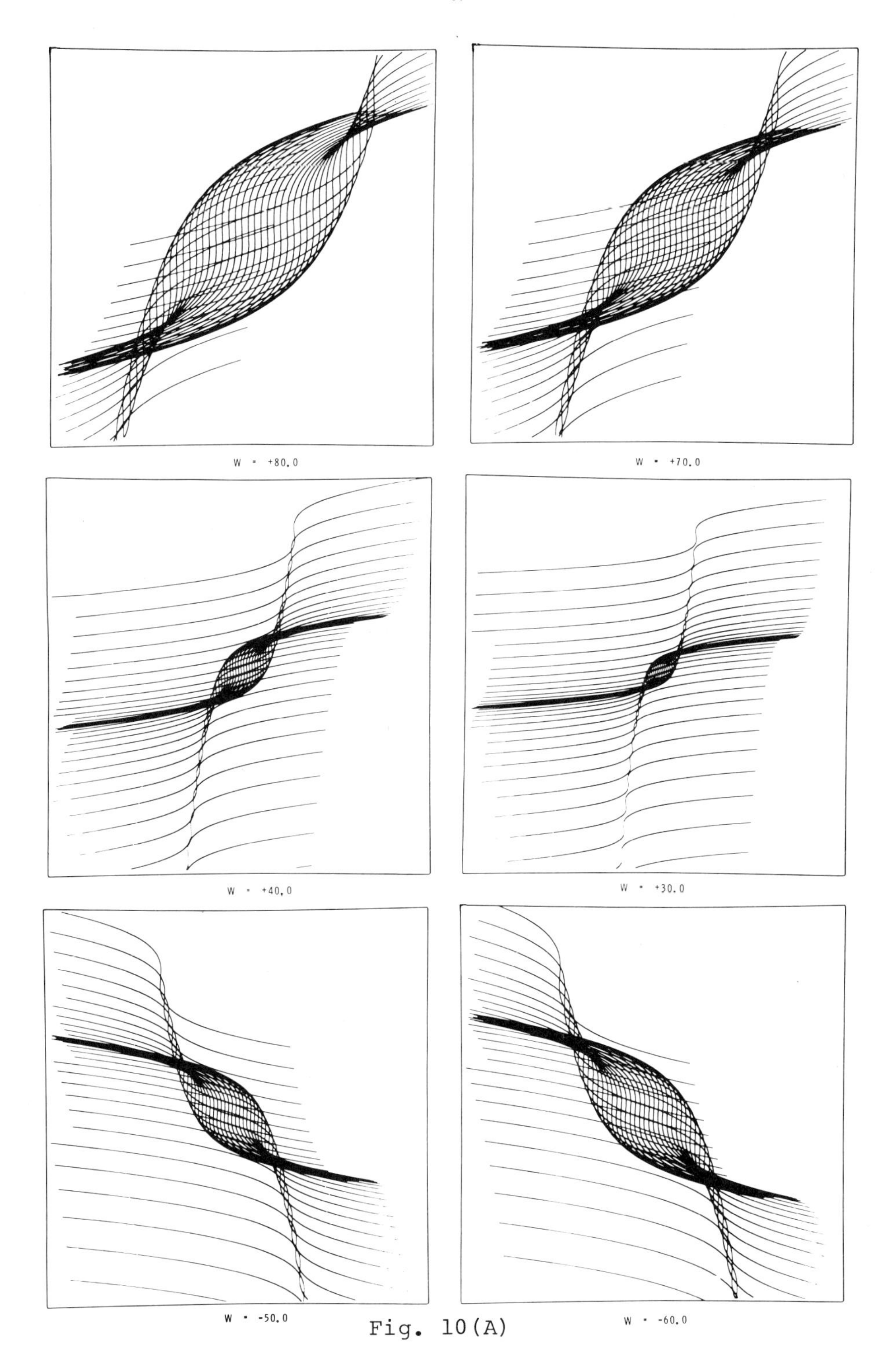

Fig. 10(A)

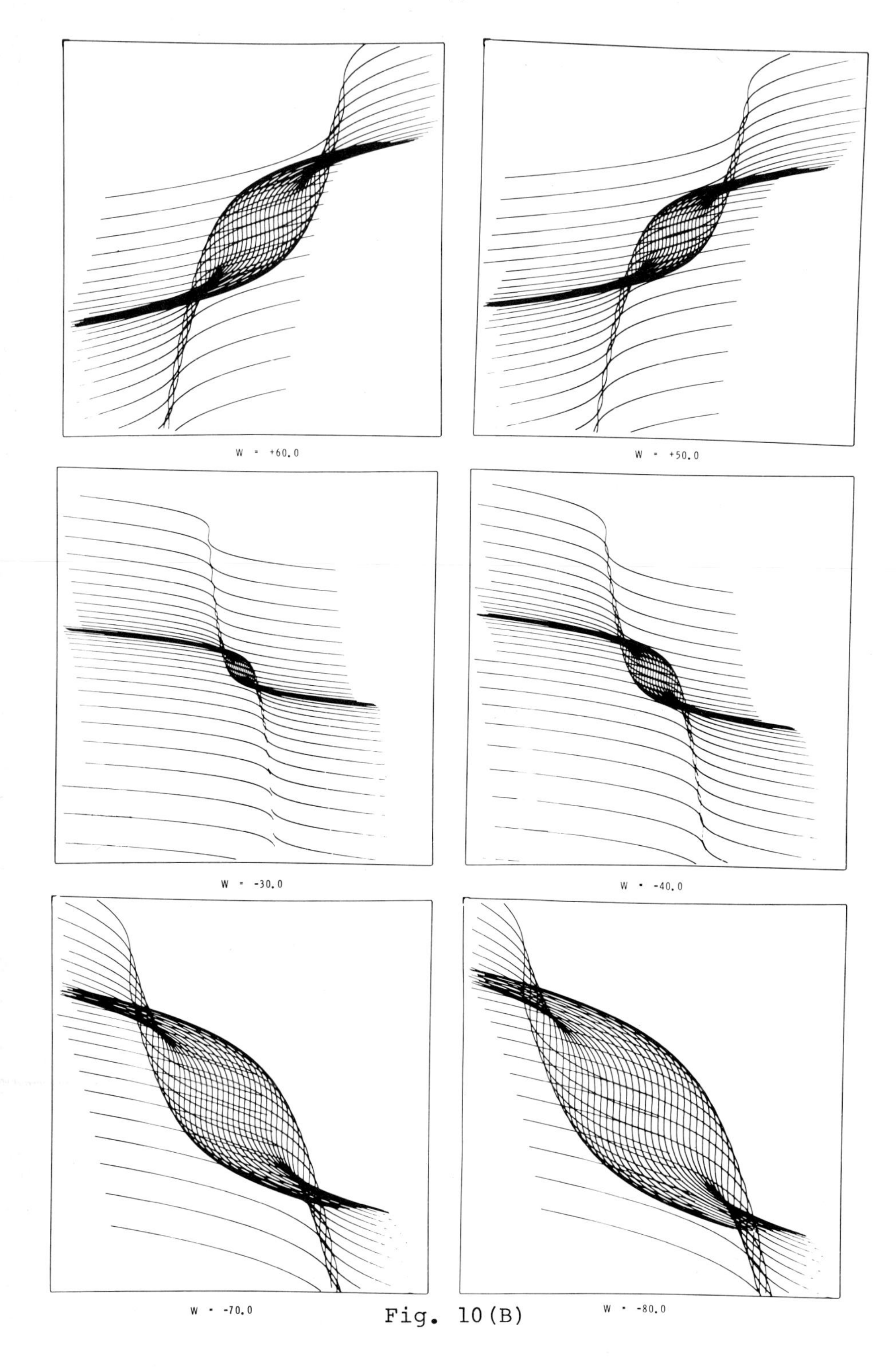
W = +60.0
W = +50.0
W = -30.0
W = -40.0
W = -70.0
W = -80.0

Fig. 10(B)

THE GEOMETRY OF THE ELEMENTARY CATASTROPHES:

(3). The Parabolic Umbilic.

by

A. E. R. Woodcock
IBM Thomas J. Watson Research Center
Yorktown Heights, N.Y. 10598, U.S.A.

and

T. Poston
Instituto de Mathemática Pura e Aplicada
Rio-de-Janeiro, Brazil

ABSTRACT: The Behavior Space of the Parabolic Umbilic is two dimensional and the full geometry of this Umbilic Catastrophe is essentially six-dimensional. This paper describes the geometry of both the non-compact and compact versions of the Parabolic Umbilic.

This work was begun when both authors were at the Institute of Mathematics, University of Warwick, Coventry CV4 7AL, Warwickshire, England.

Some Sections of the Noncompact version of the Parabolic Umbilic:

$V = x^2y + y^4 + ty^2 + wx^2 + ux + vy$

Projected onto the (u,v) plane.

Godwin (1) has investigated the nature of the Bifurcation Set of the Parabolic Umbilic. Figs. 1,2,3 and 4 are drawn with the same values of t and w as chosen by Godwin, and thus permit a direct comparison between the results obtained by the different methods. (The only, superficial, difference being that Godwins' Equation for V was written with (-ux) and (-vy); this results only in a rotation of the pictures through 180°). As mentioned in our earlier papers (1,2) use of our plotting technique gives rise to not only a picture of the Bifurcation Set but also an insight into the nature of the hyper-surface in (Control Space) x (Behavior Space) giving rise to the Bifurcation Set. Thus, we produce essentially the same outlines for the different Bifurcation Sets and also retain information about the surface from which it arises; this information is lost in Godwin's approach to the problem. It is instructive to make a comparison between our Fig. 1 and Fig. 5 of Godwin (1); Fig. 2 with Figs. 6 and 8 (Godwin); Fig. 3 with Fig. 9 (Godwin-This figure in Godwin has an incorrect value for w) and Fig. 4 with Fig. 11 (Godwin).

An instructive sequence of sections can be constructed in which the magnitudes of t and w are related. This has been done in Fig. 5 (A and B) where t and w are given by: -

$$t^2 + w^2 = R^2 = (0.5)^2.$$

In this case, the pictures complete a cycle and demonstrate rather well the types of transitions necessary to achieve any particular morphology.

The cycle, taken in the direction (0.46, -0.19) → (0.35, -0.35) → ... has been suggested by Thom to represent the growth and reproduction cycle of living organisms.

It is appropriate here to mention another aspect of the geometry of the Elementary Catastrophes, which can aid both in their description - since it permits a dimensional reduction - and in considering their possible applications; in each case the Bifurcation Set is topologically a cone on its intersection with a sphere in the Control Space, centered on the point of singularity. This means that the phenomenology due to a particular Catastrophe is effectively described by this intersection: for example (Fig. 6A) the behavior of a system governed by the Cusp Catastrophe does not depend significantly on the fact that the two curves of the Bifurcation Set meet tangentially. The latter is a differential condition in the immediate locality of the singularity itself, and a changing control will, generically, never pass through this point. Its importance lies in the geometry it imposes on its surroundings: the way, for example, that a path in the control space that goes around the cusp point will produce a steady change/jump back/steady change/jump back cycle behavior. This is perfectly captured by the intersection of the diagram with the unit circle in C and its preimage by χ. The strict cone, as opposed to the topological one, is illustrated in Fig. 6B (with rays cut off at the points fixing them, for pictorial convenience); the reader should convince himself that, for each of the applications of the Cusp Catastrophe he knows, Fig. 6B gives as good a fit to behavior as Fig. 6A. They differ in fact only by the (experimentally untestable) condition that χ be differentiable. (N.B. This property of conicality does _not_ apply globally to 'compactified' versions of the Catastrophe.)

The phenomenology of each Catastrophe, then, is given by the intersection of its Bifurcation Set with a sphere: this is illustrated for the Swallowtail, Elliptic Umbilic and Hyperbolic Umbilic in Fig. 7, and for the Parabolic Umbilic in Fig. 8, using the usual presentation of S^3 minus the N. Pole as R^3.

Note that in each of these pictures the Catastrophe itself does not and cannot occur, because the sphere as Control Space is of one too low a dimension; in each, however, all the Catastrophes of lower order than the one described occur (Cuspoids, however, only have Cuspoids as sections: thus the similar picture for the Butterfly does not involve the,dimensionally allowable, Elliptic and Hyperbolic Umbilics.) This is the way that a Catastrophe of higher dimension than permitted by the control space can act, not as a local structure at some one point in the control space but as an 'organising centre' controlling the relationship of several lower order catastrophes that do occur as point structures. The latter control the geometry of the situation in their immediately locality, while the organising center - perhaps inaccessible in the normal working of the system - controls their relationship. Thus for example a biological system governed by the cusp [4] may exhibit behavior which in healthy conditions involves only a plane or cylindrical section of the full diagram, as in Fig. 6A, and thus directly involves only the Fold Catastrophe. The Cusp Catastrophe remains central to its explanation.

This phenomenon of the relationship of Catastrophes at different points being governed by a higher catastrophe which does not appear directly gives importance to the study of characteristic or special sections of such higher catastrophes, even in applications where dimensional considerations would appear to rule them out. Hence the interest in, for instance, the Reduced Double Cusp [2], which is precisely such a section (the Double Cusp itself being 7 or 8 dimensional, the Reduced version a 5-dimensional slice).

The concept of 'organizing centre' is due to Thom [3].

Figs. 9(A, B, C, D, E, F, G and H) constitute a matrix of pictures of the noncompact Parabolic Umbilic with the values of t and w varying in steps of 0.1.

Figs. 9A and E with t positive and w either positive, negative or zero show a Cusp-like Bifurcation Set. As t is reduced to zero, to Cusp becomes either more rounded (Fig. 9B) for w positive or appears to penetrate part of the (x,y) hypersurface. (Fig. 9F). as t is further reduced, the Cusp "passes through" the (x,y) surface and forms an Elliptic Umbilic (Figs. 9C, D) with w-positive. The most interesting region occurs when t and w are both negative. In this case it is possible to go from a Cusp (Fig. 9E) to an Elliptic Umbilic-like region through a Hyperbolic Umbilic-like transition (Fig. 9, G and H). In Fig. 9H, t = 0.5 sequence of pictures in which the value of w runs from -0.4 to -0.1 clearly shows this transition. In t = -0.5, w = -0.4, the cusp is sitting inside a triangular region, at w = -0.3, the cusp has broadened and at w = -0.2 has "passed through" (or more correctly, has appeared to pass through) the boundary of the triangular region and now lies outside (t = -0.2, w= -0.3); the internal triangular region for t= -0.4, w= -0.1, has a profile characteristic of the Elliptic Umbilic. (See Woodcock and Poston (2)).

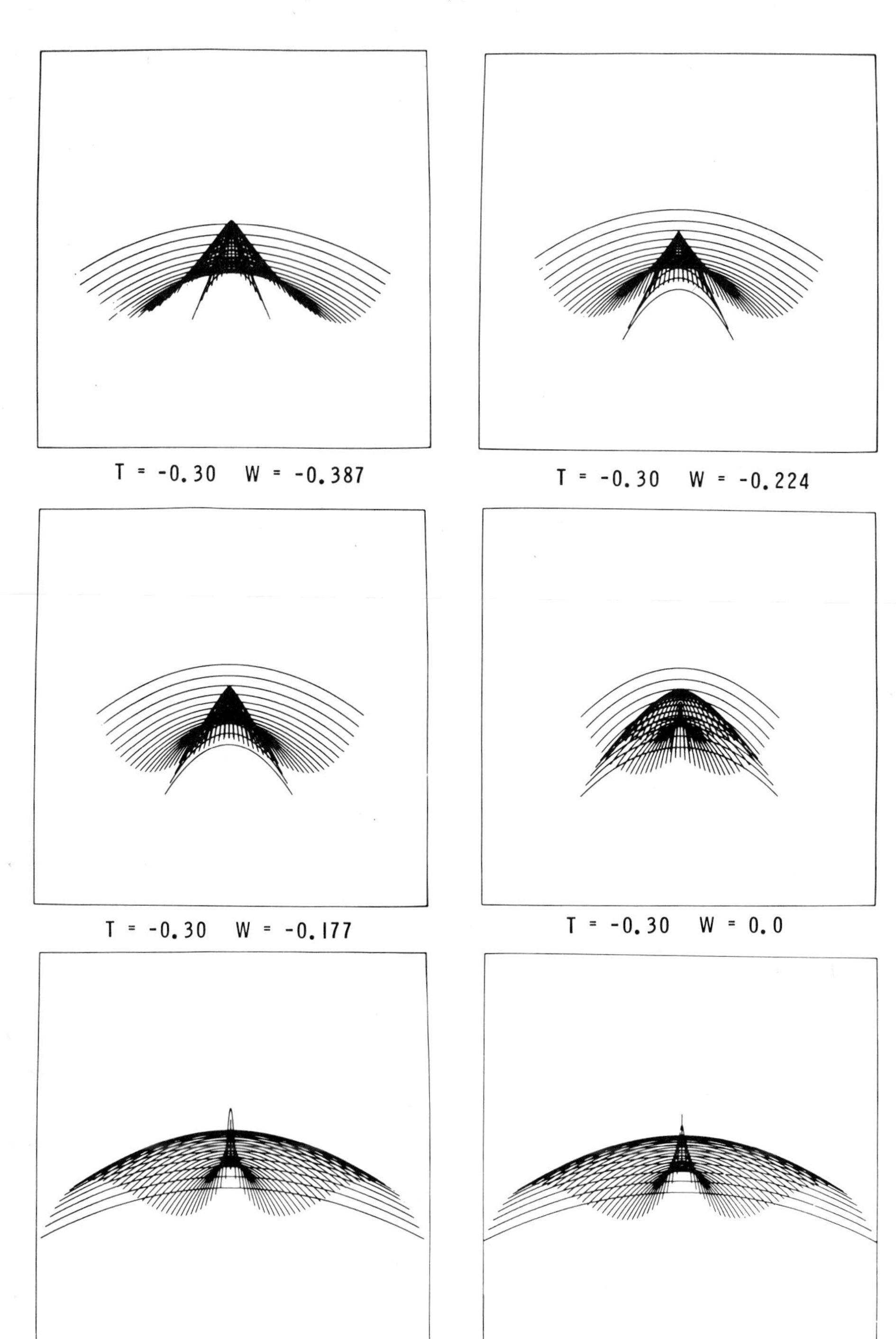
T = -0.30 W = -0.387
T = -0.30 W = -0.224
T = -0.30 W = -0.177
T = -0.30 W = 0.0
T = -0.30 W = +0.477
T = -0.30 W = +0.50

Fig. 1

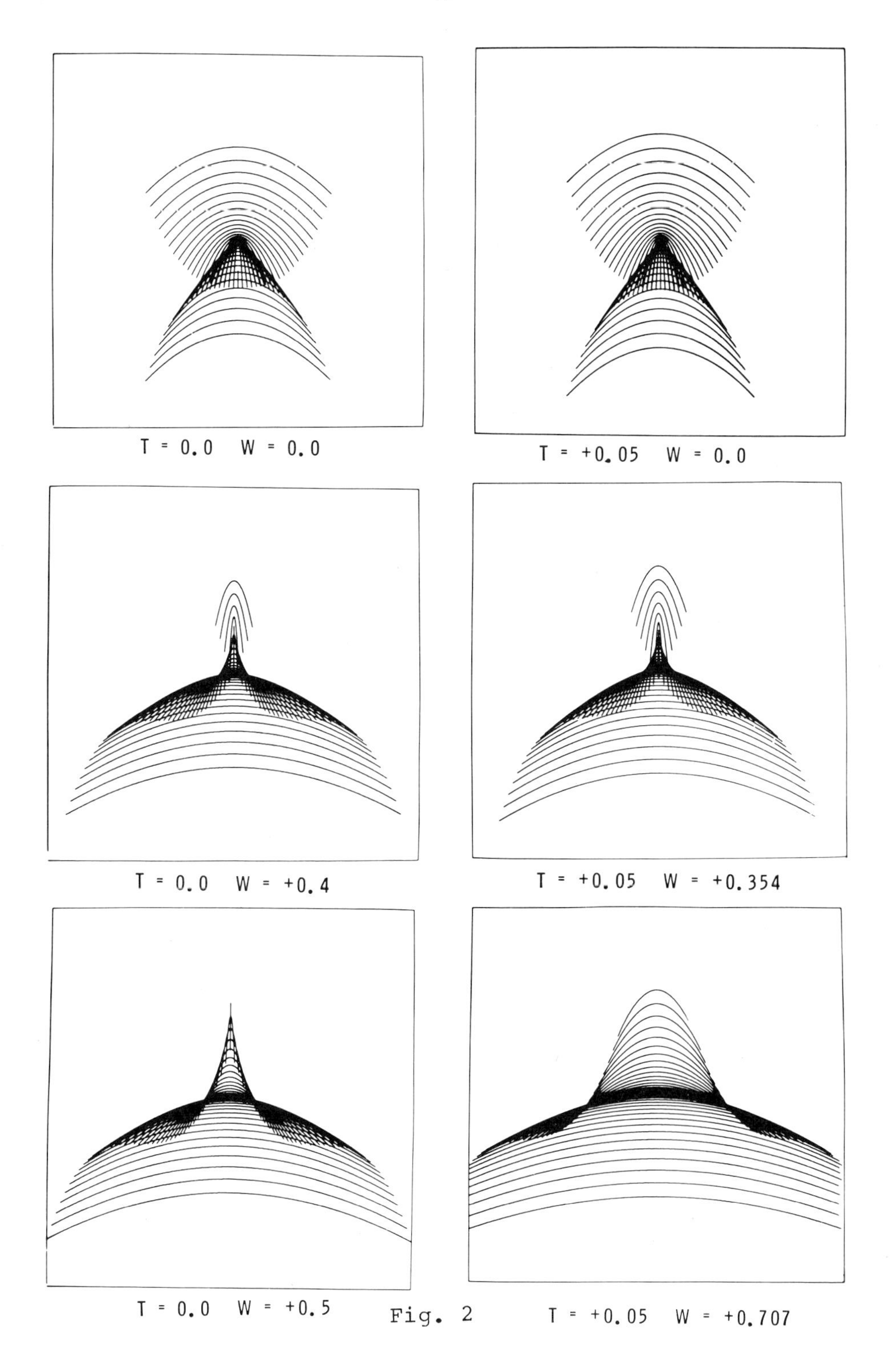
T = 0.0 W = 0.0
T = +0.05 W = 0.0
T = 0.0 W = +0.4
T = +0.05 W = +0.354
T = 0.0 W = +0.5
T = +0.05 W = +0.707

Fig. 2

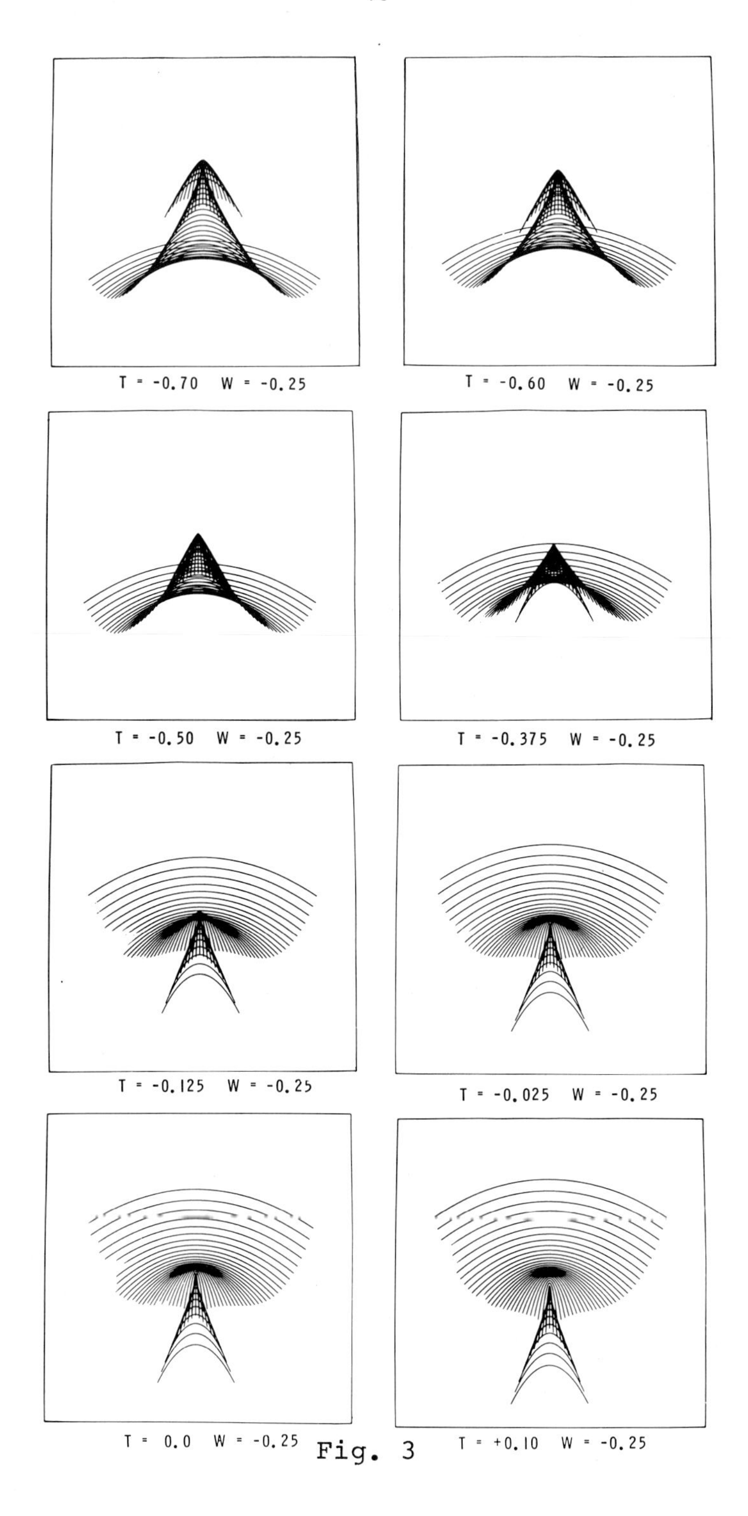

T = -0.70 W = -0.25
T = -0.60 W = -0.25
T = -0.50 W = -0.25
T = -0.375 W = -0.25
T = -0.125 W = -0.25
T = -0.025 W = -0.25
T = 0.0 W = -0.25
T = +0.10 W = -0.25

Fig. 3

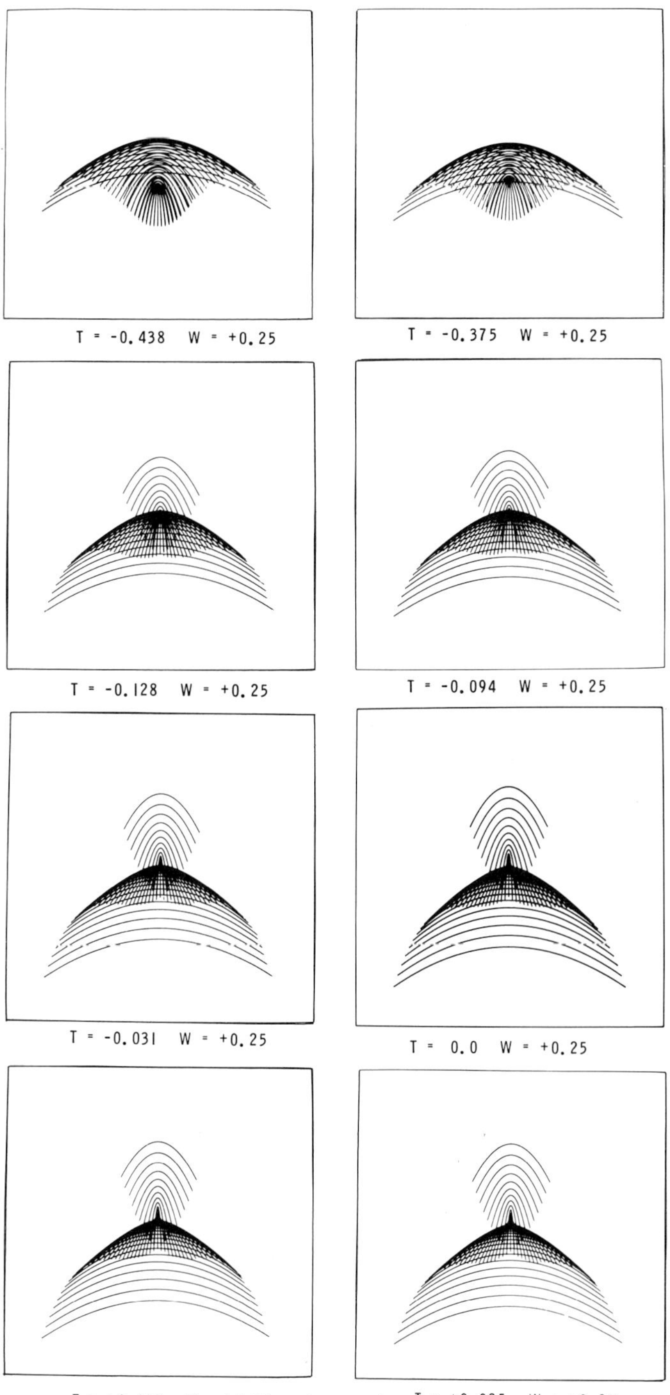
T = -0.438 W = +0.25
T = -0.375 W = +0.25
T = -0.128 W = +0.25
T = -0.094 W = +0.25
T = -0.031 W = +0.25
T = 0.0 W = +0.25
T = +0.013 W = +0.25
T = +0.025 W = +0.25

Fig. 4

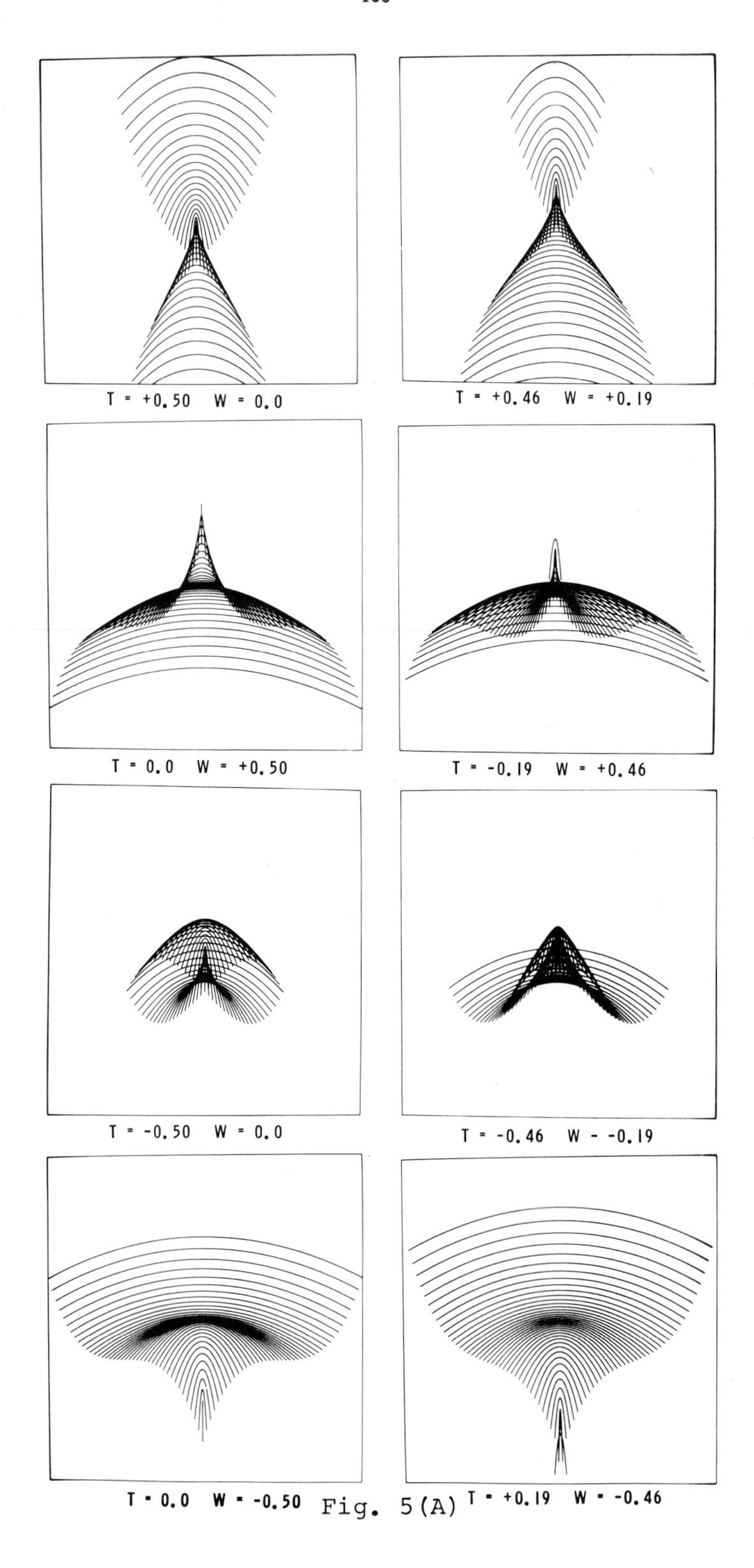

T = +0.50 W = 0.0
T = +0.46 W = +0.19
T = 0.0 W = +0.50
T = -0.19 W = +0.46
T = -0.50 W = 0.0
T = -0.46 W - -0.19
T - 0.0 W - -0.50
T - +0.19 W - -0.46

Fig. 5(A)

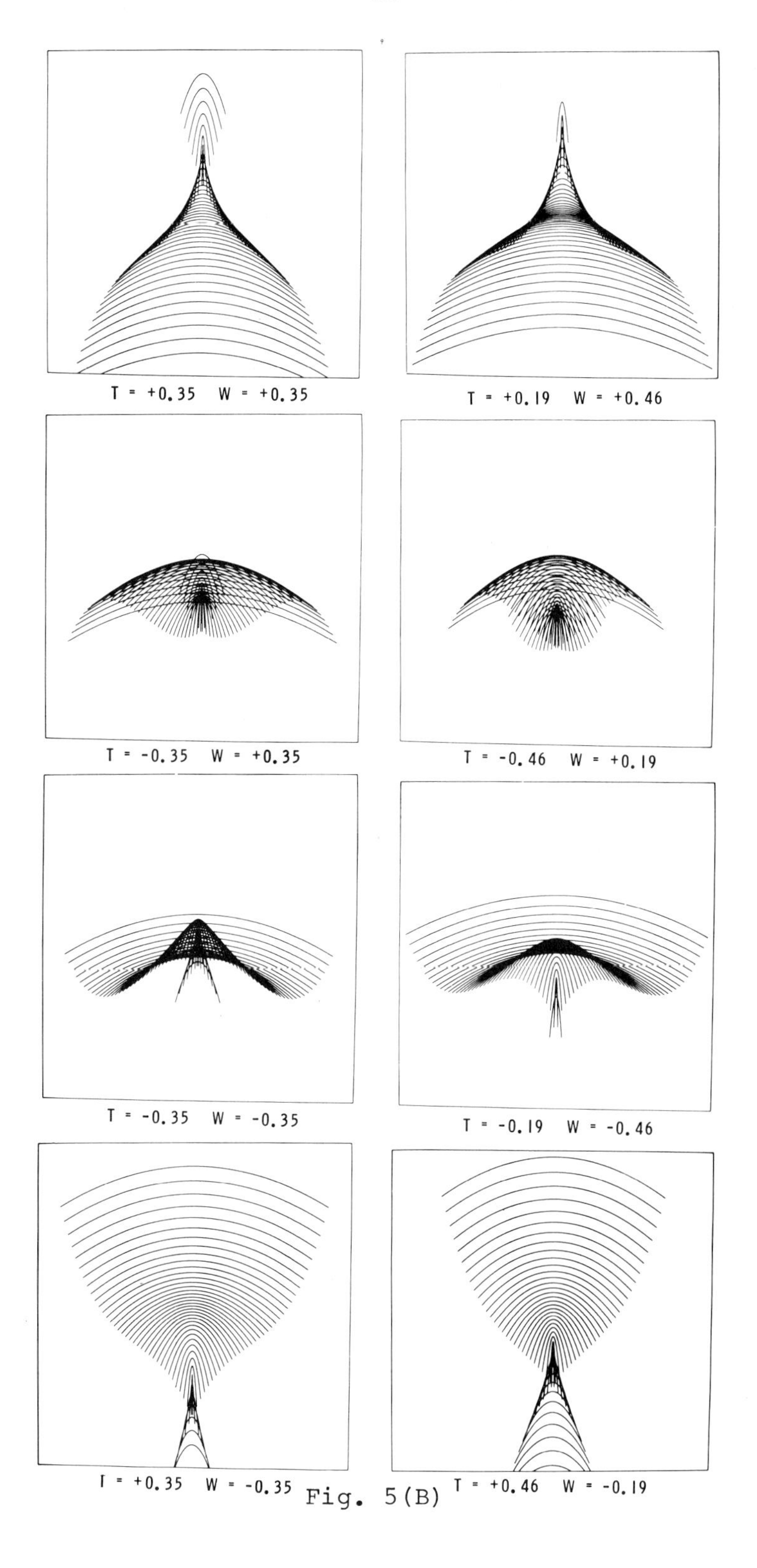
T = +0.35 W = +0.35
T = +0.19 W = +0.46
T = -0.35 W = +0.35
T = -0.46 W = +0.19
T = -0.35 W = -0.35
T = -0.19 W = -0.46
T = +0.35 W = -0.35
T = +0.46 W = -0.19

Fig. 5(B)

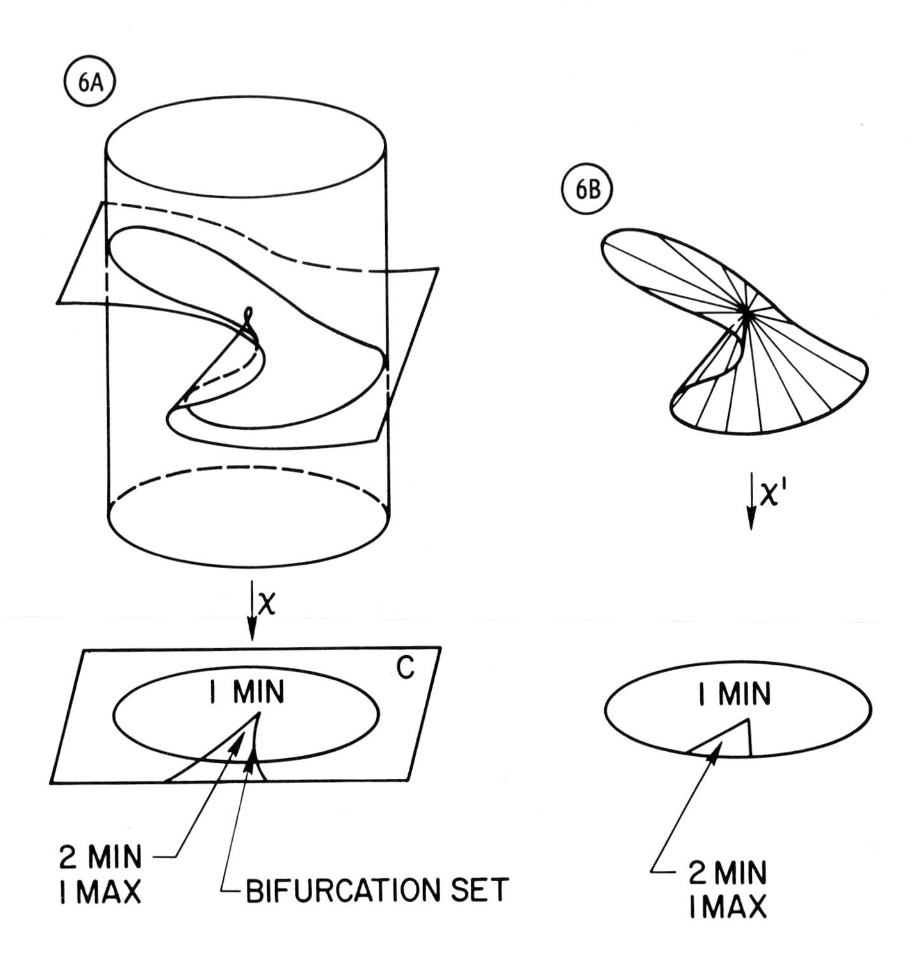

Fig. 6

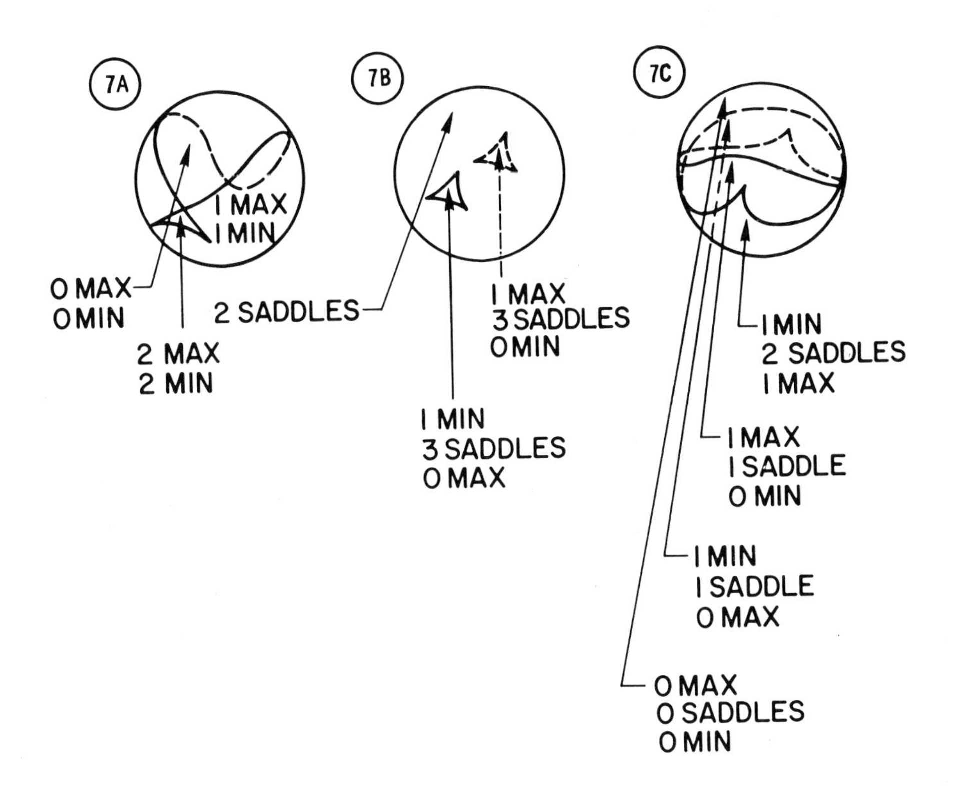
7A
1 MAX
1 MIN
0 MAX
0 MIN
2 MAX
2 MIN
7B
2 SADDLES
1 MAX
3 SADDLES
0 MIN
1 MIN
3 SADDLES
0 MAX
7C
1 MIN
2 SADDLES
1 MAX
1 MAX
1 SADDLE
0 MIN
1 MIN
1 SADDLE
0 MAX
0 MAX
0 SADDLES
0 MIN

Fig. 7

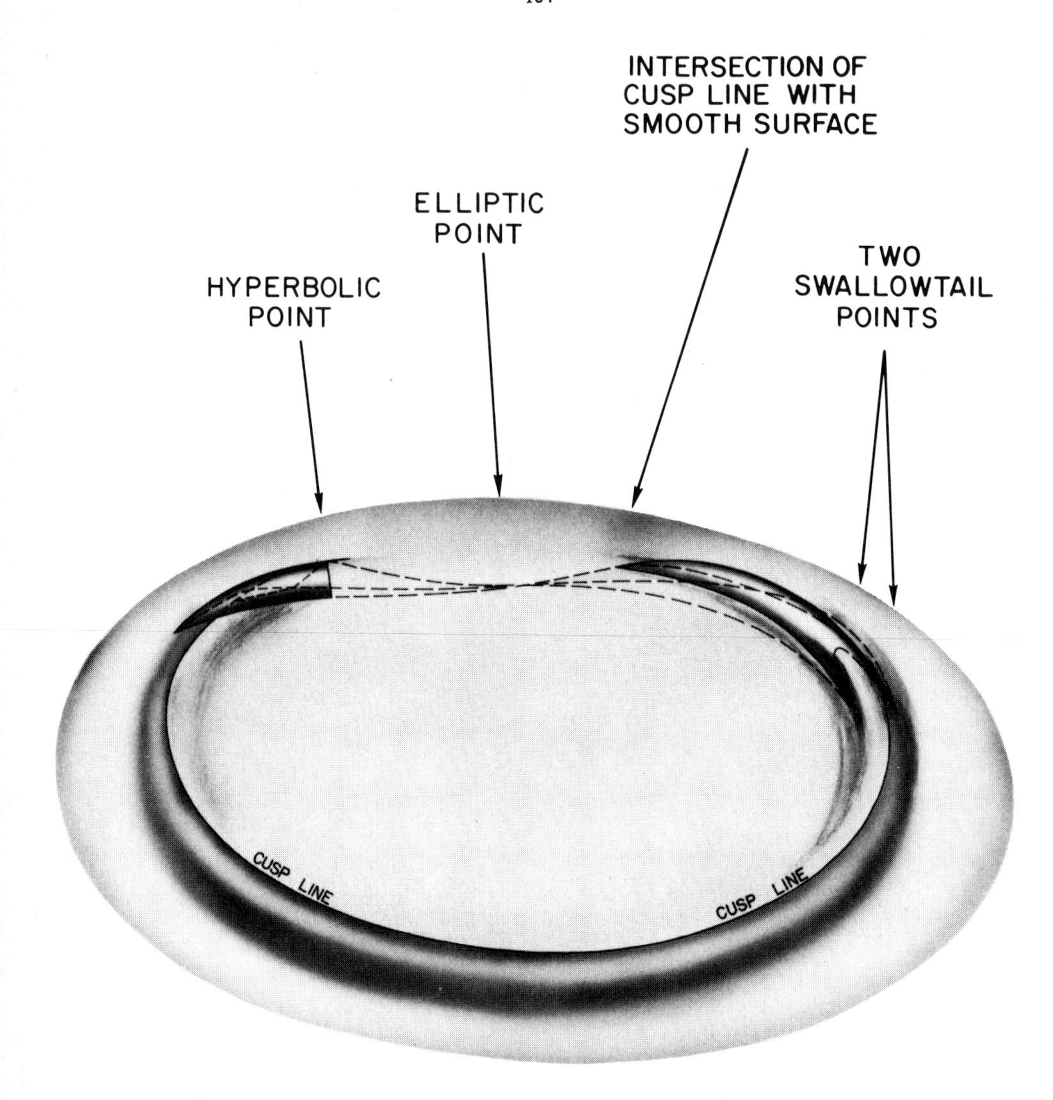
INTERSECTION OF
CUSP LINE WITH
SMOOTH SURFACE
ELLIPTIC
POINT
HYPERBOLIC
POINT
TWO
SWALLOWTAIL
POINTS
CUSP LINE
CUSP LINE

Fig. 8

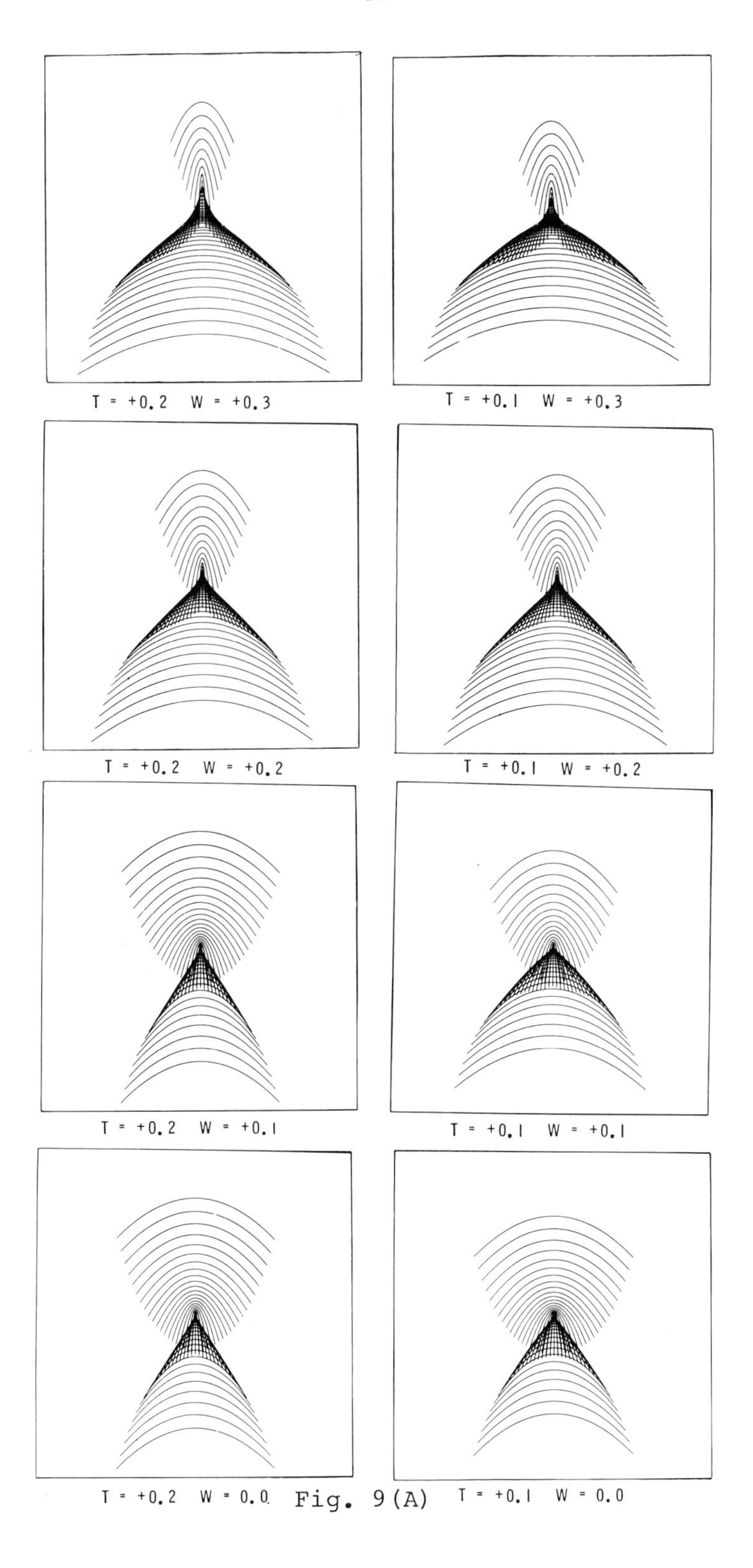
T = +0.2 W = +0.3
T = +0.1 W = +0.3
T = +0.2 W = +0.2
T = +0.1 W = +0.2
T = +0.2 W = +0.1
T = +0.1 W = +0.1
T = +0.2 W = 0.0
T = +0.1 W = 0.0

Fig. 9(A)

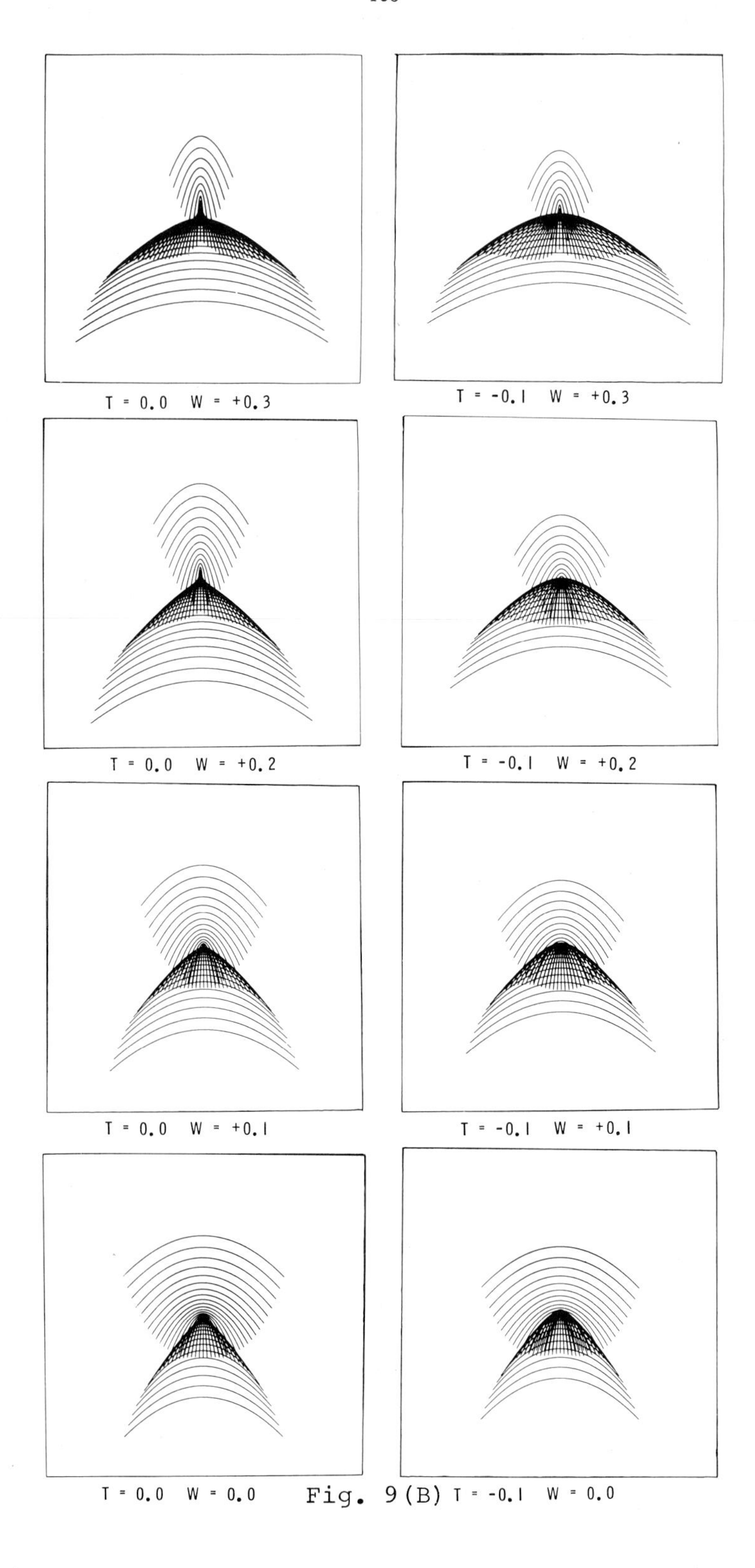
T = 0.0 W = +0.3
T = -0.1 W = +0.3
T = 0.0 W = +0.2
T = -0.1 W = +0.2
T = 0.0 W = +0.1
T = -0.1 W = +0.1
T = 0.0 W = 0.0
T = -0.1 W = 0.0

Fig. 9(B)

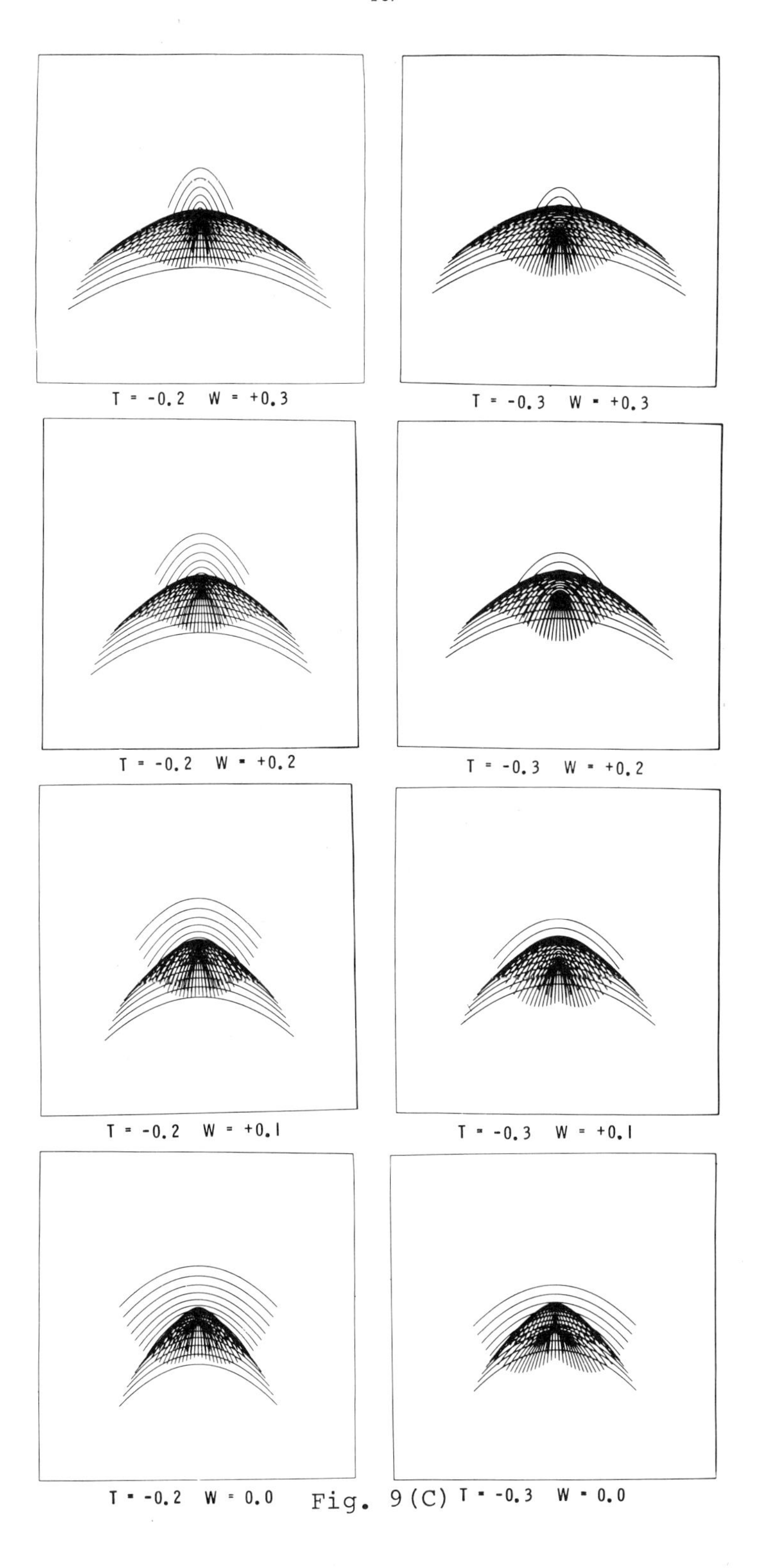
T = -0.2 W = +0.3
T = -0.3 W = +0.3
T = -0.2 W = +0.2
T = -0.3 W = +0.2
T = -0.2 W = +0.1
T = -0.3 W = +0.1
T = -0.2 W = 0.0
T = -0.3 W = 0.0

Fig. 9(C)

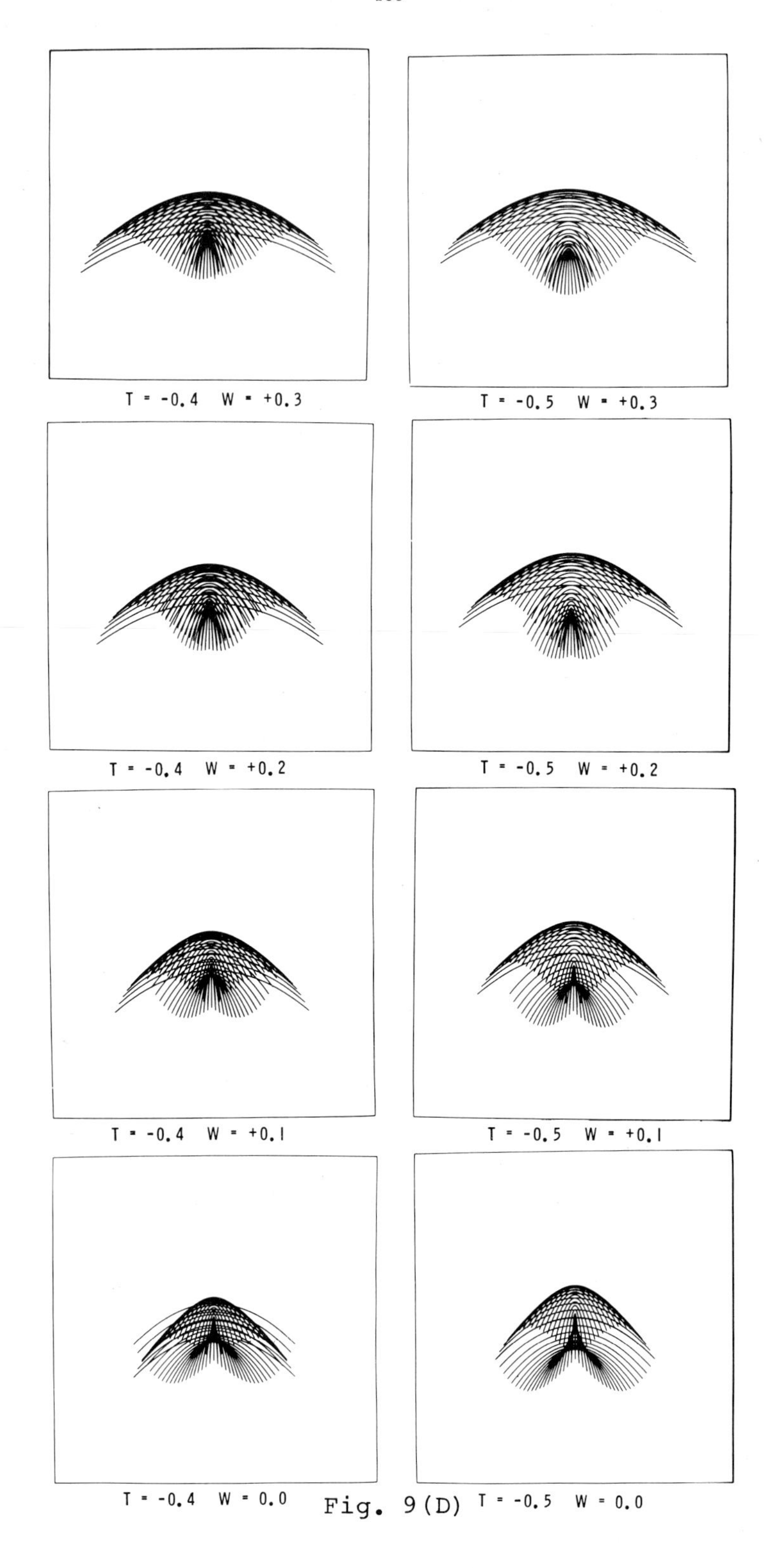
T = -0.4 W = +0.3
T = -0.5 W = +0.3
T = -0.4 W = +0.2
T = -0.5 W = +0.2
T = -0.4 W = +0.1
T = -0.5 W = +0.1
T = -0.4 W = 0.0
T = -0.5 W = 0.0

Fig. 9(D)

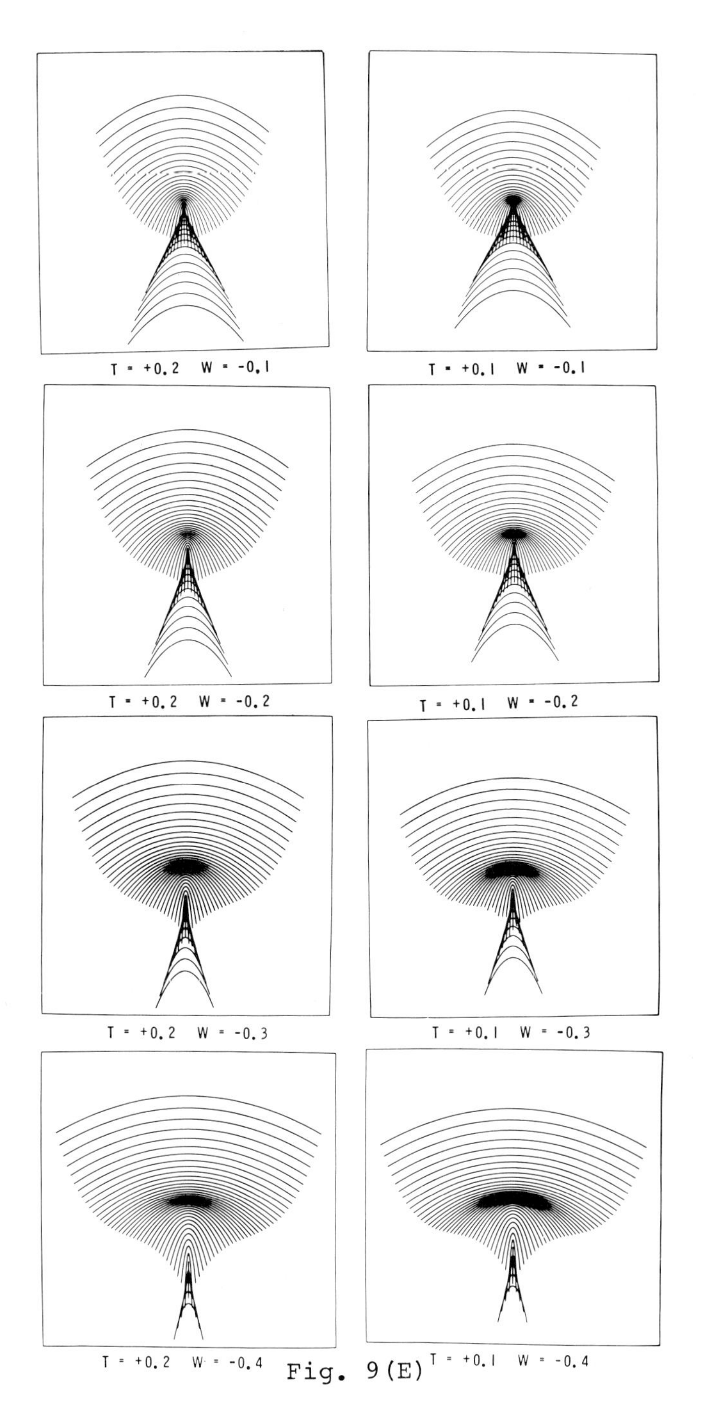

T = +0.2 W = -0.1
T = +0.1 W = -0.1
T = +0.2 W = -0.2
T = +0.1 W = -0.2
T = +0.2 W = -0.3
T = +0.1 W = -0.3
T = +0.2 W = -0.4
T = +0.1 W = -0.4

Fig. 9(E)

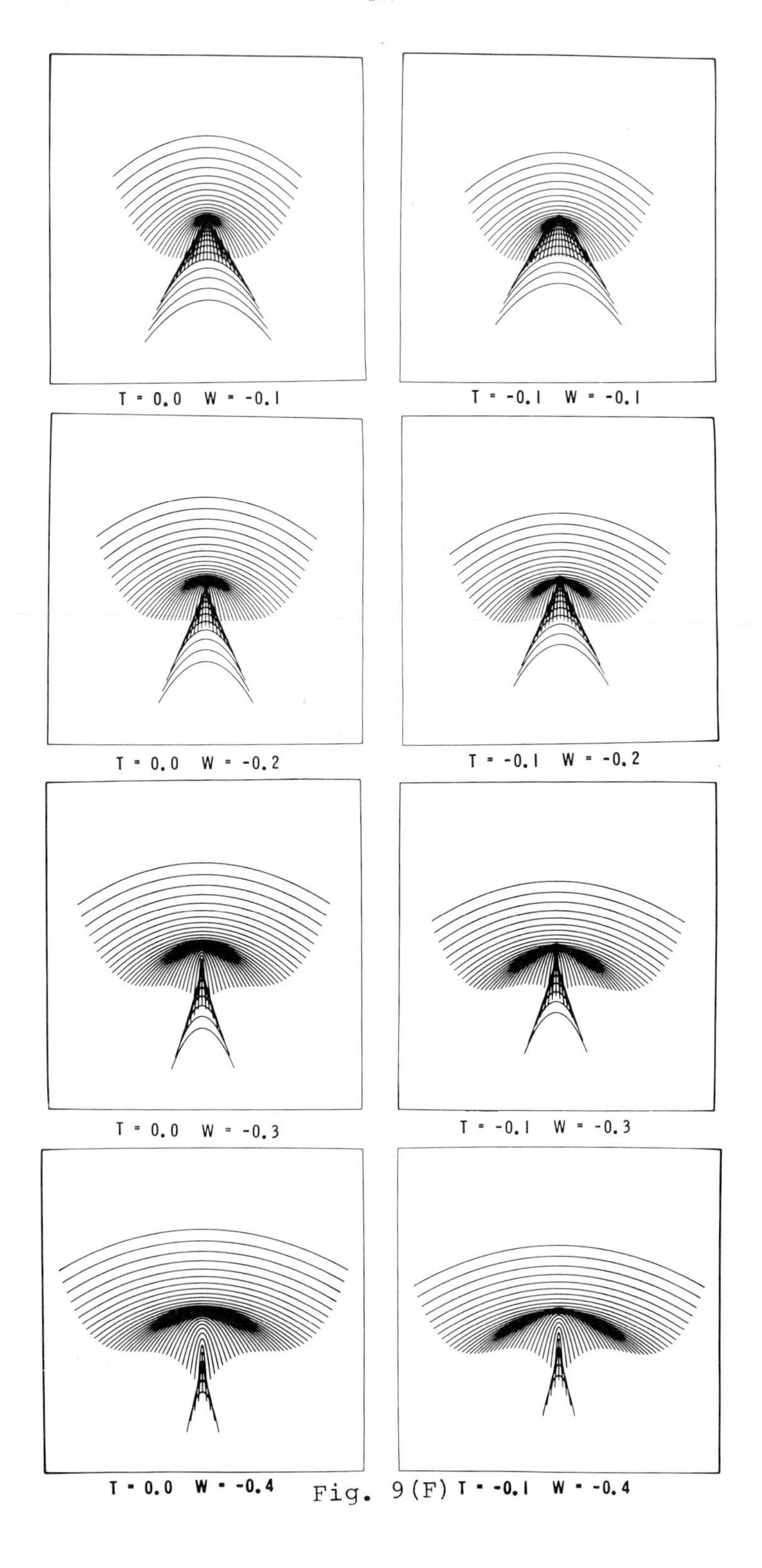
T = 0.0 W = -0.1
T = -0.1 W = -0.1
T = 0.0 W = -0.2
T = -0.1 W = -0.2
T = 0.0 W = -0.3
T = -0.1 W = -0.3
T = 0.0 W = -0.4
T = -0.1 W = -0.4

Fig. 9(F)

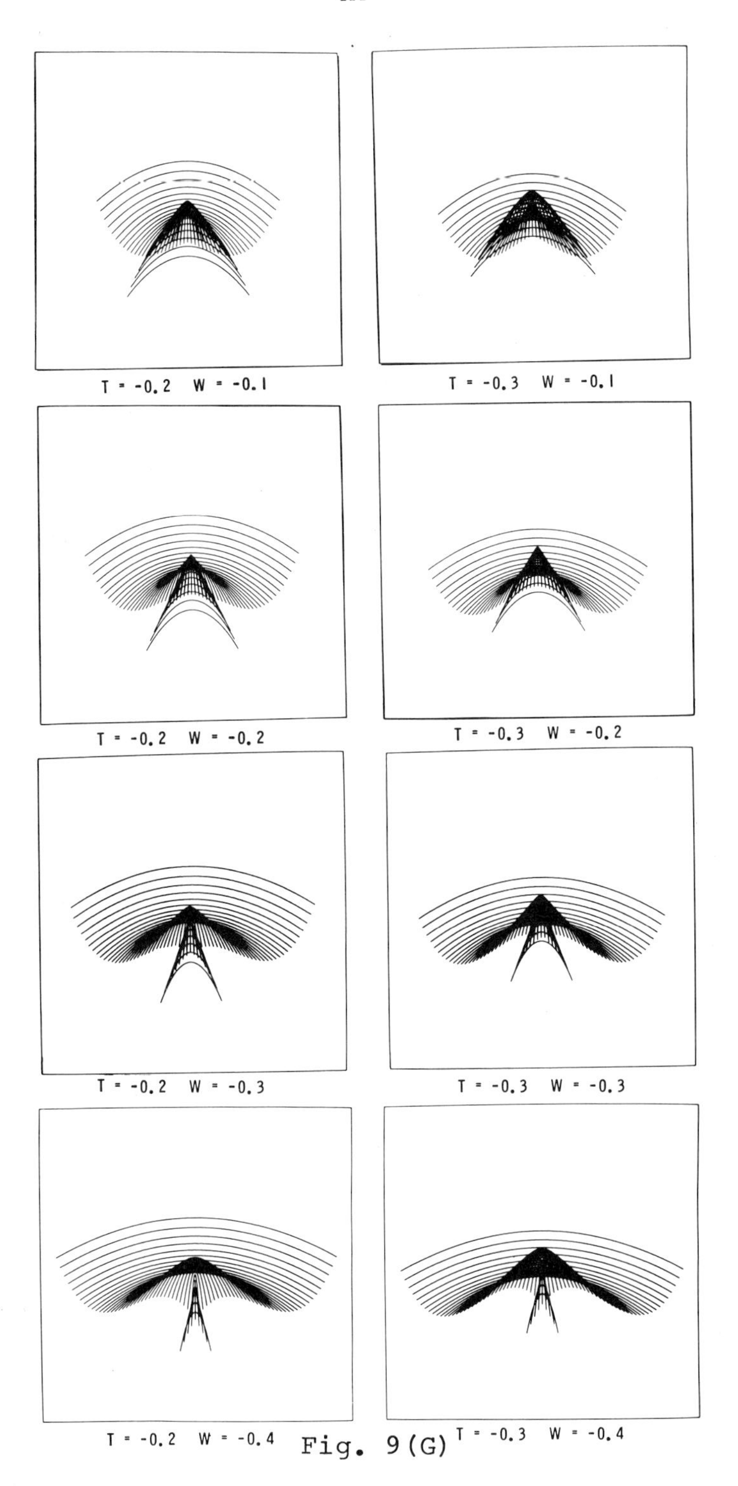
T = -0.2 W = -0.1
T = -0.3 W = -0.1
T = -0.2 W = -0.2
T = -0.3 W = -0.2
T = -0.2 W = -0.3
T = -0.3 W = -0.3
T = -0.2 W = -0.4
T = -0.3 W = -0.4

Fig. 9(G)

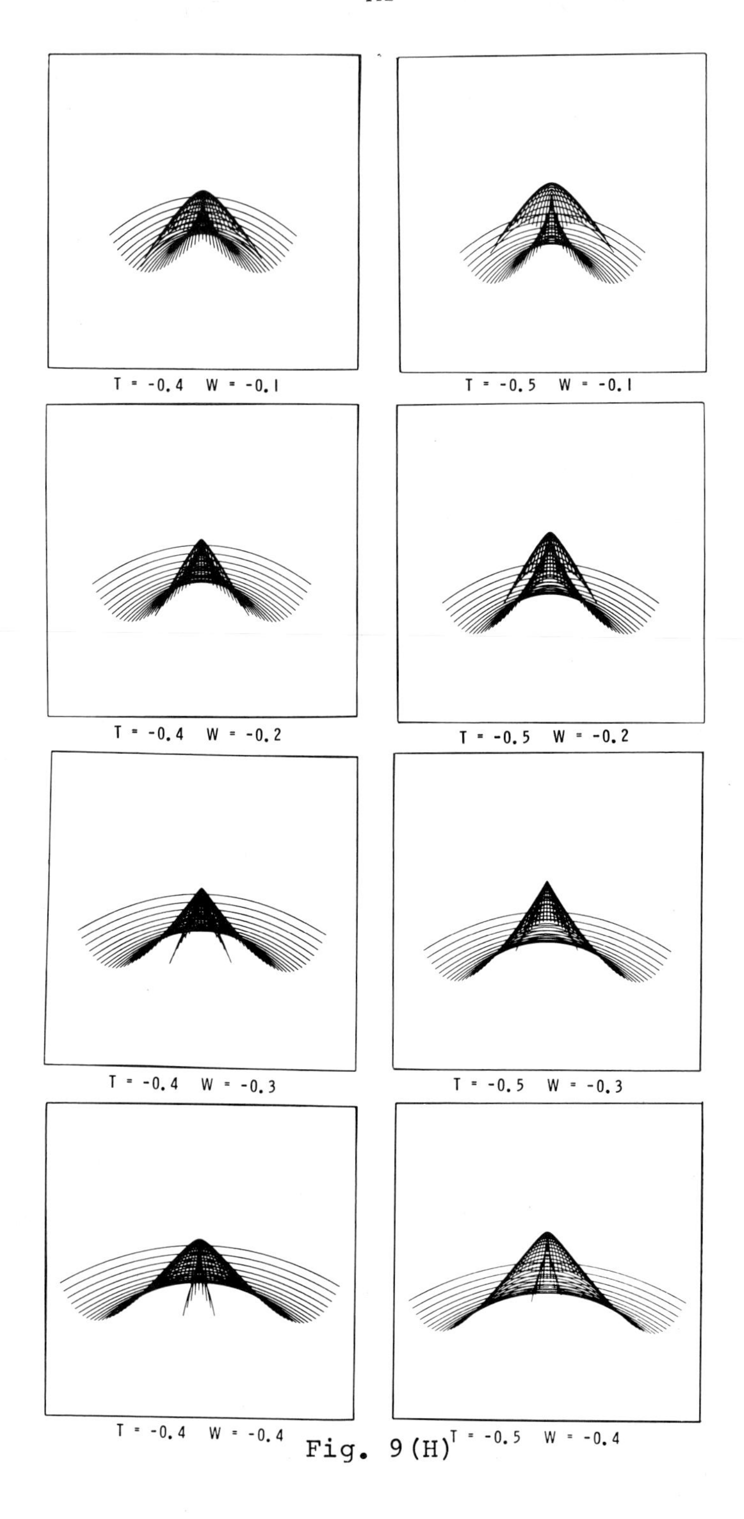
T = -0.4 W = -0.1
T = -0.5 W = -0.1
T = -0.4 W = -0.2
T = -0.5 W = -0.2
T = -0.4 W = -0.3
T = -0.5 W = -0.3
T = -0.4 W = -0.4
T = -0.5 W = -0.4

Fig. 9(H)

Some Sections of the Compactified Parabolic Umbilic

$V = x^2y + x^4 + y^4 + ty^2 + wx^2 + ux + vy$

projected onto the plane (u,v).

Figs. 10 (A, B, C, D, E, F, G and H) are a matrix of pictures of sections of the compact version of the Parabolic Umbilic, with t and w varying in steps of 0.1 Fig. 10 (A and E) show essentially a ridged surface with a cusp-like region growing out of the u = v = zero region of the ridge. This cusp region broadens as w is reduced from +0.3 to -0.4, and is wider but less elongated for smaller values of t.

As t is reduced the end of the cusp region advances on the u-negative region of the (u,v) plane and at the same time the edge of the (x,y) surface begins to roll over (Figs. 10 (B, C, F, G)).

The edge of the (xy) surface progressively rolls over as t is made more negative and, for w negative, an Elliptic Umbilic-like region develops inside the cusp. (Fig. 10, C. D, G, H). a Hyperbolic Umbilic-like transition (similar to that in Fig. 9) may be seen in this sequence (see, for example Fig. 10. t = -0.4, w = - 0.4 to -0.1). (The appearance of the Bifurcation Set at t = -0.3, w = 0.4 led S. H. W. to call this "Koala Bear" Catastrophe.)

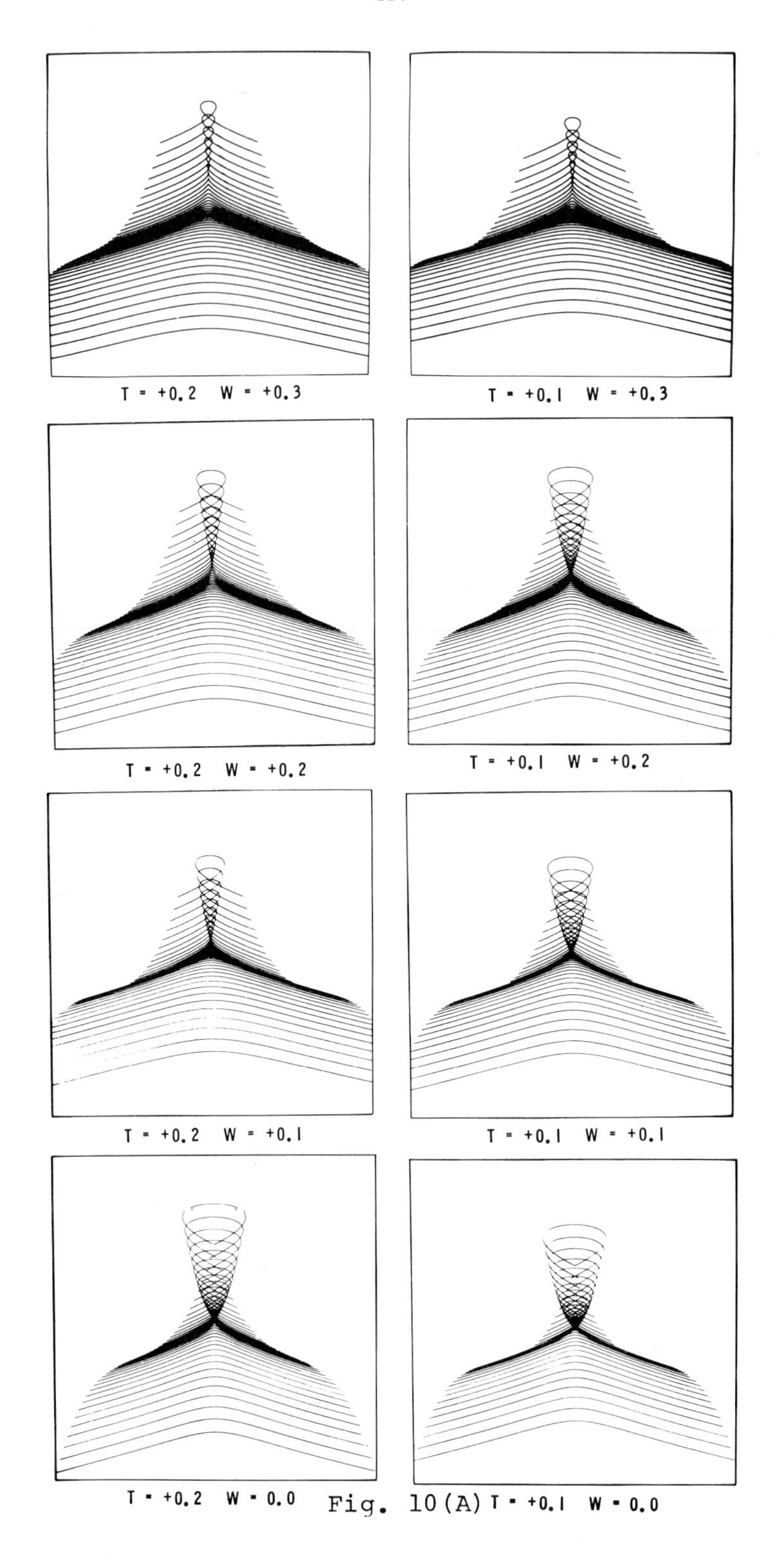
T = +0.2 W = +0.3
T = +0.1 W = +0.3
T = +0.2 W = +0.2
T = +0.1 W = +0.2
T = +0.2 W = +0.1
T = +0.1 W = +0.1
T = +0.2 W = 0.0
T = +0.1 W = 0.0

Fig. 10(A)

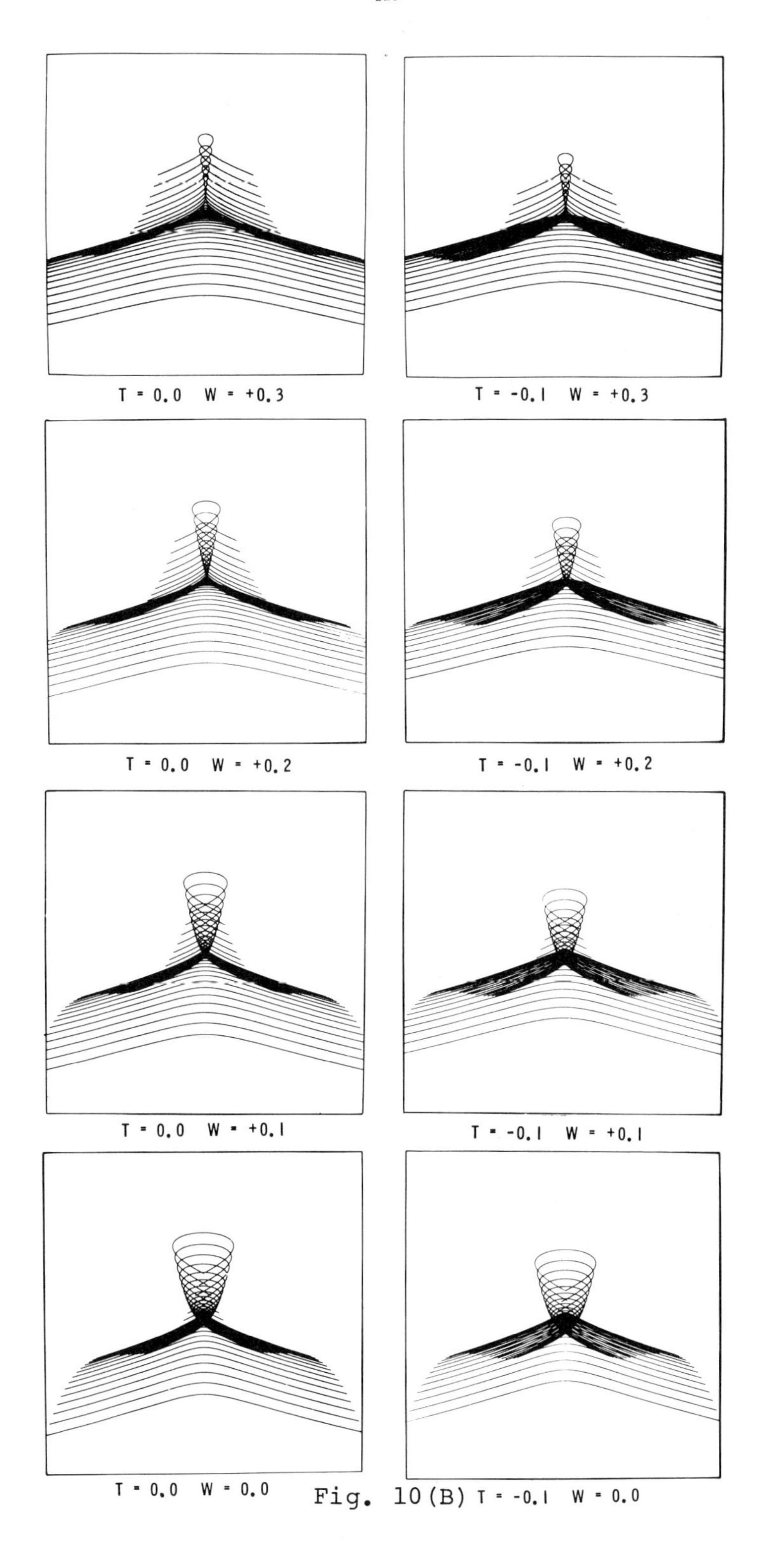

T = 0.0 W = +0.3
T = -0.1 W = +0.3
T = 0.0 W = +0.2
T = -0.1 W = +0.2
T = 0.0 W = +0.1
T = -0.1 W = +0.1
T = 0.0 W = 0.0
T = -0.1 W = 0.0

Fig. 10(B)

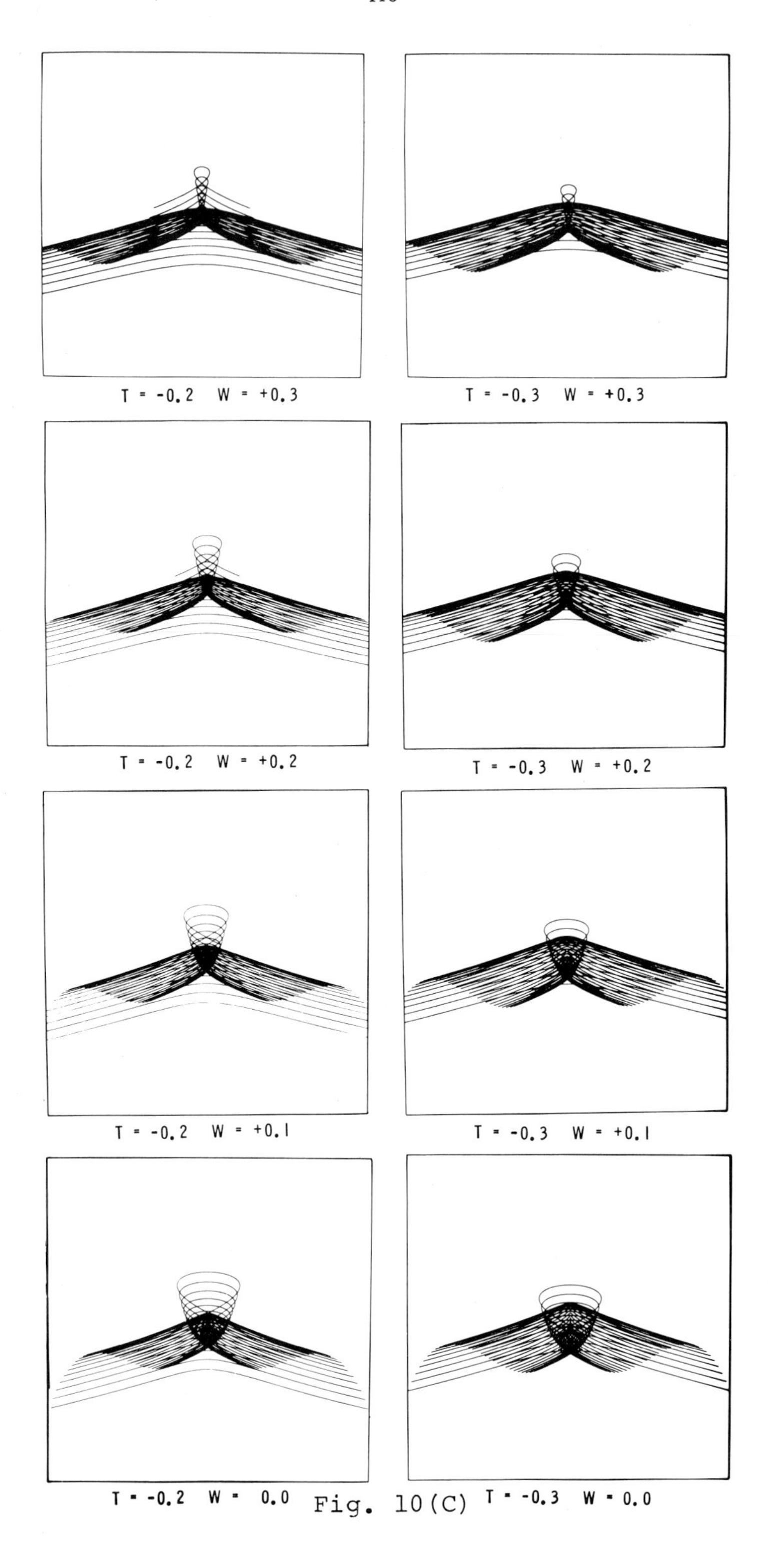
T = -0.2 W = +0.3
T = -0.3 W = +0.3
T = -0.2 W = +0.2
T = -0.3 W = +0.2
T = -0.2 W = +0.1
T = -0.3 W = +0.1
T = -0.2 W = 0.0
T = -0.3 W = 0.0

Fig. 10(C)

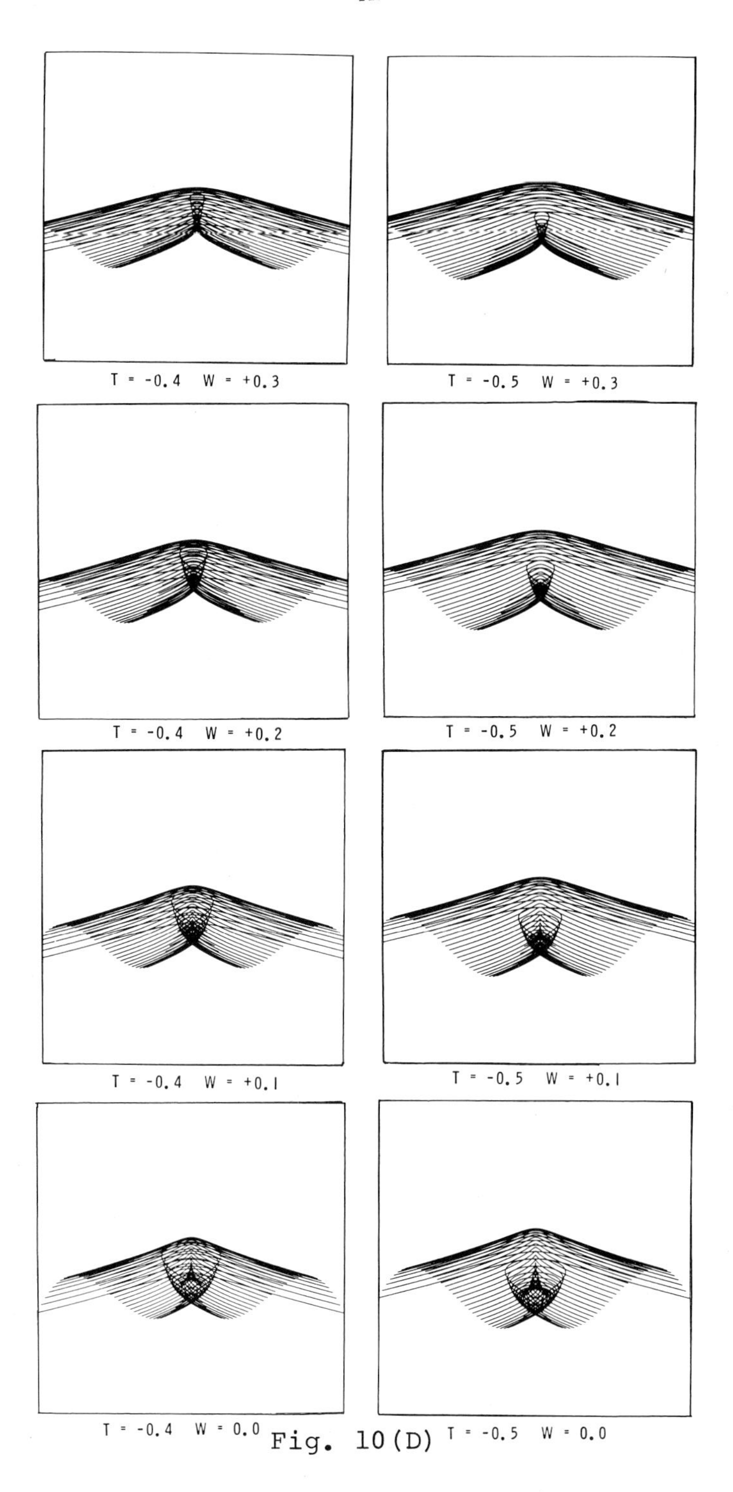

Fig. 10(D)

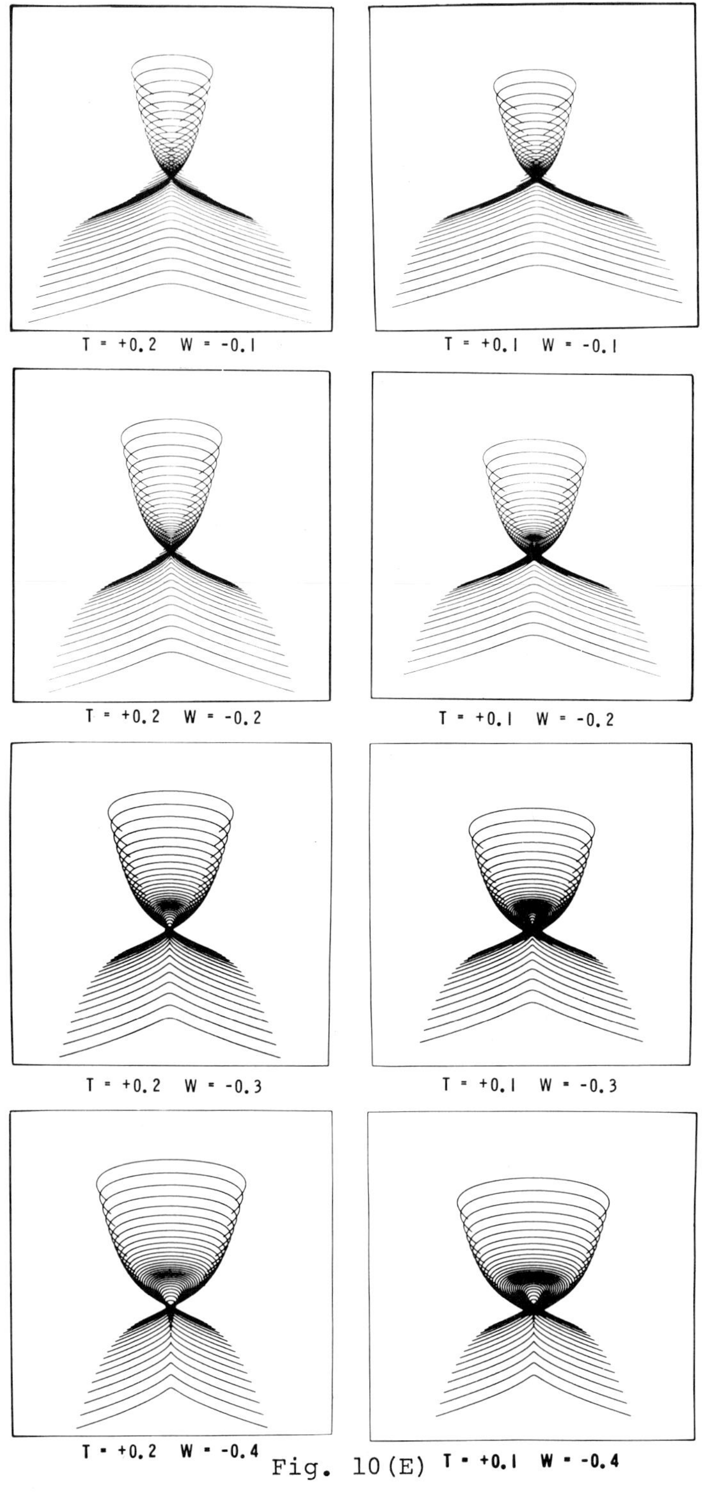
T = +0.2 W = -0.1
T = +0.1 W = -0.1
T = +0.2 W = -0.2
T = +0.1 W = -0.2
T = +0.2 W = -0.3
T = +0.1 W = -0.3
T = +0.2 W = -0.4
T = +0.1 W = -0.4

Fig. 10(E)

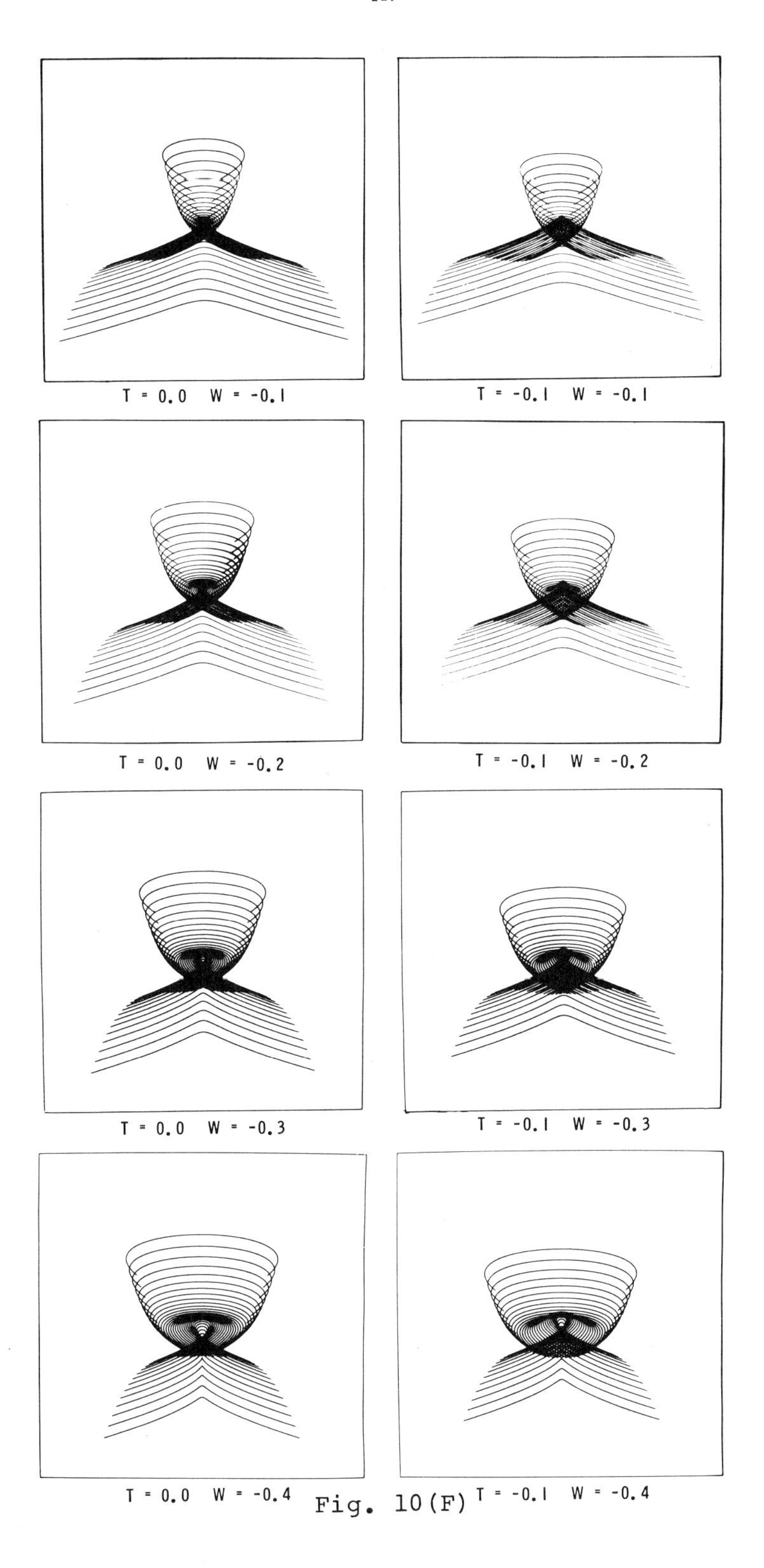
T = 0.0 W = -0.1
T = -0.1 W = -0.1
T = 0.0 W = -0.2
T = -0.1 W = -0.2
T = 0.0 W = -0.3
T = -0.1 W = -0.3
T = 0.0 W = -0.4
T = -0.1 W = -0.4

Fig. 10(F)

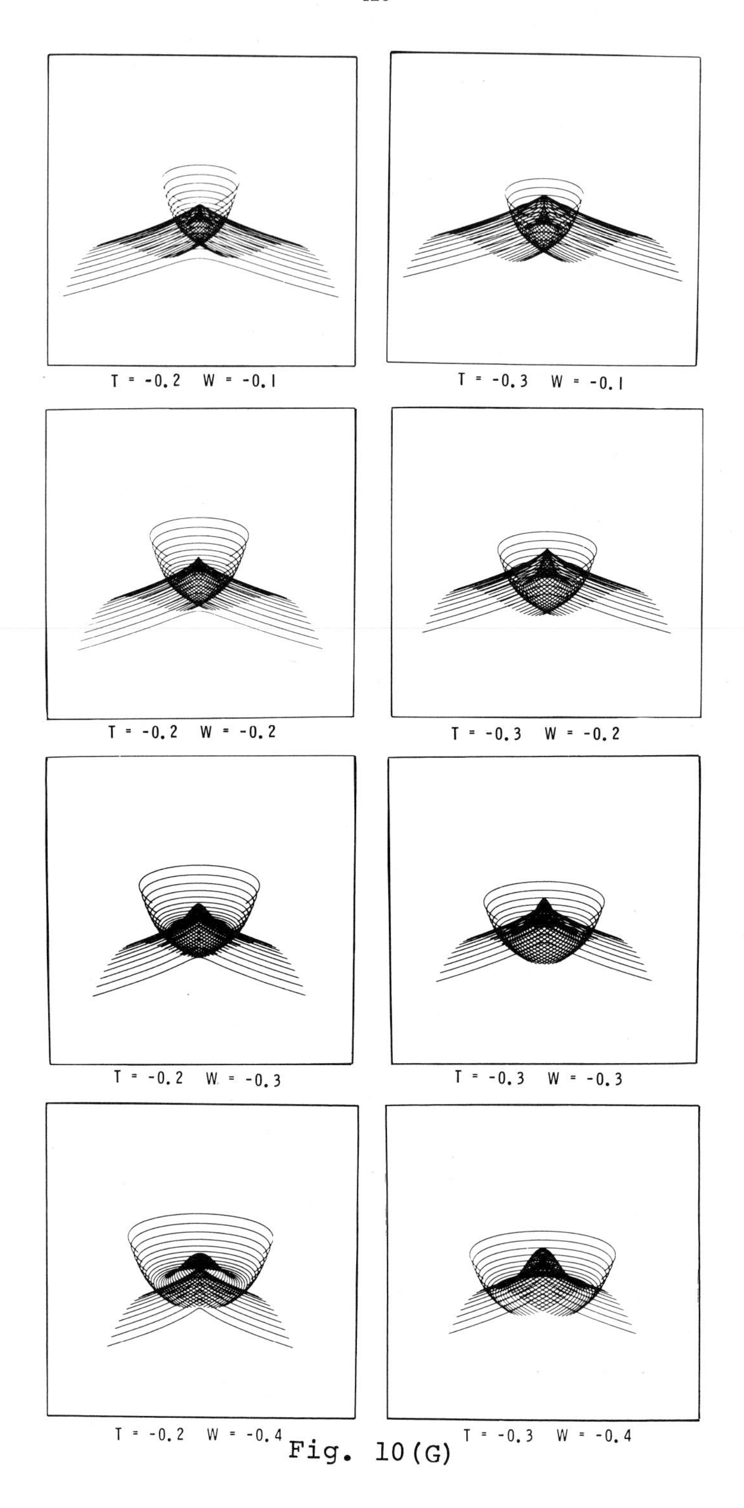

Fig. 10(G)

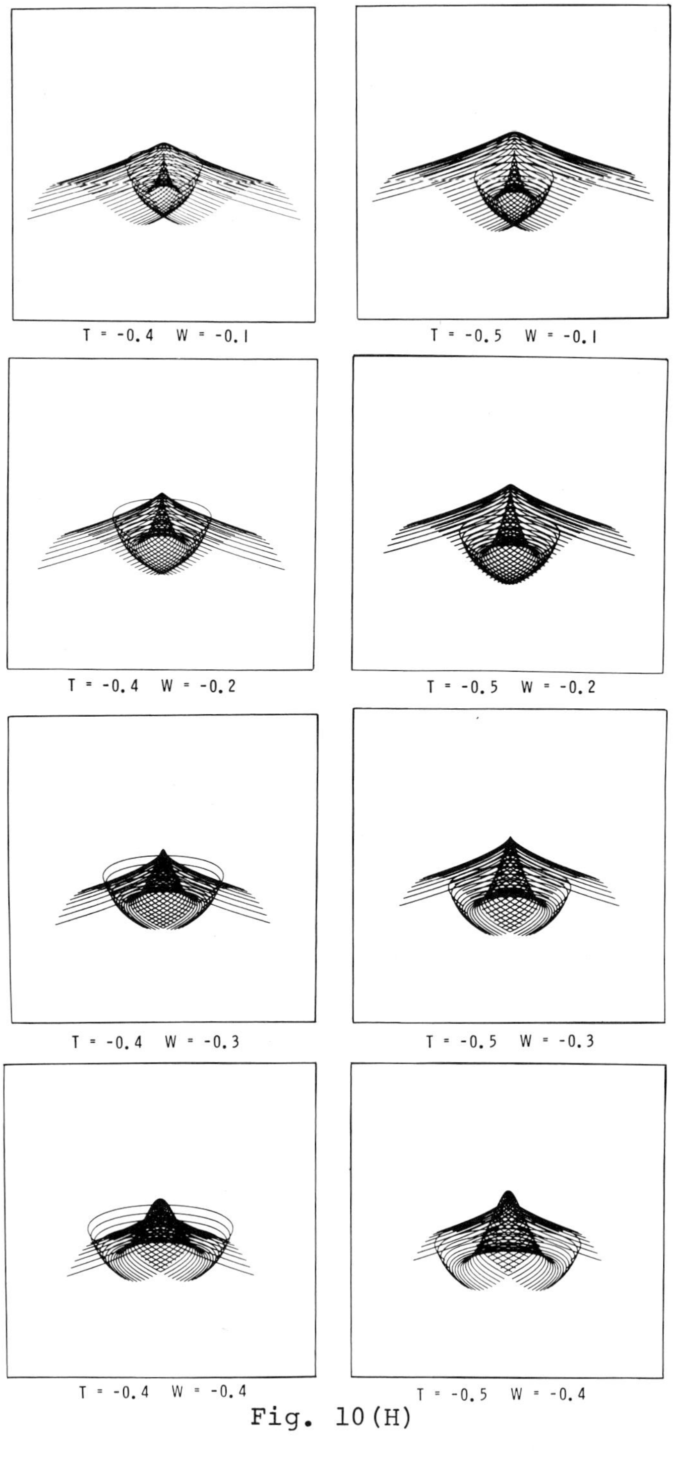
T = -0.4 W = -0.1
T = -0.5 W = -0.1
T = -0.4 W = -0.2
T = -0.5 W = -0.2
T = -0.4 W = -0.3
T = -0.5 W = -0.3
T = -0.4 W = -0.4
T = -0.5 W = -0.4

Fig. 10(H)

Some Sections of the weakly compactified Version of the Parabolic Umbilic

$$V = x^2y + \frac{y^4}{4} + \frac{x^4}{4} + ty^2 + wx^2 + ux + vy$$

Projected onto the (u,v) plane

In this example of the Parabolic Umbilic the relative influence of both the y^4 and the x^4 ("compactifying") terms is reduced. The matrix of pictures (Fig. 11, A, B, C, D, E, F, G and H) has been drawn in a similar way to those of Figs. 9 and 10.

With t positive the sections are of a rounded cusp nature (Figs. 11 (A,E)). With t negative, the cusp sinks into the round shouldered region and passes through the (x,y) surface to form an Elliptic Umbilic-like region. (Figs. 11 (B, C, D, F, G, H).

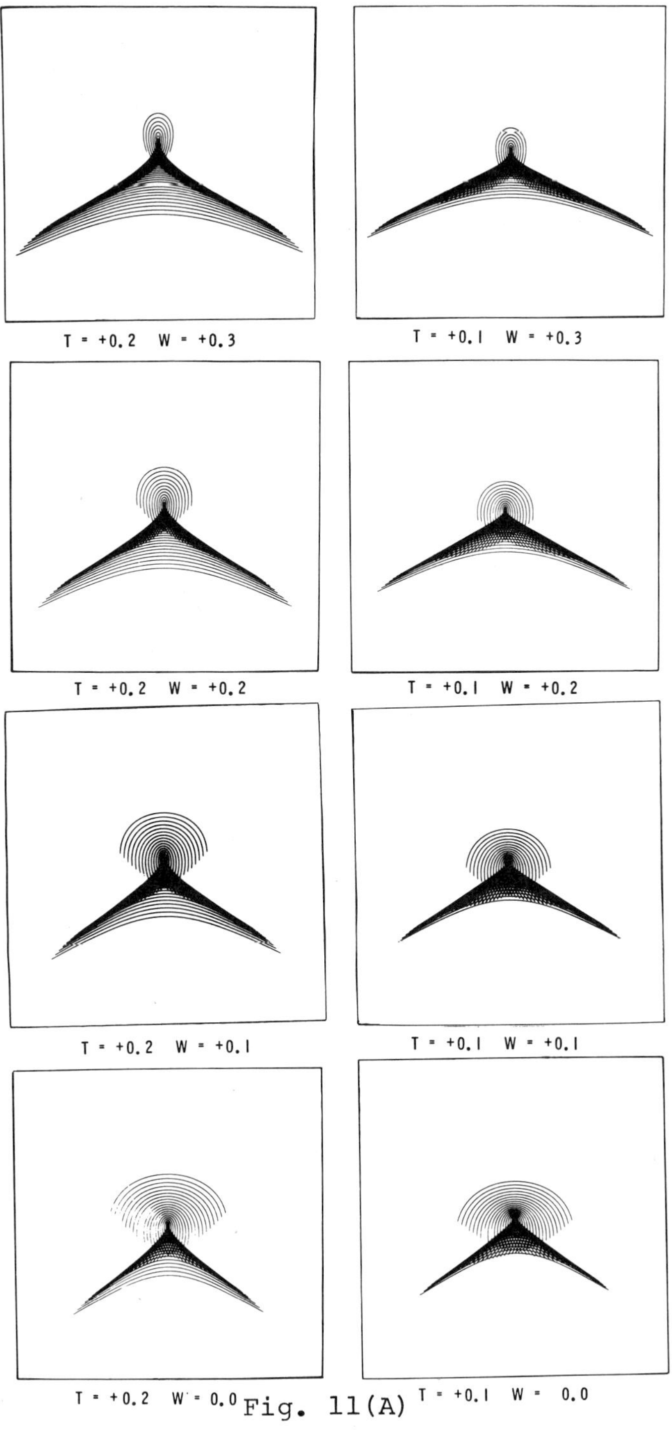
T = +0.2 W = +0.3
T = +0.1 W = +0.3
T = +0.2 W = +0.2
T = +0.1 W = +0.2
T = +0.2 W = +0.1
T = +0.1 W = +0.1
T = +0.2 W = 0.0
T = +0.1 W = 0.0

Fig. 11(A)

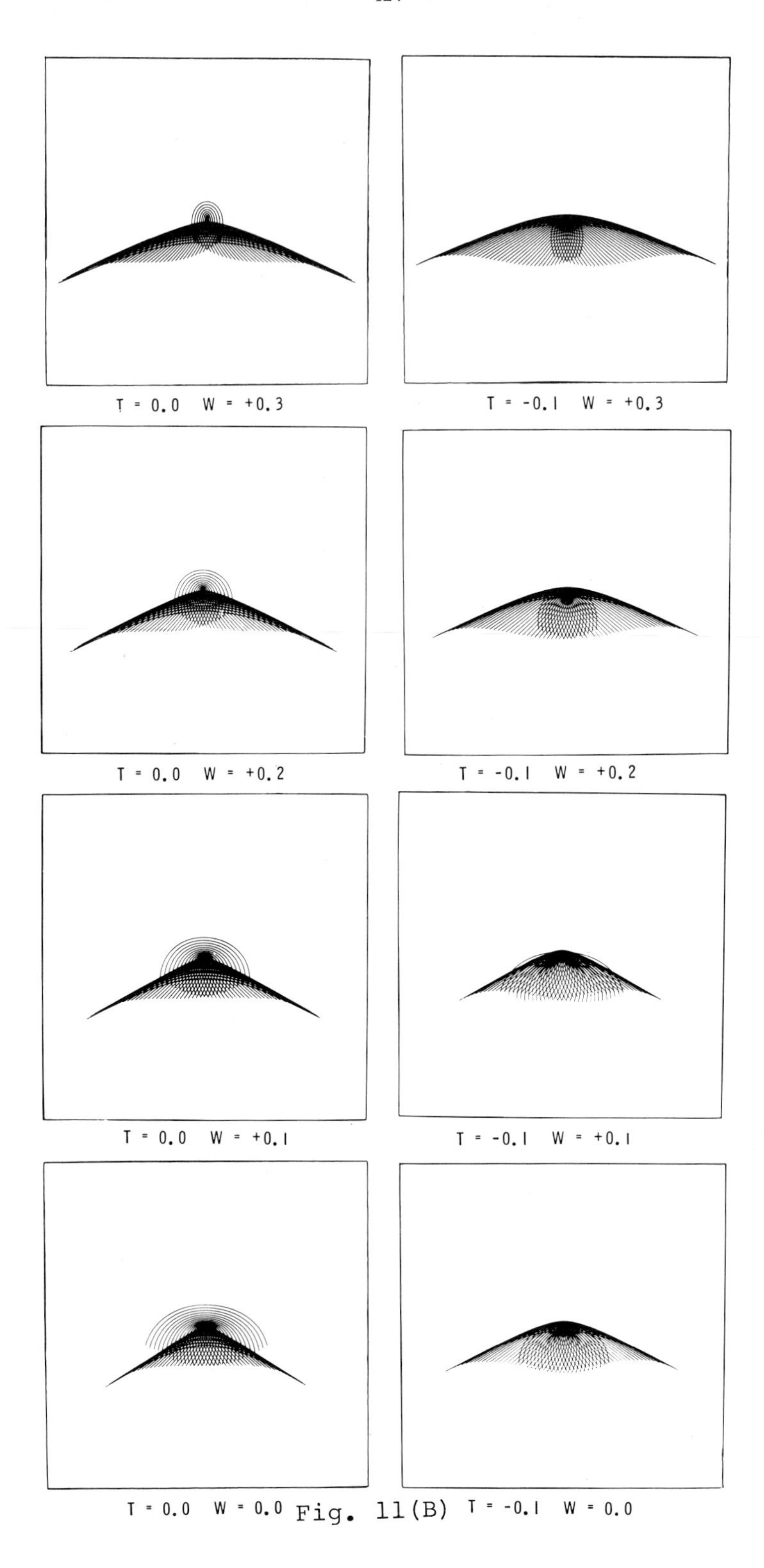

Fig. 11(B)

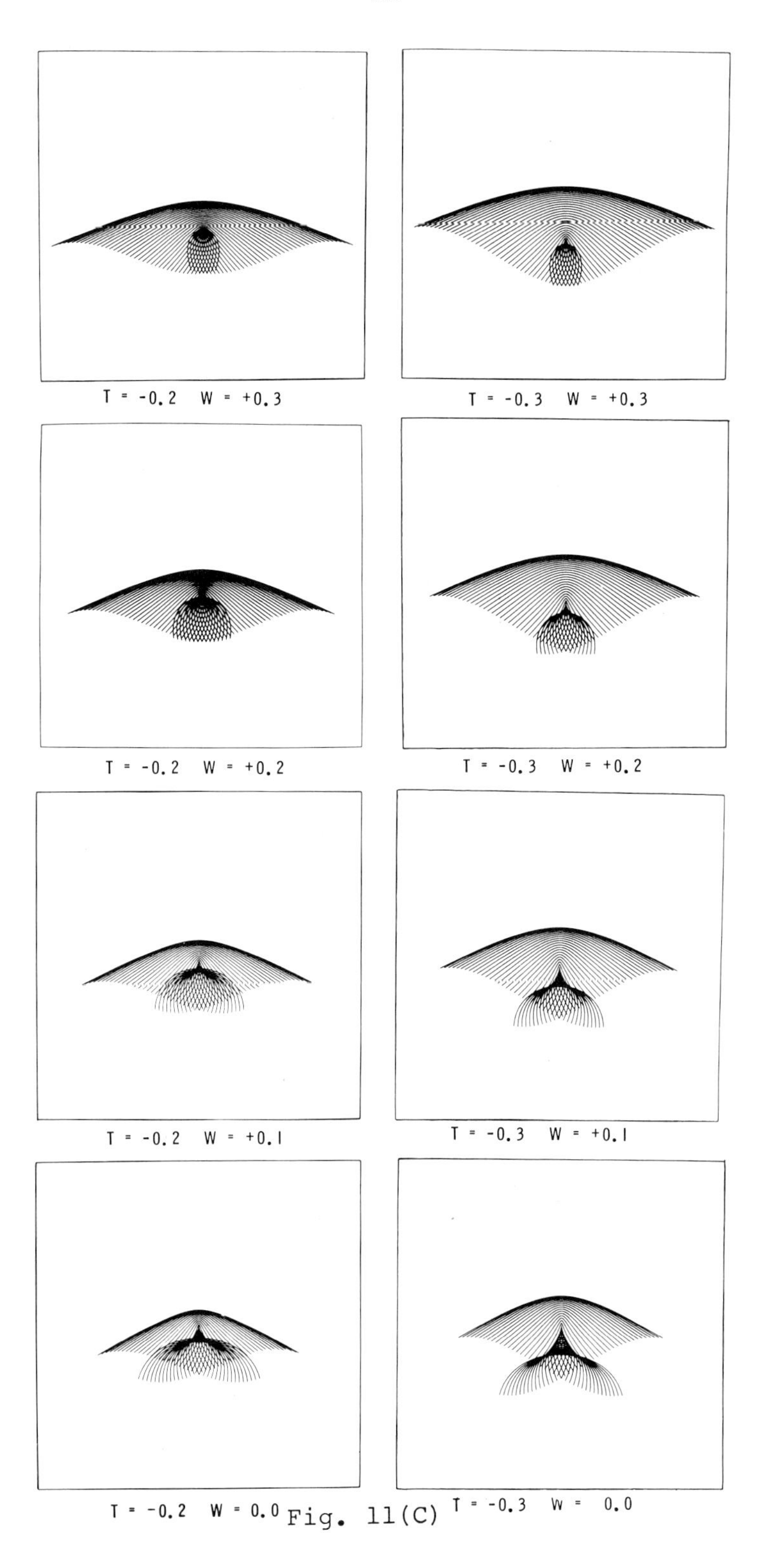
T = -0.2 W = +0.3
T = -0.3 W = +0.3
T = -0.2 W = +0.2
T = -0.3 W = +0.2
T = -0.2 W = +0.1
T = -0.3 W = +0.1
T = -0.2 W = 0.0
T = -0.3 W = 0.0

Fig. 11(C)

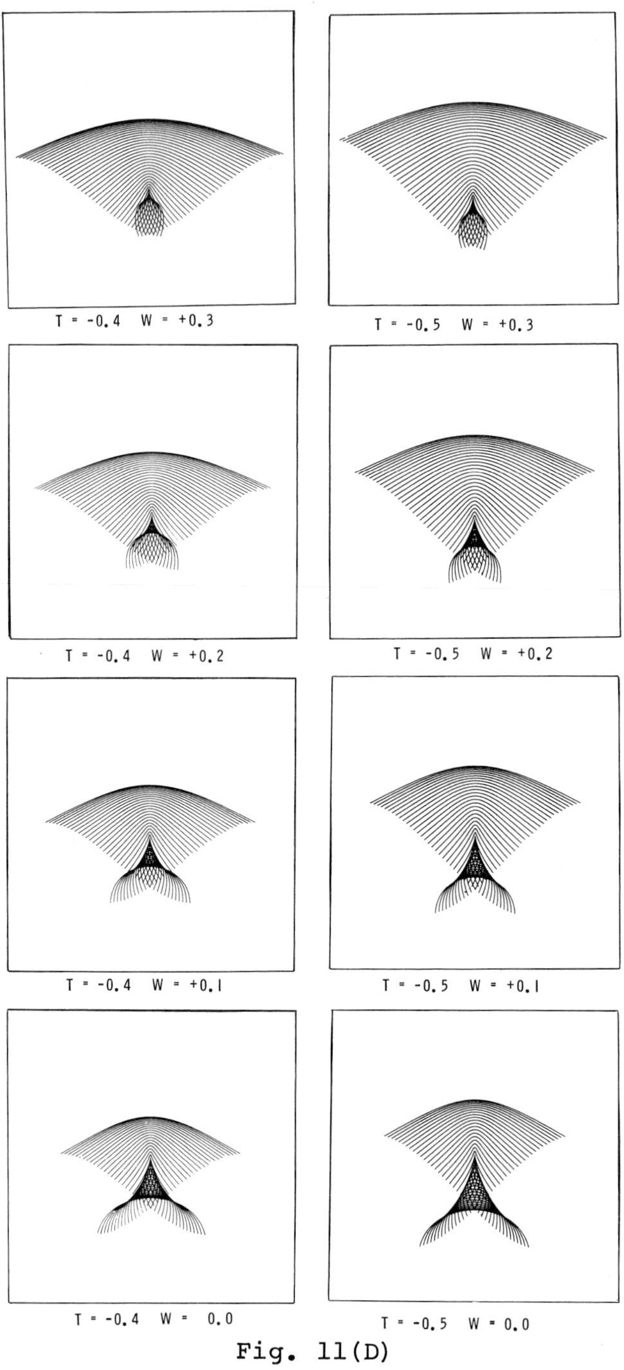
T = -0.4 W = +0.3
T = -0.5 W = +0.3
T = -0.4 W = +0.2
T = -0.5 W = +0.2
T = -0.4 W = +0.1
T = -0.5 W = +0.1
T = -0.4 W = 0.0
T = -0.5 W = 0.0

Fig. 11(D)

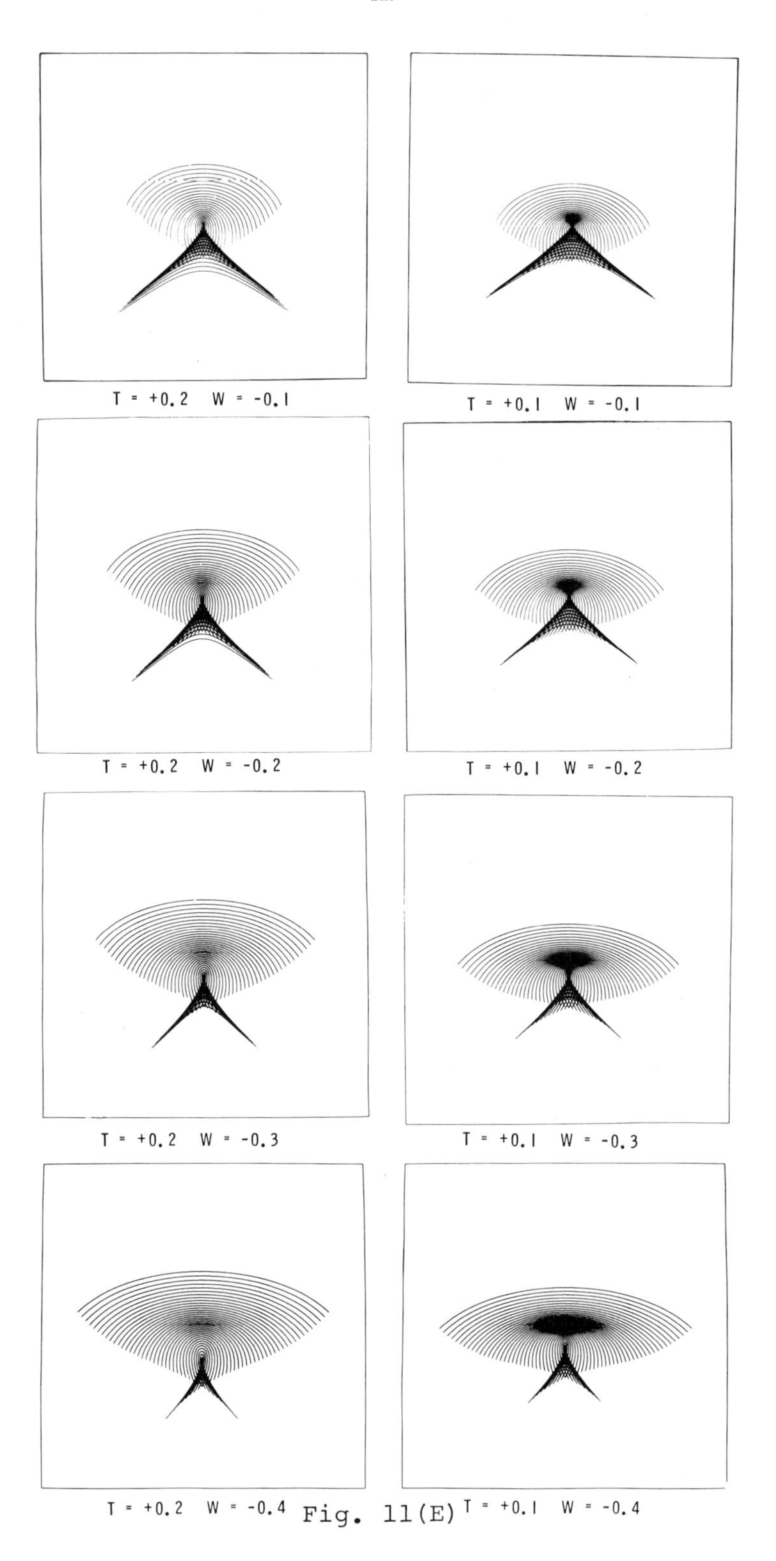
T = +0.2 W = -0.1
T = +0.1 W = -0.1
T = +0.2 W = -0.2
T = +0.1 W = -0.2
T = +0.2 W = -0.3
T = +0.1 W = -0.3
T = +0.2 W = -0.4
T = +0.1 W = -0.4

Fig. 11(E)

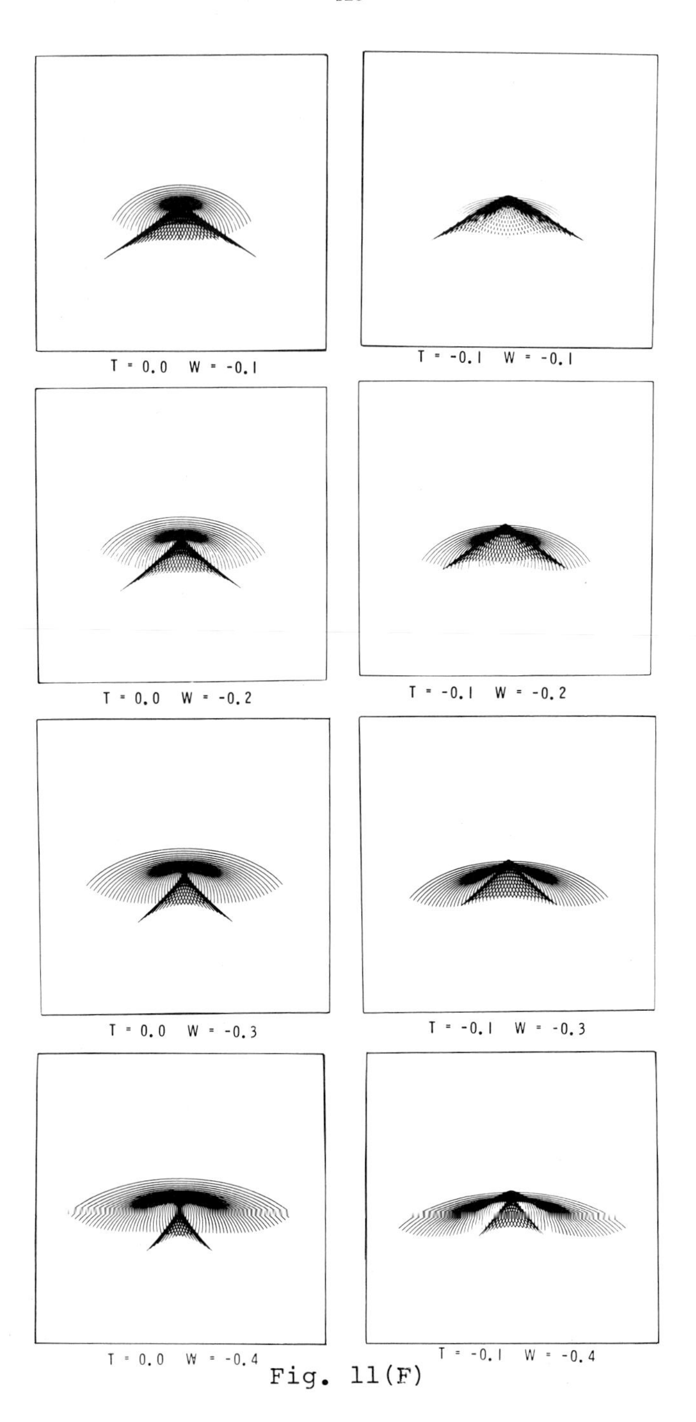
T = 0.0 W = -0.1
T = -0.1 W = -0.1
T = 0.0 W = -0.2
T = -0.1 W = -0.2
T = 0.0 W = -0.3
T = -0.1 W = -0.3
T = 0.0 W = -0.4
T = -0.1 W = -0.4

Fig. 11(F)

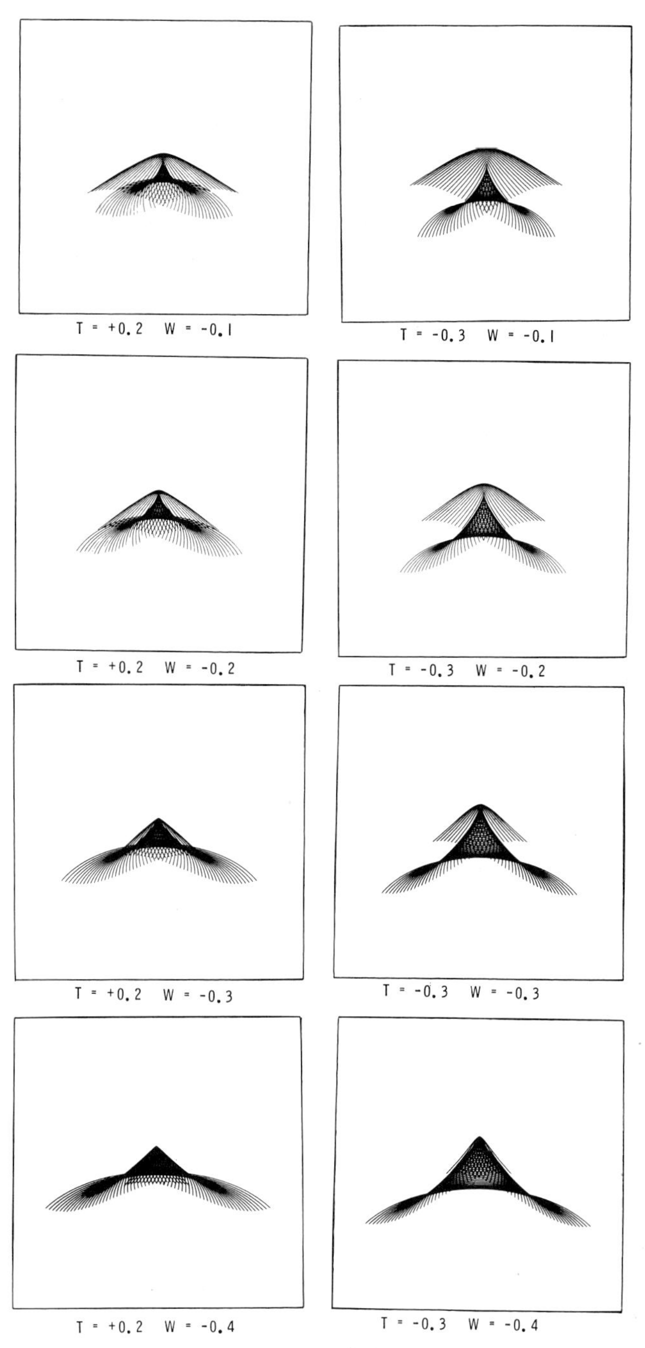
T = +0.2 W = -0.1
T = -0.3 W = -0.1
T = +0.2 W = -0.2
T = -0.3 W = -0.2
T = +0.2 W = -0.3
T = -0.3 W = -0.3
T = +0.2 W = -0.4
T = -0.3 W = -0.4

Fig. 11(G)

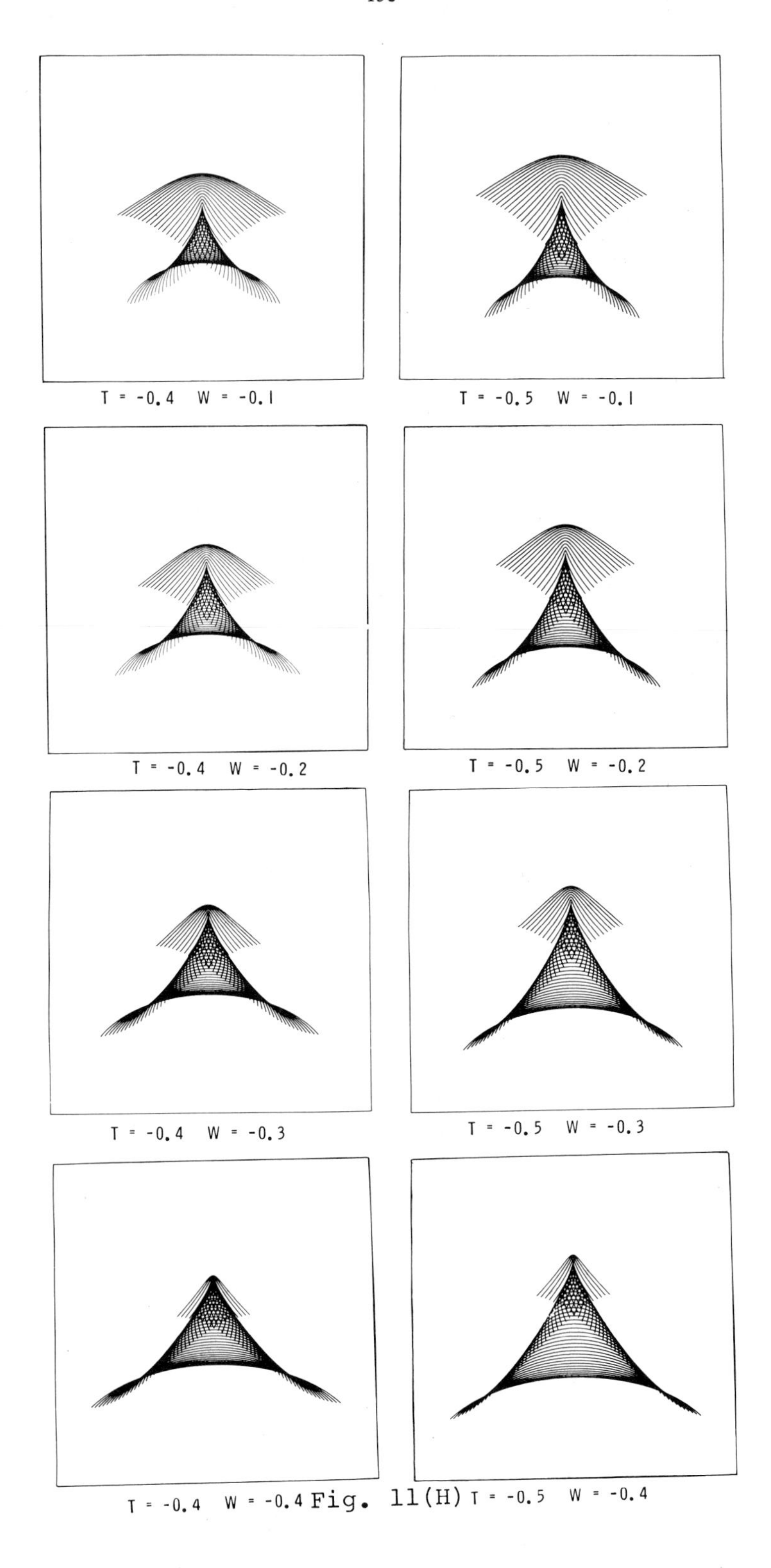
T = -0.4 W = -0.1
T = -0.5 W = -0.1
T = -0.4 W = -0.2
T = -0.5 W = -0.2
T = -0.4 W = -0.3
T = -0.5 W = -0.3
T = -0.4 W = -0.4
T = -0.5 W = -0.4

Fig. 11(H)

Bibliography

[1] Godwin, N., Three Dimensional Pictures for Thom's Parabolic Umbilic, 1972. I.H.E.S. No. 40, p.117-138.

[2] Woodcock, A. E. R. & Poston, T., Geometrical Properties of the Reduced Double Cusp (this volume).

[3] Thom, R., Stabilité Structurelle et Morphogénèse, Published by Benjamin - Addison Wesley, 1973.

[4] Zeeman, E. C., Differential Equations for Heartbeat and Nervous Impulse, In: Towards a Theoretical Biology IV ed. by C. H. Waddington.

STEREOGRAPHIC RECONSTRUCTIONS OF THE CATASTROPHE MANIFOLDS OF THE CUSPOID CATASTROPHES

by

A. E. R. Woodcock

IBM T. J. Watson Research Center
Yorktown Heights, New York

ABSTRACT: The Catastrophe Manifolds of the Cuspoid Catastrophes are generated as surfaces by the loci of maximum and minimum values of a Generalized Potential Energy Function of the form:

$$V = x^n + A \frac{x^{n-2}}{n-2} + \ldots + P x.$$

The present series of papers present a series of sterographic-pair photographs and therefore permit the three-dimensional reconstruction of the Catastrophe Manifold by the optical fusion of the separate pairs of the photographs.

STEREOGRAPHIC RECONSTRUCTIONS OF THE CATASTROPHE MANIFOLD OF THE CUSPOIDS: (1) THE SIMPLE CUSP

Introduction:

In a previous paper, Woodcock and Poston (1) have described the nature of the Catastrophe Manifolds of the Cuspoid Catastrophes in terms of their projection onto planes in the Control Spaces of the Catastrophes. In that presentation, the three-dimensional nature of the Catastrophe Manifold was destroyed by projection onto a two-dimensional plane. The present papers will present reconstructions of the three-dimensionality of the Catastrophe Manifolds by drawing them as warped surfaces in three dimensions.

Method:

The coordinates of the three-dimensional warped surface were input to a computer which was programmed to draw two-dimensional projections of the three-dimensional cube at any given series of angular rotations from the initial direction of projection. In this way, it is possible to give the same impression that one would obtain from looking through the actual three-dimensional surface after such a rotation, along the direction of projection. The new two-dimensional projections were made after rotations of approximately 6 degrees in the "T" and "P" directions (see Fig. 1). The (projected) figures were photographed by a Stromberg-Carlson camera attachment to the computer. The projections of the Catastrophe Manifold are presented in this, and subsequent papers, in the form of sterographic pairs. When each pair of these photographs is viewed through a stereographic viewer, the complete three-dimensional nature of the Catastrophe Manifold is reconstructed, giving a much more realistic description of the three-dimensional surface. Fig. 1 shows the criteria for the generation of the stereographic illusion. The three-dimensional object (in the present case, the surface of the Catastrophe Manifold) is photographed from viewpoints separated by about 6 degrees of arc. The two two-dimensional photographs are placed side-by-side with a separation of about two to three inches between the centers of the photographs (the optimal separation depending upon the actual type of viewer used to form the illusion) so that the picture presented to the right eye was photographed with a greater angular displacement than the one presented to the left eye. The criteria for the creation of the stereographic illusion are summarized for rotations in the "T" and "P" directions in Fig. 1.

Observations:

After the Fold, the Simple (or Riemann-Hugoniot) Cusp is the next most complicated of the Cuspoids, and the first one in which a three-dimensional reconstruction of the curved (x,A,B) Catastrophe Manifold becomes a valid exercise. Creation of the stereographic illusion by the fusion of the stereographic pair of pictures reveals the true S-shaped nature of the (x,A,B) surface. Furthermore, as the (X,A,B) cube is rotated in the P direction, one can look further and further over the edge of the folded surface until, at P = 90 degrees, (Fig. 5) one is looking in a direction perpendicular to the (x,A) plane of the Catastrophe Manifold and the lower edge of the figure appears to be S-shaped. For rotations beyond P = 90 degrees, the two limbs of the surface again overlie the middle region. Rotations of the (x,A,B) cube in the T direction also dramatically show the S-shaped nature of the surface; this is, again, most striking at projections with values of T about 90 degrees.

The Generalized Potential Energy Function for the Simple Cusp is of the form:

$$V = \frac{x^4}{4} + A\,\frac{x^2}{2} + B\,x \qquad (1)$$

Stationary (or maxima and minima) values of the equation (1) occur when:

$$\frac{dV}{dx} = x^3 + A\,x + B = 0 \qquad (2)$$

In general, there will be three real roots of the equation (2) and, for negative values of the parameter A, the equivalent potential energy well will be of the form shown in Fig. 2(A); the three real roots corresponding to the existence of two minima separated by a maxima. However, the Potential Energy Function could be of the form:

$$V = -\left(\frac{x^4}{4} + A\,\frac{x^2}{2} + B\,x\right) \qquad (3)$$

and so:

$$\frac{dV}{dx} = -(x^3 + A\,x + B) = 0 \qquad (4)$$

The roots of the equation (4) now correspond to two maxima of the function, separated by a minima (Fig. 2(B)). The equation (3), as pointed out by Woodcock and Poston (1), is the Dual-Cusp. The Simple Cusp is not Self-Dual since it does not possess an equal number of maxima and minima. However, the Fold, Swallowtail and Wigwam Catastrophes are Self-Dual, since they have an even number of possible real roots; while the Simple Cusp, Butterfly and Star Catastrophes are all Non-Self-Dual Catastrophes. Throughout the presentation we will bear the possibility of Self-Duality in mind, but will only consider Potential Energy Functions of the type described by equation (1) in our analysis.

The pictures presented in this paper show the surface of the Catastrophe Manifold of the Simple-Casp to be simply a plane surface with a twist imposed upon it. The twist arises from the existence of more than one real root of the equation (1) (or, of equation (3)) which thus forces the surface to overlie itself for the relevant values of A and B in the equations (1) or (3), as shown in Figs. 6, 7, 8 and 9.

Criteria for generation of the Stereographic illusion:

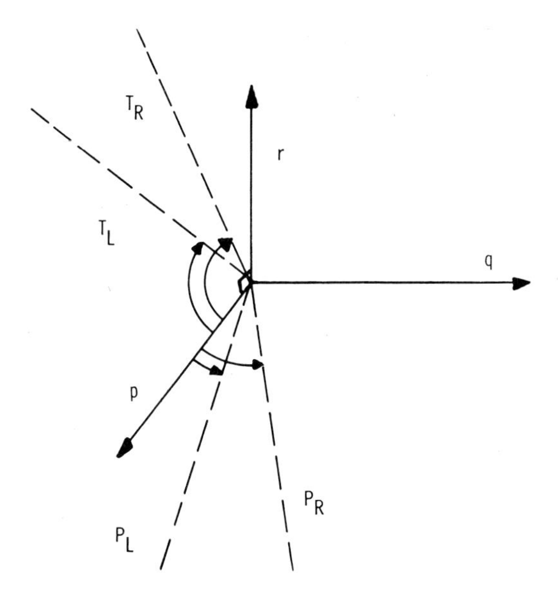

Either: (1) T_R greater than T_L if $P_R = P_L$

Or: (2) P_R greater than P_L if $T_R = T_L$

Difference in T or P must be about 6 degrees of arc.

Fig. 1

The Non-Self-Duality of the Simple Cusp.

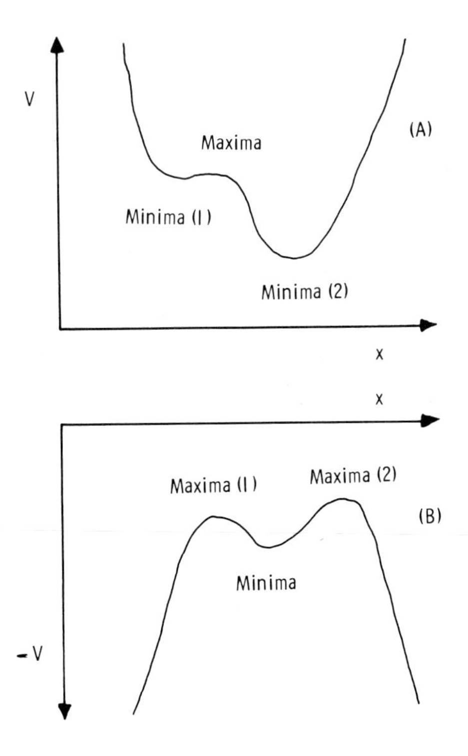

Fig. 2

Figs. 3, 4 and 5 show the rotation of the cube (x,A,B) in the P direction from P = 0 to P = 113 degrees; with T = 0 throughout the sequence. The sequence of pictures from P = 45 degrees through P = 113 degrees show dramatically that the Catastrophe Mfd is, for the Simple Cusp, an S-shaped warped surface (Figs. 4 and 5). The pictures placed side-by-side are stereographic pairs and, as such, permit the generation of an illusion of three-dimensionality when the separate images are fused using a stereographic viewer.

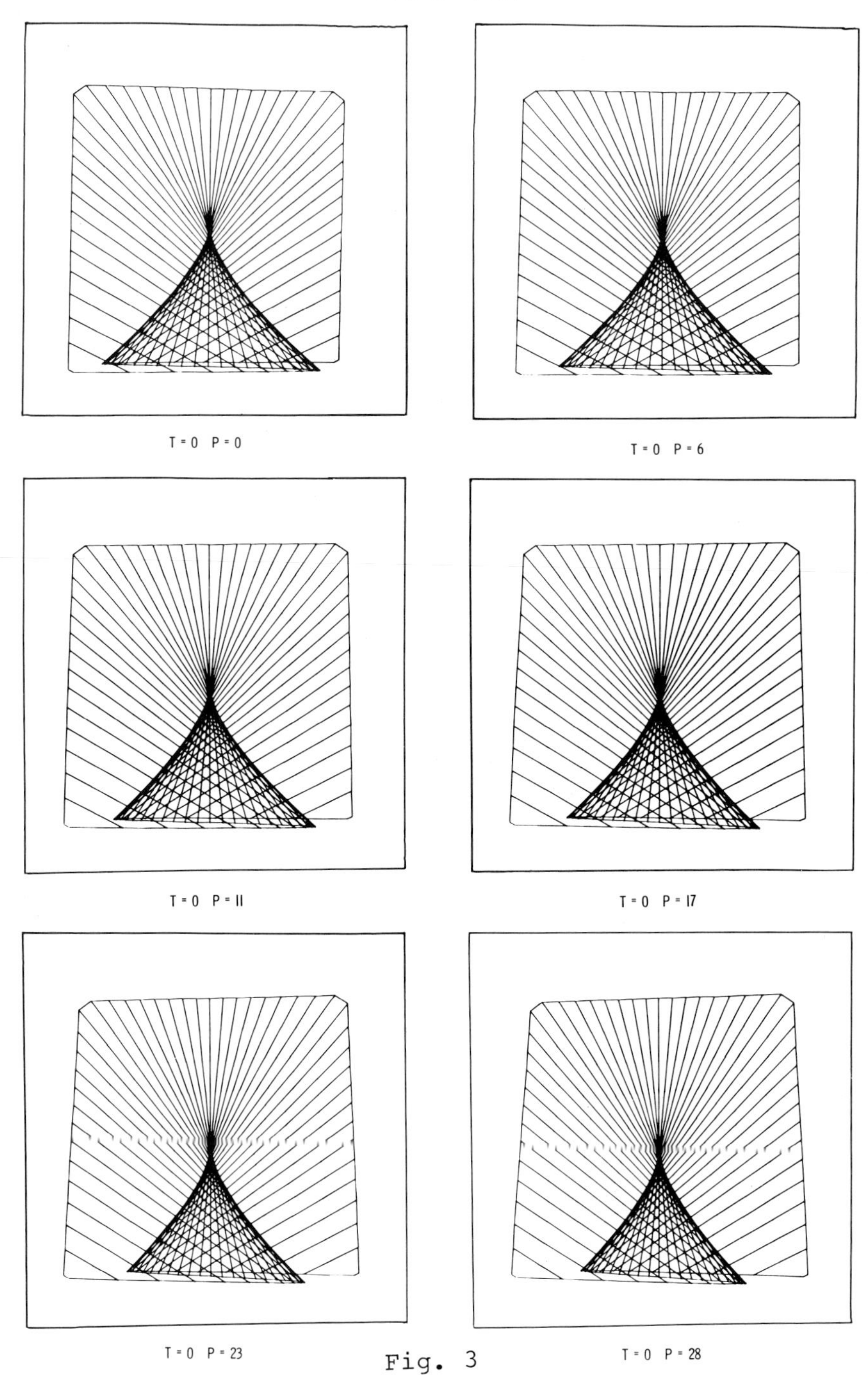
SIMPLE CUSP
T = 0 P = 0
T = 0 P = 6
T = 0 P = 11
T = 0 P = 17
T = 0 P = 23
T = 0 P = 28

Fig. 3

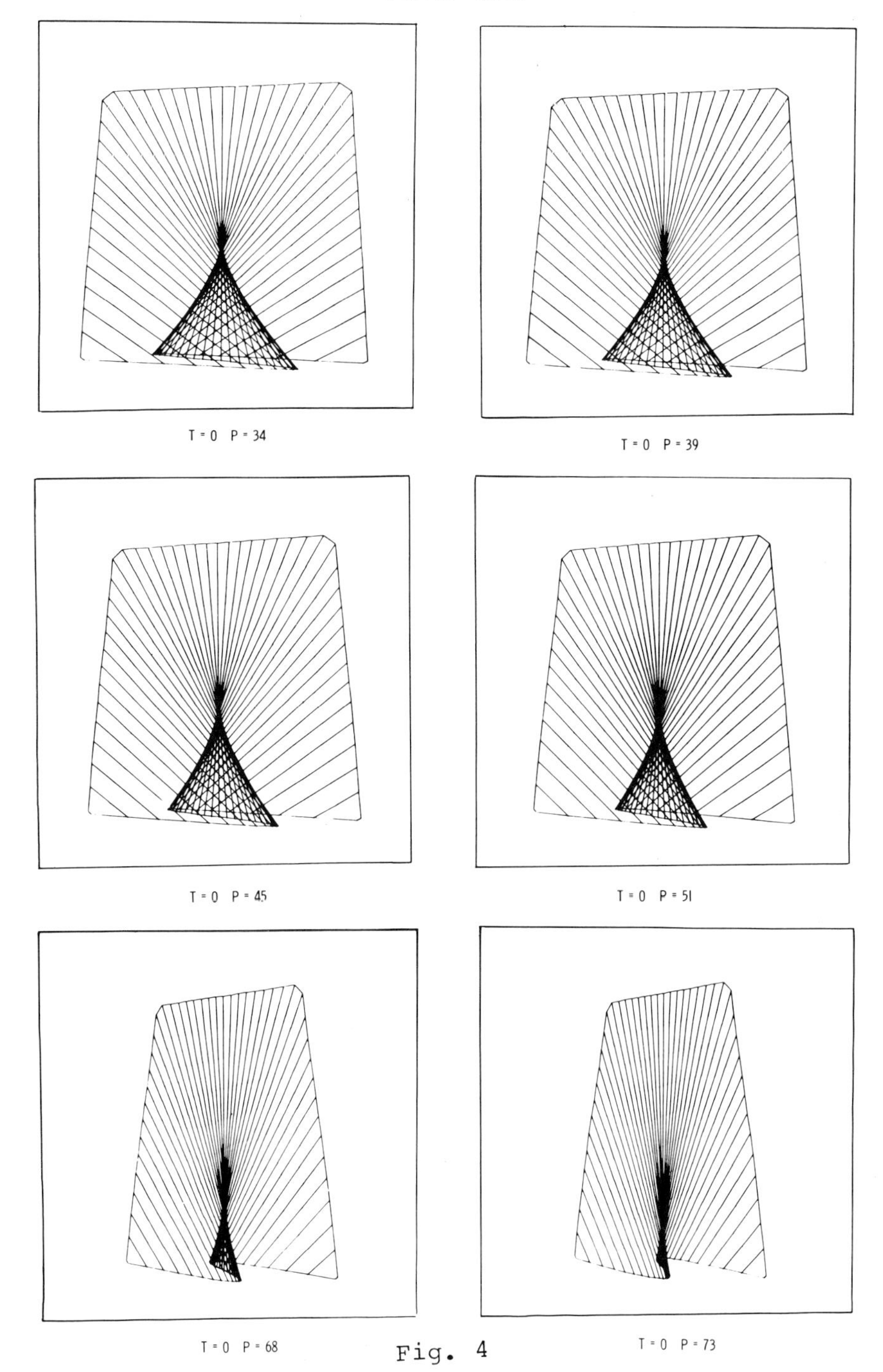
SIMPLE CUSP
T = 0 P = 34
T = 0 P = 39
T = 0 P = 45
T = 0 P = 51
T = 0 P = 68
T = 0 P = 73

Fig. 4

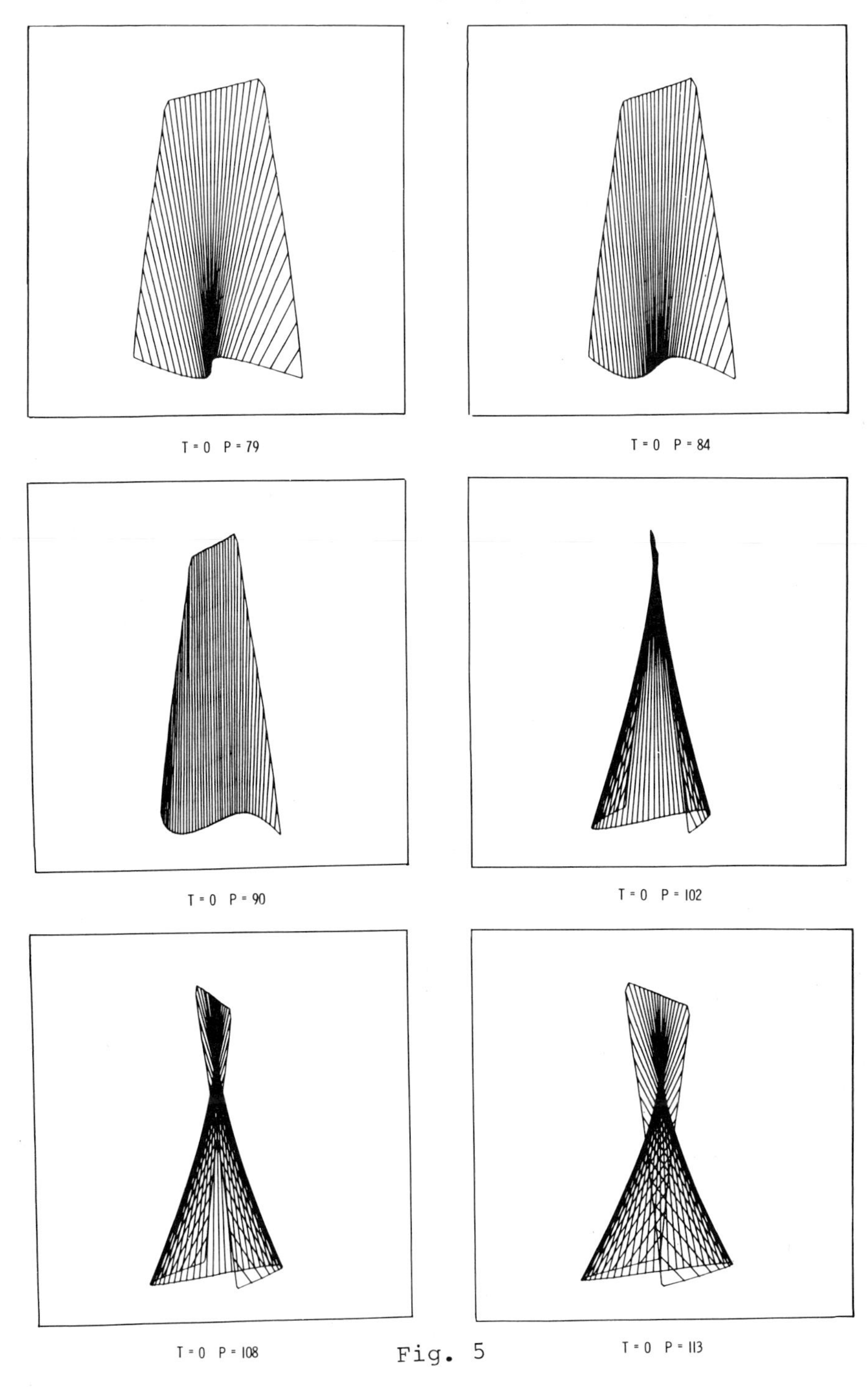

Fig. 5

Figs. 6, 7, 8 and 9 show rotations in the T direction for P = 0 from T = 0 degrees to T = 137 degrees. The sequence from T = 56 degrees to T = 125 degrees again shows the S-shaped nature of the surface (Figs. 6 and 7). The pairs of pictures are stereo-pairs and, as such, permit the generation of an illusion of three-dimensionality when the two pictures are fused using a stereographic viewer.

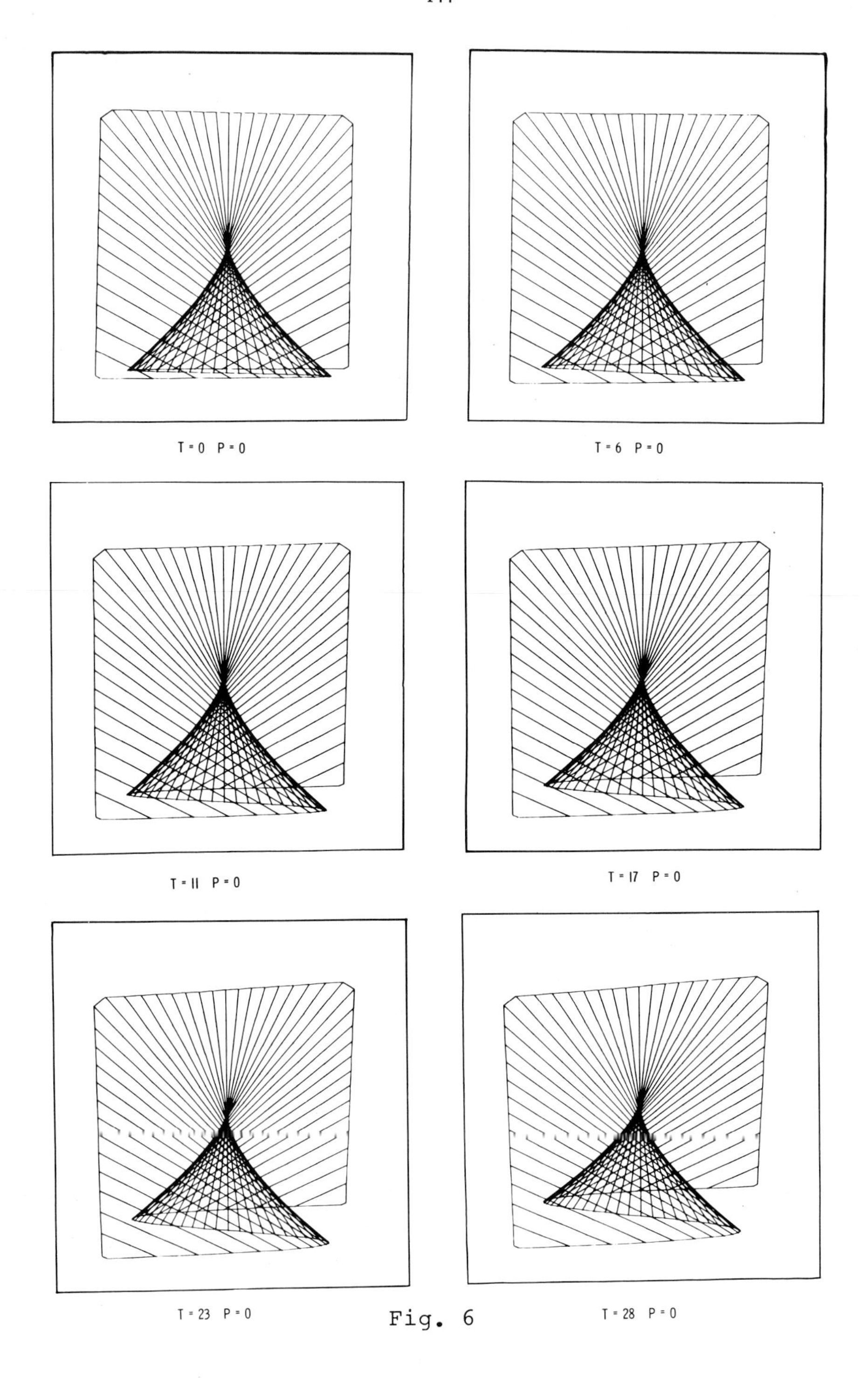
T = 0 P = 0
T = 6 P = 0
T = 11 P = 0
T = 17 P = 0
T = 23 P = 0
T = 28 P = 0

Fig. 6

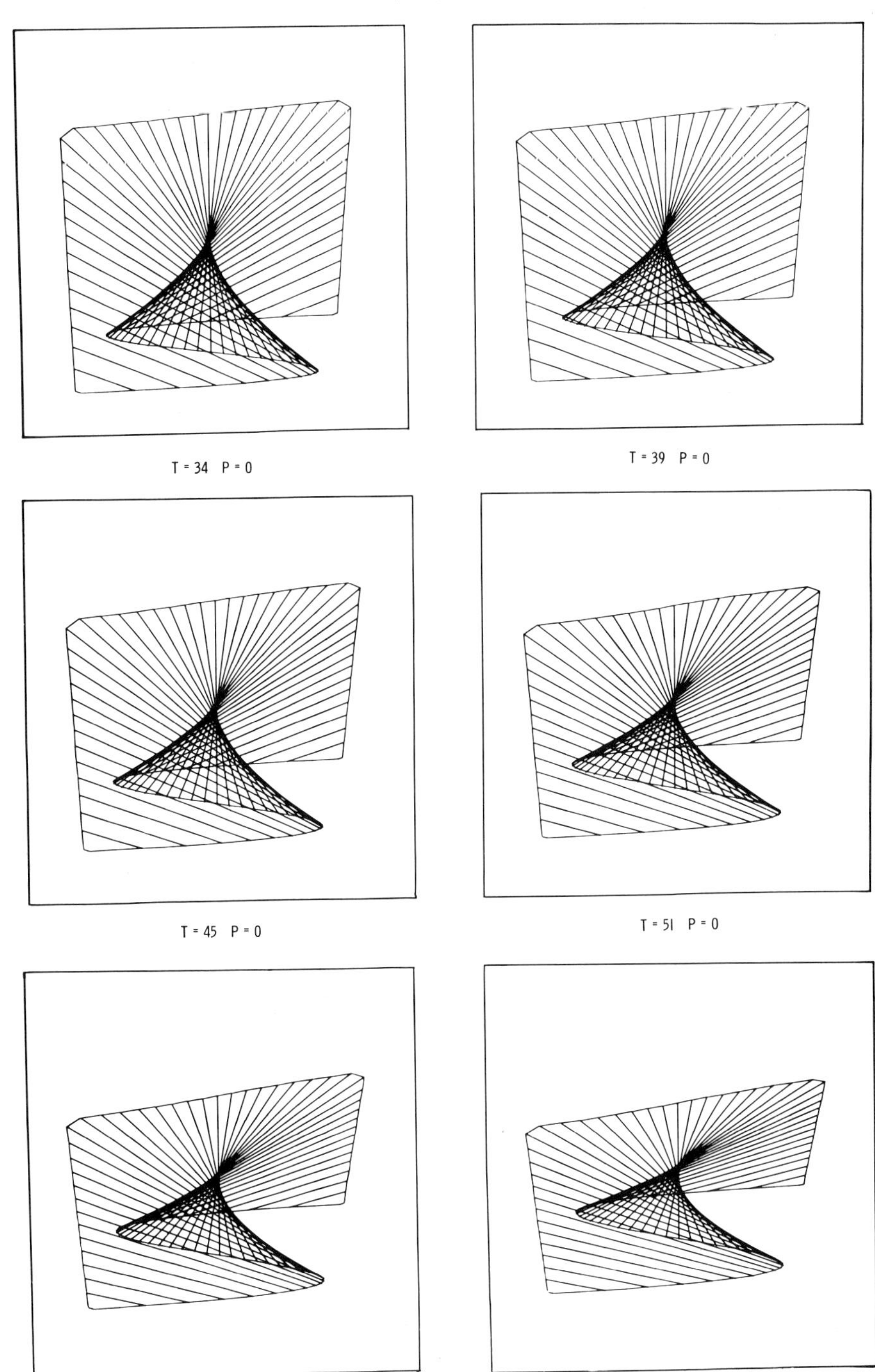
SIMPLE CUSP
T = 34 P = 0
T = 39 P = 0
T = 45 P = 0
T = 51 P = 0
T = 56 P = 0
T = 62 P = 0

Fig. 7

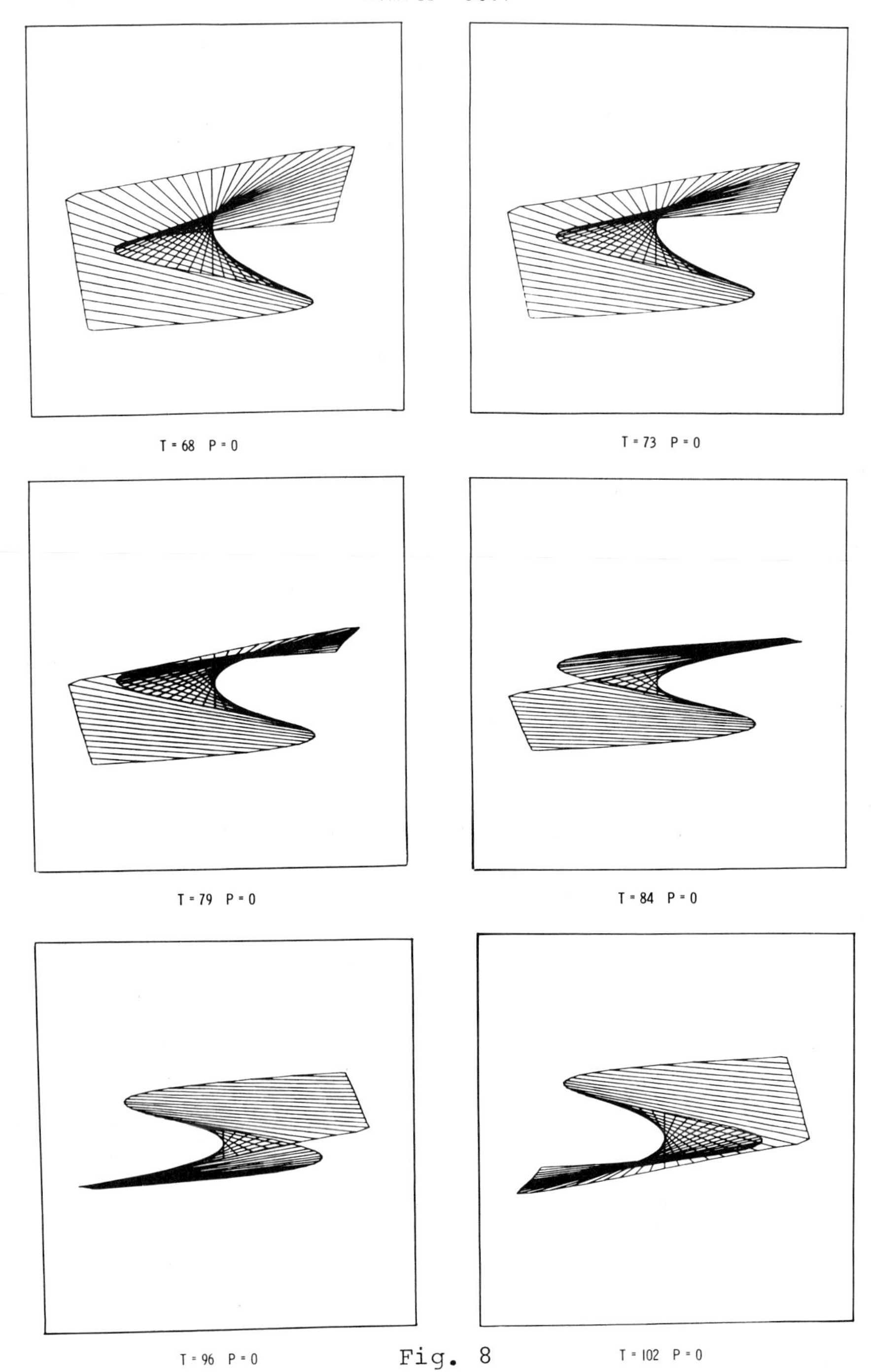

Fig. 8

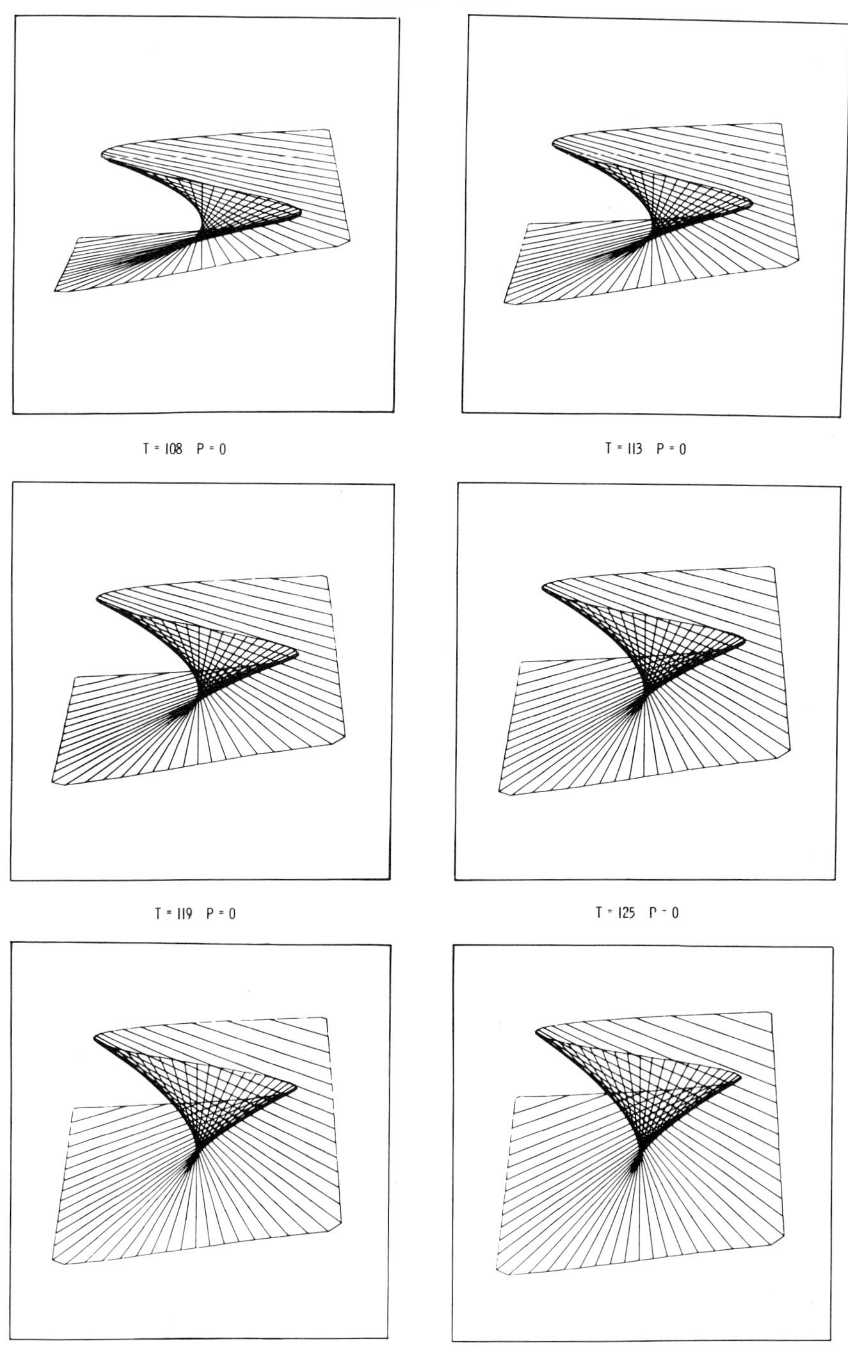
SIMPLE CUSP
T = 108 P = 0
T = 113 P = 0
T = 119 P = 0
T = 125 P = 0
T = 131 P = 0
T = 137 P = 0

Fig. 9

ACKNOWLEDGEMENT

I am grateful to Arthur Appel for permission to use one of his Computer Graphics Programs and for his interest and encouragement in this study.

Bibliography:

(1) Woodcock, A.E.R. and Poston, T., The Geometrical Properties of the Elementary Catastrophes, (1) The Cuspoids. (this volume.)

STEREOGRAPHIC RECONSTRUCTIONS OF THE CATASTROPHE MANIFOLDS OF THE CUSPOIDS: (2) THE SWALLOWTAIL.

Introduction:

The Swallowtail Catastrophe is derived from a Generalized Potential Energy Function of the following form:

$$V = \frac{x^5}{5} + A\frac{x^3}{3} + B\frac{x^2}{2} + C\,x \qquad (1)$$

The Control Space of the Swallowtail Catastrophe (A,B,C) is three-dimensional. The stationary values of equation (1) occur when:

$$\frac{dV}{dx} = x^4 + A\,x^2 + B\,x + C = 0 \qquad (2)$$

The equation (2) is a quartic equation and as such has, at most, four real roots. Thus the Catastrophe Mfd (x,A,B,C) of the Swallowtail Catastrophe overlies itself at most four times for the relevant values of the parameters A, B and C in equation (1). The existence of an even number of real roots of equation (2) determines that the Swallowtail Catastrophe is a Self-Dual Catastrophe (see, 1). Woodcock and Poston (2) have described the development of the characteristic Swallowtail bifurcation set and show that the richest regions of the Catastrophe Mfd occur for negative values of the parameter A in equation (1). A section of the Catastrophe Mfd , taken with A = -10.0, was therefore chosen as the region of interest for the present study. The figures were drawn as stereographic pairs and presented as described in (1).

Observations:

The essential four-sheet (folded) nature of the Catastrophe Mfd is shown, for example, in Fig. 1. In this and subsequent figures, increasing P from P = 0 degrees to P = 90 degrees (Figs. 1, 2 and 3) affords a view of the Catastrophe Mfd at right angles to the x-axis. These pictures (especially when viewed stereoscopically) demonstrate that the characteristic Swallowtail-like sections arise from a projection onto the Control Space of a Catastrophe Mfd consisting of two Simple Cusps with a common limb. In other words, the Swallowtail may be generated from the Simple Cusp by folding one of its surfaces of minima so that it comes to lie underneath the other surfaces. This operation creates another surface of maxima and forces the Catastrophe Mfd to have no lower minima. This behavior is summarized in Fig. 10. In this figure, the sheets of maxima and minima are simply shown without the twist in the surface that is responsible for producing the characteristic triangular area of overlap of the Swallowtail. Fig. 10 demonstrates the essential similarity between the Catastrophe Mfds of the Fold Catastrophe ($V = \frac{x^3}{3} + A\,x$) and the Swallowtail ($V = \frac{x^5}{5} + A\,\frac{x^3}{3} + B\,\frac{x^2}{2} + D\,x$) in that both have an equal number of maxima and minima of the Generalized Potential Energy Function, and both have no lower minimum value for that function. Fig. 11 shows a reconstruction of the Catastrophe Mfd showing the folded nature of the surface with the twist present, and thus the triangular area of overlap of the various sheets of maxima and minima of the surface.

It is interesting to note that in the sequence of pictures with T varying from T = 0 degrees to T = 73 degrees (P = 0 degrees throughout), the envelope generated by the ruled lines (Figs. 5, 6 and 7) varies in a manner very reminiscent of the variations reported by Woodcock and Poston (2; Fig. 7) for variations of A alone, the projections always being along the x-axis. Fig. 7 shows that the Catastrophe Mfd of the Swallowtail is simply a plane with an imposed twist and Figs. 8 and 9 show very clearly the generation of typical Swallowtail morphologies by the apparent folding over of the surface as the Catastrophe Mfd is rotated about the T axis.

Figs. 1, 2, 3 and 4 show, for the Catastrophe Mfd of the Swallowtail Catastrophe ($V = \frac{x^5}{5} + A\frac{x^3}{3} + B\frac{x^2}{2} + C\,x$); with A = -10.0, a sequence of projections from the (x,B,C) space with T = 0 degrees and P ranging from P = 34 degrees to P = 131 degrees in increments of 6 degrees.

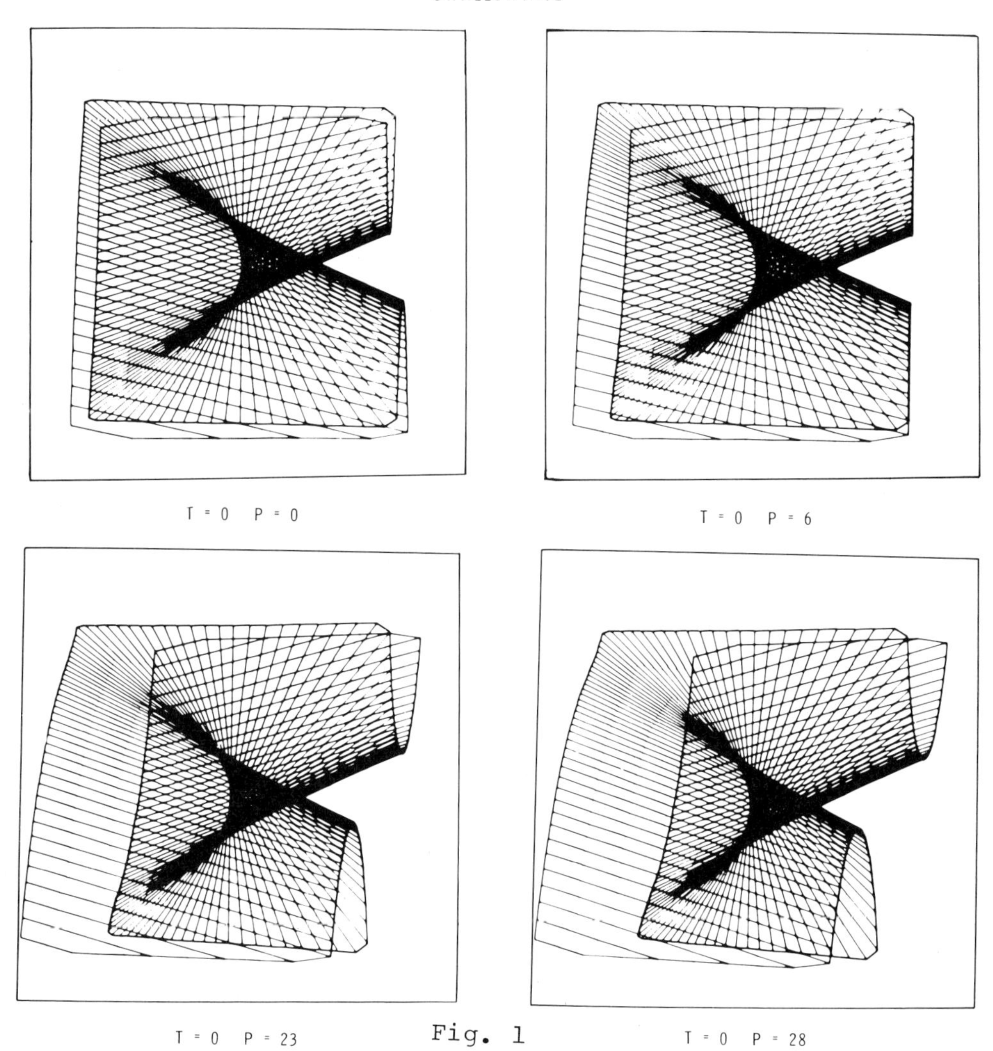

Fig. 1

SWALLOWTAIL

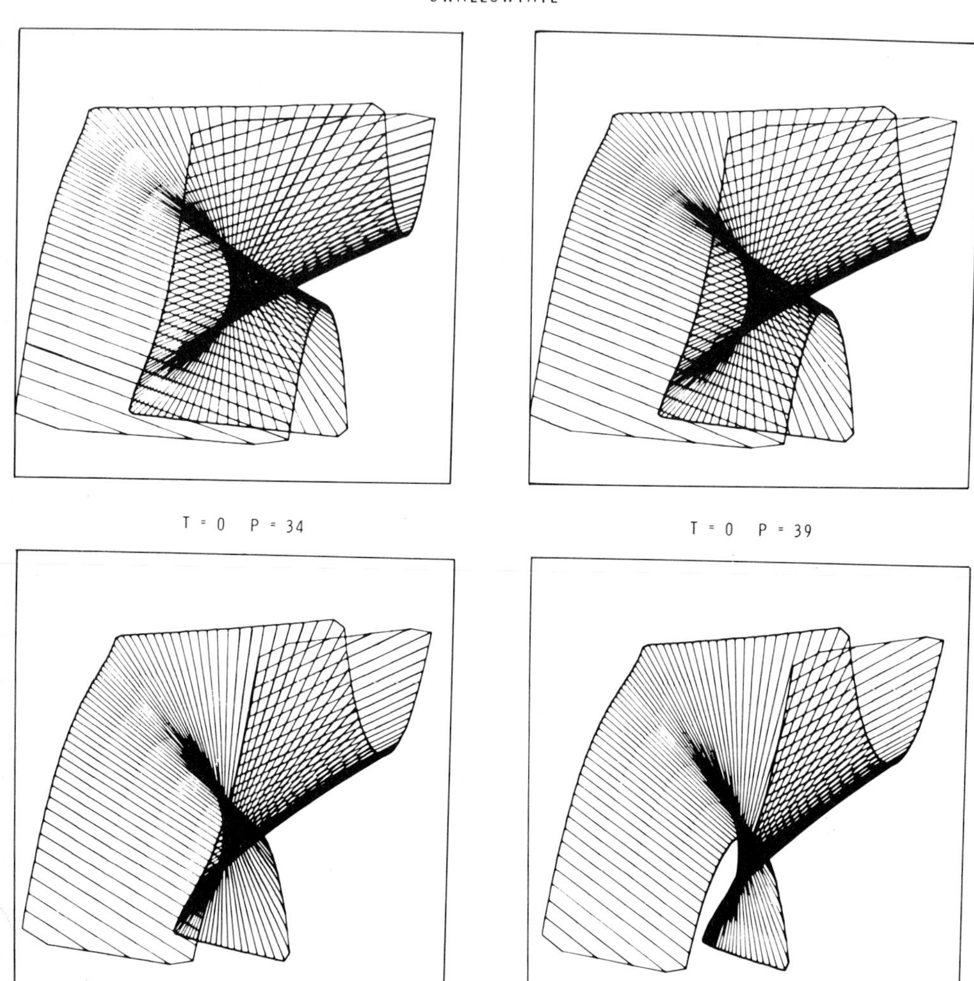

Fig. 2

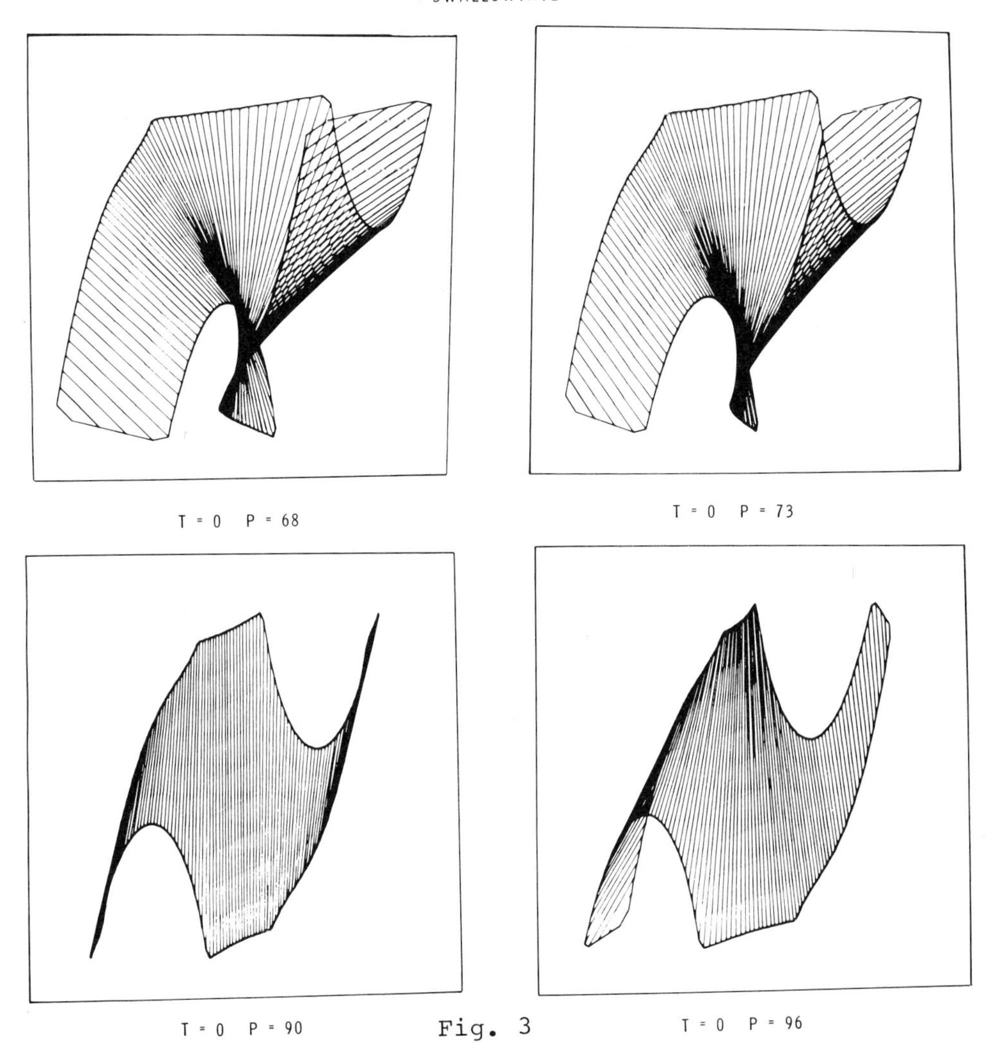

Fig. 3

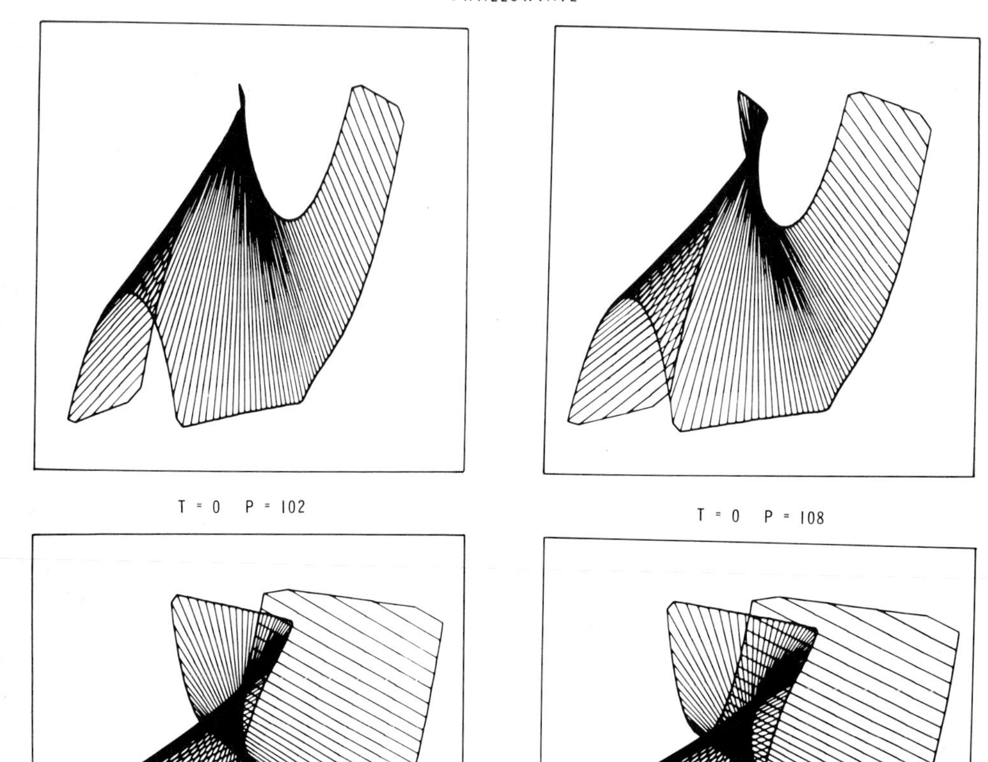

Fig. 4

Figs. 5, 6, 7, 8 and 9 show the sequence of projections of the Catastrophe Mfd of the Swallowtail Catastrophe ($V = \frac{x^5}{5} + A\frac{x^3}{3} + B\frac{x^2}{2} + C\,x$) drawn with A = -10.0 with P = 0 degrees and T varying in 6 degree steps from T = 0 to T = 172 degrees.

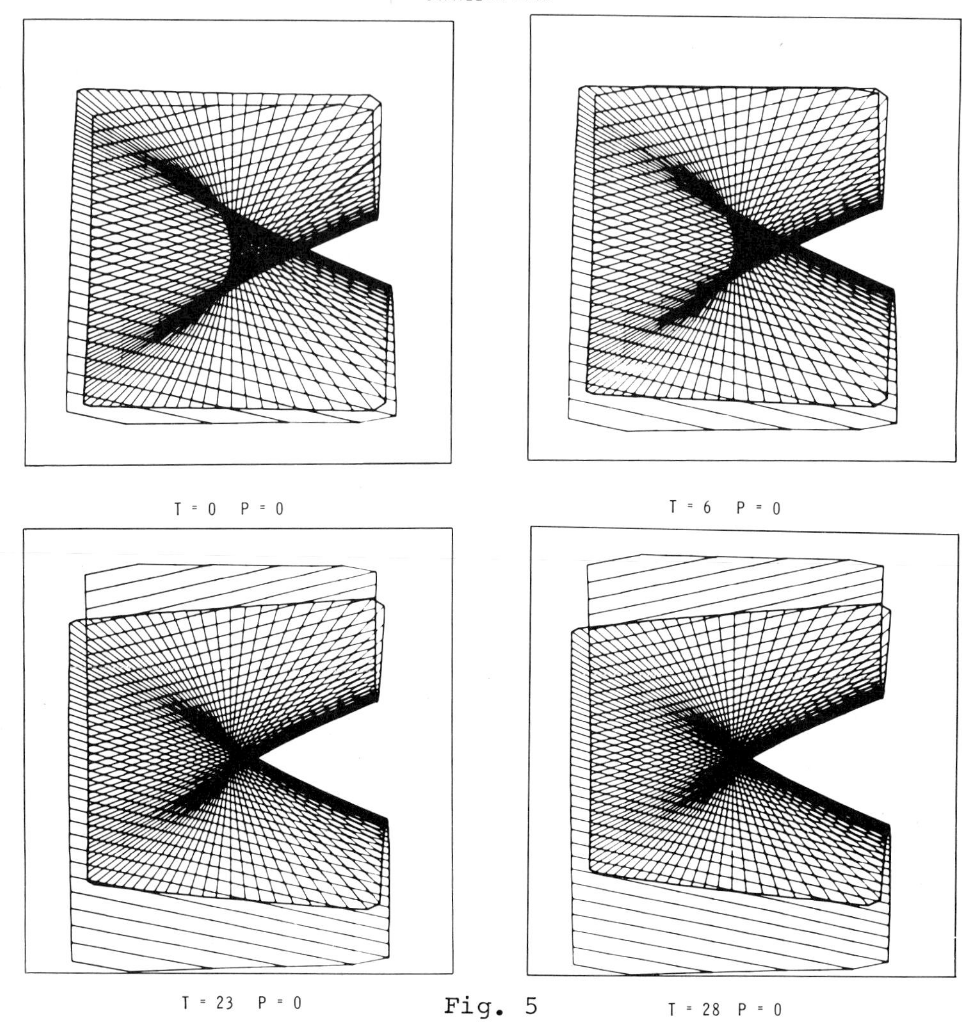

Fig. 5

SWALLOWTAIL

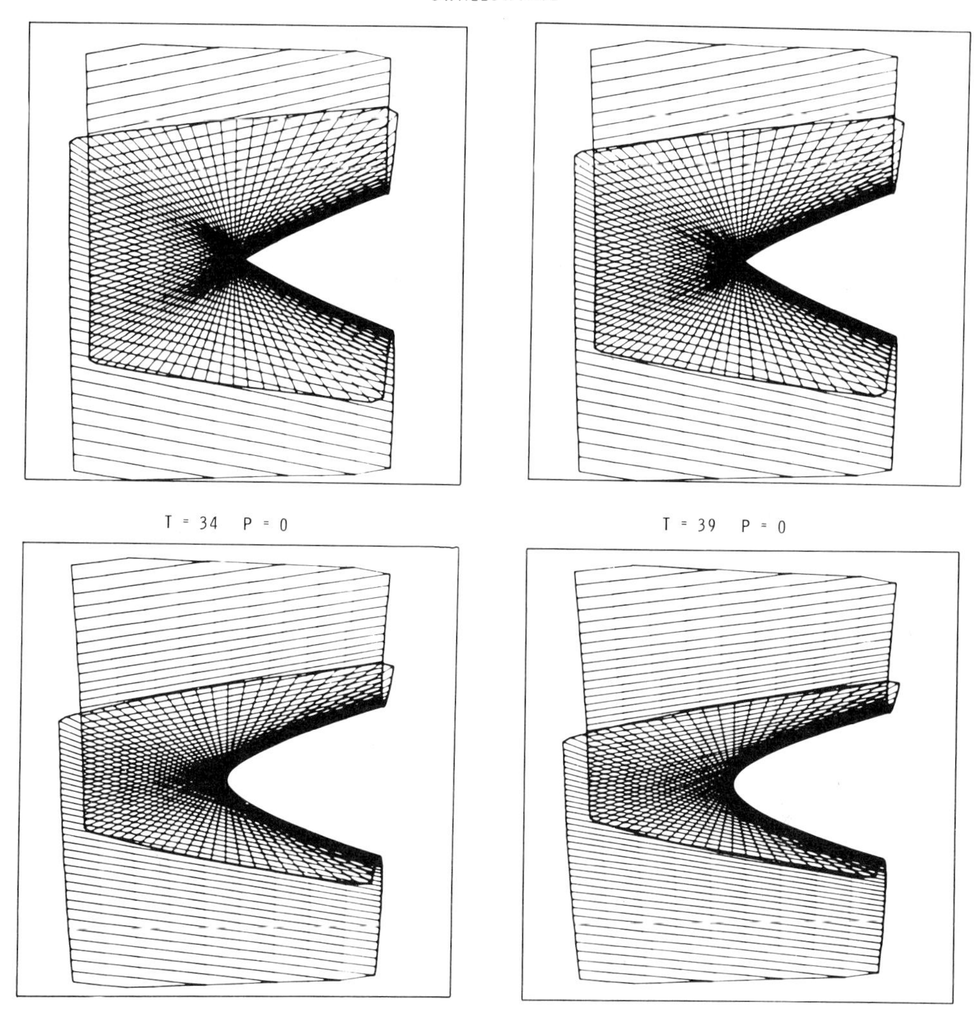

Fig. 6

SWALLOWTAIL

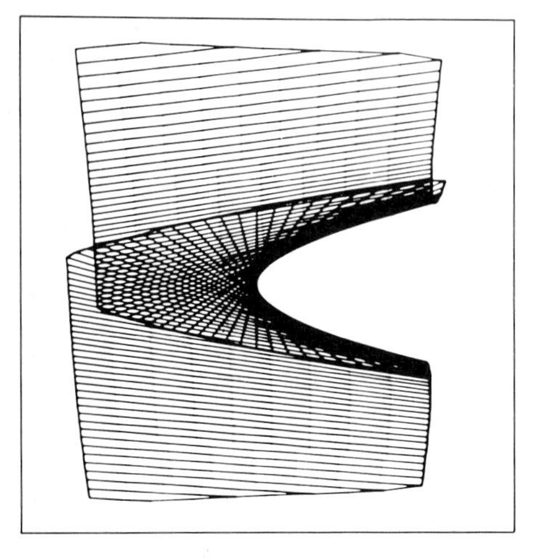

T = 68 P = 0

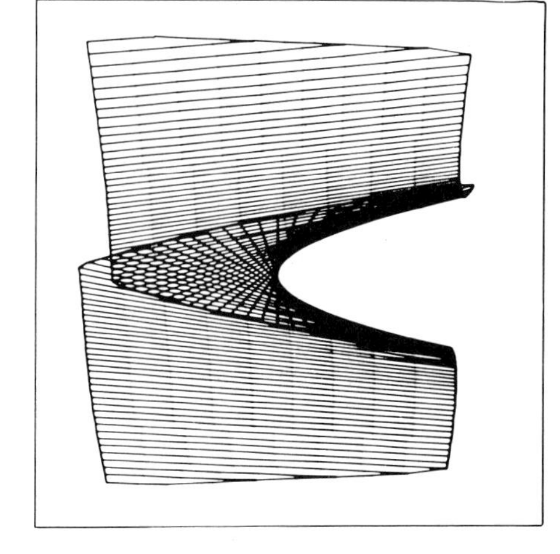

T = 73 P = 0

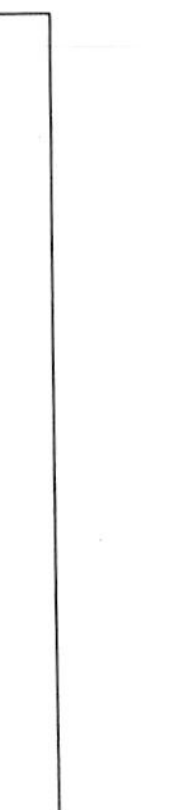

T = 96 P = 0

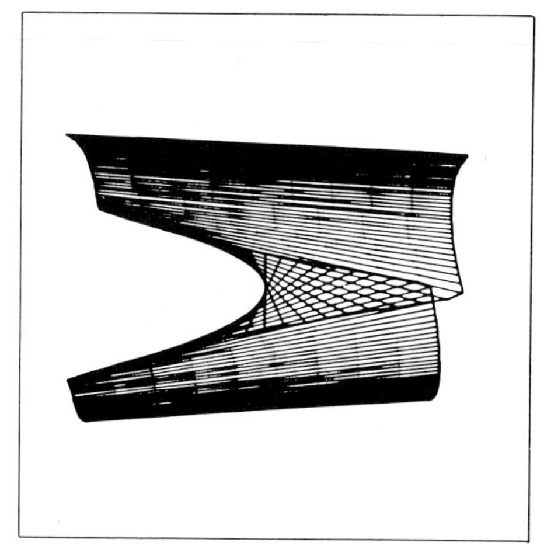

T = 102 P = 0

Fig. 7

SWALLOWTAIL

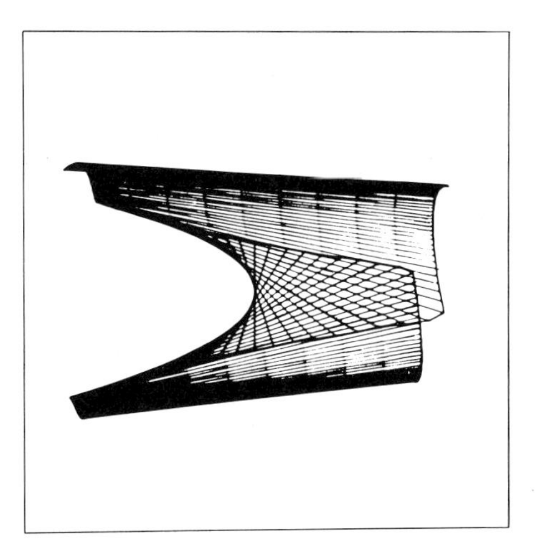

T = 108 P = 0

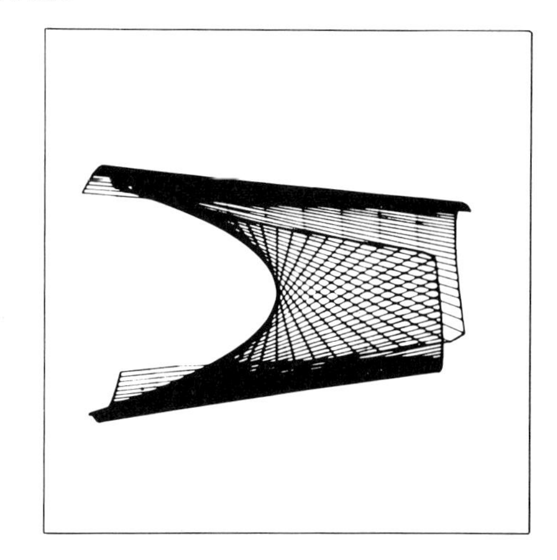

T = 113 P = 0

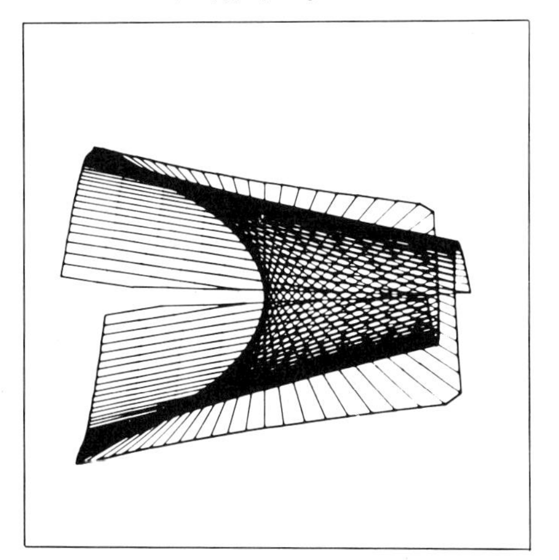

T = 131 P = 0

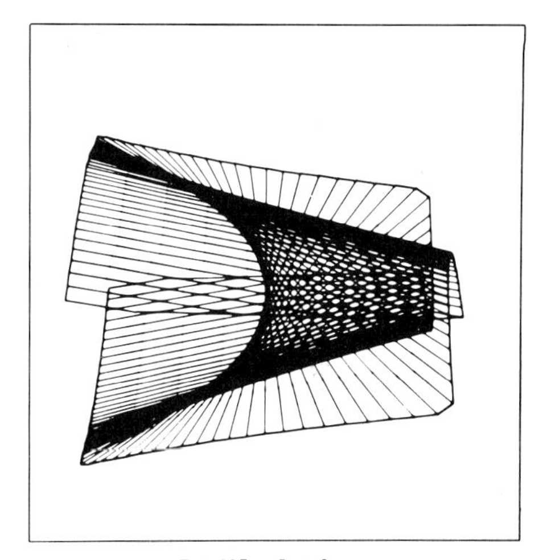

T = 137 P = 0

Fig. 8

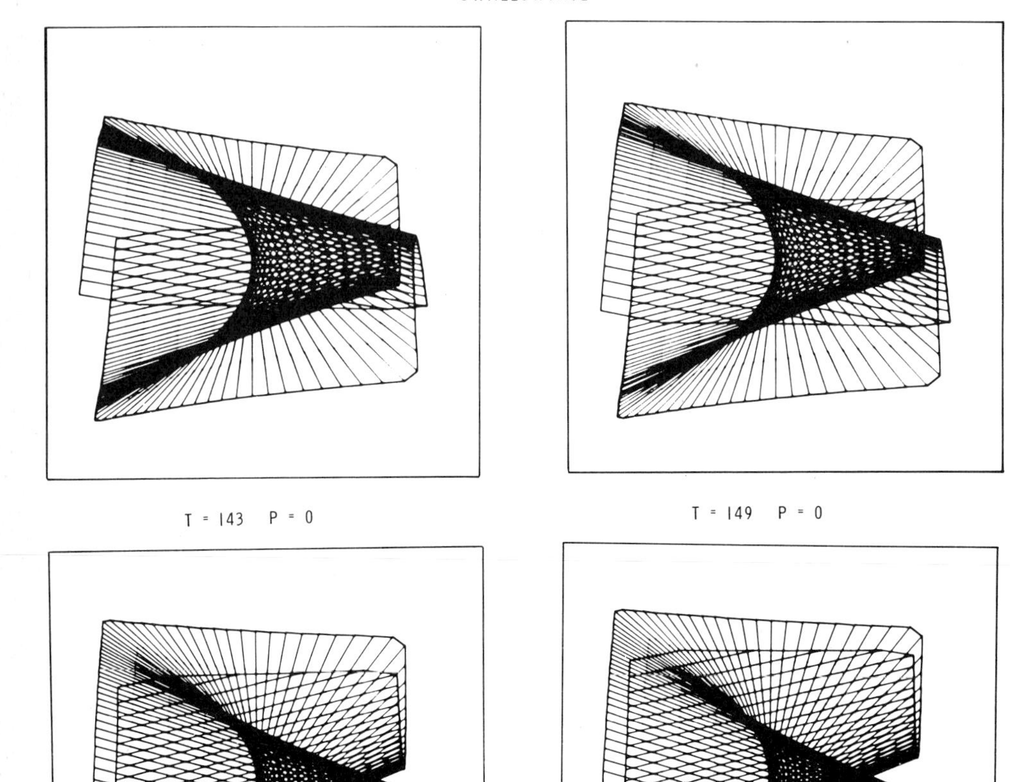

Fig. 9

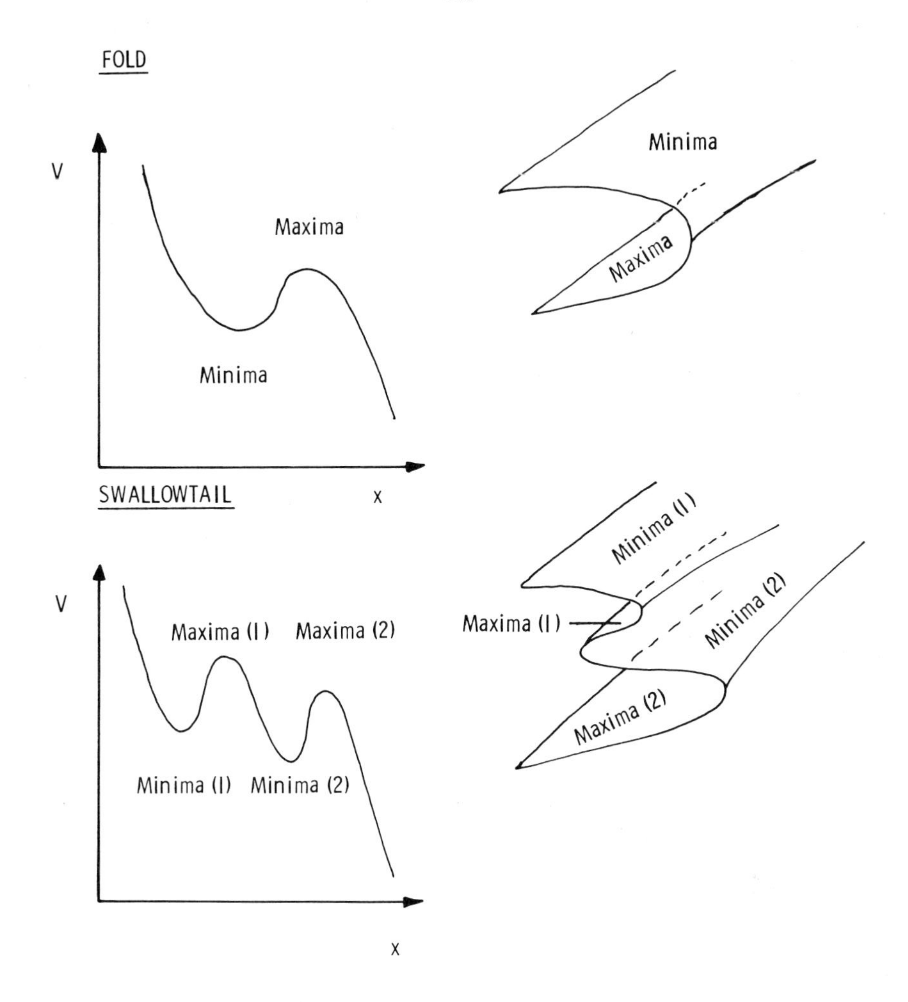
FOLD
V
Maxima
Minima
X
Minima
Maxima
SWALLOWTAIL
V
Maxima (1)
Maxima (2)
Minima (1)
Minima (2)
X
Minima (1)
Maxima (1)
Minima (2)
Maxima (2)

Fig. 10

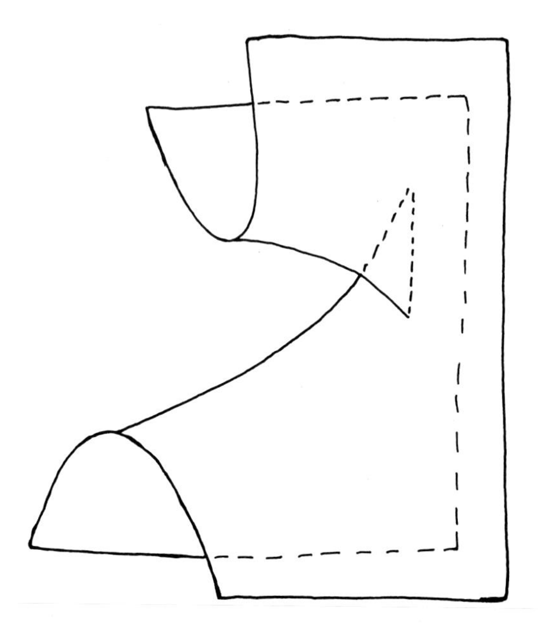

Fig. 11

Bibliography:

(1) Woodcock, A.E.R., Stereographic Reconstructions of the Catastrophe Mfd of the Cuspoids: (1) The Simple Cusp. (this volume.)

(2) Woodcock, A.E.R. and Poston, T., The Geometrical Properties of the Elementary Catastrophes. (1) The Cuspoids. (this volume.)

STEREOGRAPHIC RECONSTRUCTIONS OF THE CATASTROPHE MANIFOLDS OF THE CUSPOIDS: (3) THE BUTTERFLY

Introduction:

The Butterfly Catastrophe arises from a Generalized Potential Energy Function of the following type:

$$V = \frac{x^6}{6} + A\,\frac{x^4}{4} + B\,\frac{x^3}{3} + C\,\frac{x^2}{2} + D\,x \qquad (1)$$

Since equation (2), describing the stationary values

$$\frac{dV}{dx} = x^5 + Ax^3 + Bx^2 + Cx + D = 0 \qquad (2)$$

of the function V, with respect to changes in x, has at most five real roots, the Butterfly Catastrophe belongs to the same sub-family of cuspoids as the Simple Cusp (1) since they both have an odd number of possible real roots, and are both Non-Self-Dual Catastrophes. The potential energy well of the Butterfly Catastrophe described by equation (1) will have, in general, an odd number of minima and an even number of maxima. Therefore, in some regions of the Catastrophe Mfd (and for some orientations of the surface) there will be up to five layers of the Catastrophe Mfd overlying each other.

The figures were drawn with A = -10.0 and B = 0.0 as stereographic pairs and presented as described in (1).

Observations:

Figs. 1 and 2 show that the Butterfly Catastrophe is generated from two Swallowtail Catastrophe sections (2) with a common limb. Figs. 3 and 4 clearly show the generation of the five overlying layers of the surface as the orientation is changed. Figs. 5 and 6 show changes in the projection of the Catastrophe Mfd as T is changed from T = 0 degrees to T = 63 degrees (P = 0). The changes induced in the envelope of the lines ruled on the Catastrophe Mfd by this rotation is very similar to the effect on the morphology of this envelope when the B parameter is varied (with A fixed as negative) and the projection performed solely along the x-axis (Woodcock and Poston, 3 (Fig. 8B)). The effect is shown in reverse in Fig. 8. Fig. 9 shows the similarities between the Simple Cusp and the Butterfly Catastrophes in terms of their Catastrophe Mfds and the nature of their potential energy wells.

Swallowtail-like Sections of the Butterfly Catastrophe

Introduction:

As Woodcock and Poston (3) have pointed out, higher order Catastrophes contain as specific sections the lower order Catastrophes. So, while the projection of the Butterfly Catastrophe Mfd (x,A,B,C,D) onto the (C,D) plane of the Control Space produces Butterfly Sections, the projection onto the (B,C) plane produces Swallowtail-like sections (Woodcock and Poston (3)). These sections are also generated in the case of the true Swallowtail by projection of the (x,A,B,C) Catastrophe Mfd onto the (B,C) plane of the Control Space. Therefore, by projecting onto these different axial systems, it is possible to observe the way in which the Catastrophes of lower dimension are constructed from foldings of the Catastrophe Mfd .

Observations:

Using the approach mentioned in the Introduction, (A,D) fixed section of the Butterfly Catastrophe were drawn by projection onto the (B,C) plane (A = -10.0 and D = -2.0). The figures demonstrate that the Swallowtail is, as previously pointed out (2), the result of the apposition of two Simple Cusps which share a common limb. This becomes more clear as P is increased to about 62 degrees (T = 0 degrees), Fig. 11. Indeed, the whole sequence from P = 0 degrees to P = 131 degrees (Figs. 10 to 13) clearly show the juxtaposition of the two Simple Cusps. When D is zero, the two layers of the surface become fused (in the present example they are separated into regions with x positive and those with x negative) to form the true Swallowtail Sections. Rotations giving an increase in T (Figs. 14 and 15) produce the same changes as those reported by Woodcock and Poston (3, Fig. 9B) when D is varied for negative values of A. Figs. 14, 15, 16, 17 and 18 again show the two separate regions of the surface and the apposition of the two Simple Cusp regions of the Catastrophe Mfd .

Figs. 1, 2, 3 and 4 show projections from the Catastrophe Mfd (x,A,B,C,D) of the Butterfly Catastrophe:

$$V = \frac{x^6}{6} + A\frac{x^4}{4} + B\frac{x^3}{3} + C\frac{x^2}{2} + D\,x$$

for A = -10.0 and B = 0.0. T = 0 degrees throughout and P is varied in 6 degree increments from P = 0 degrees to P = 131 degrees.

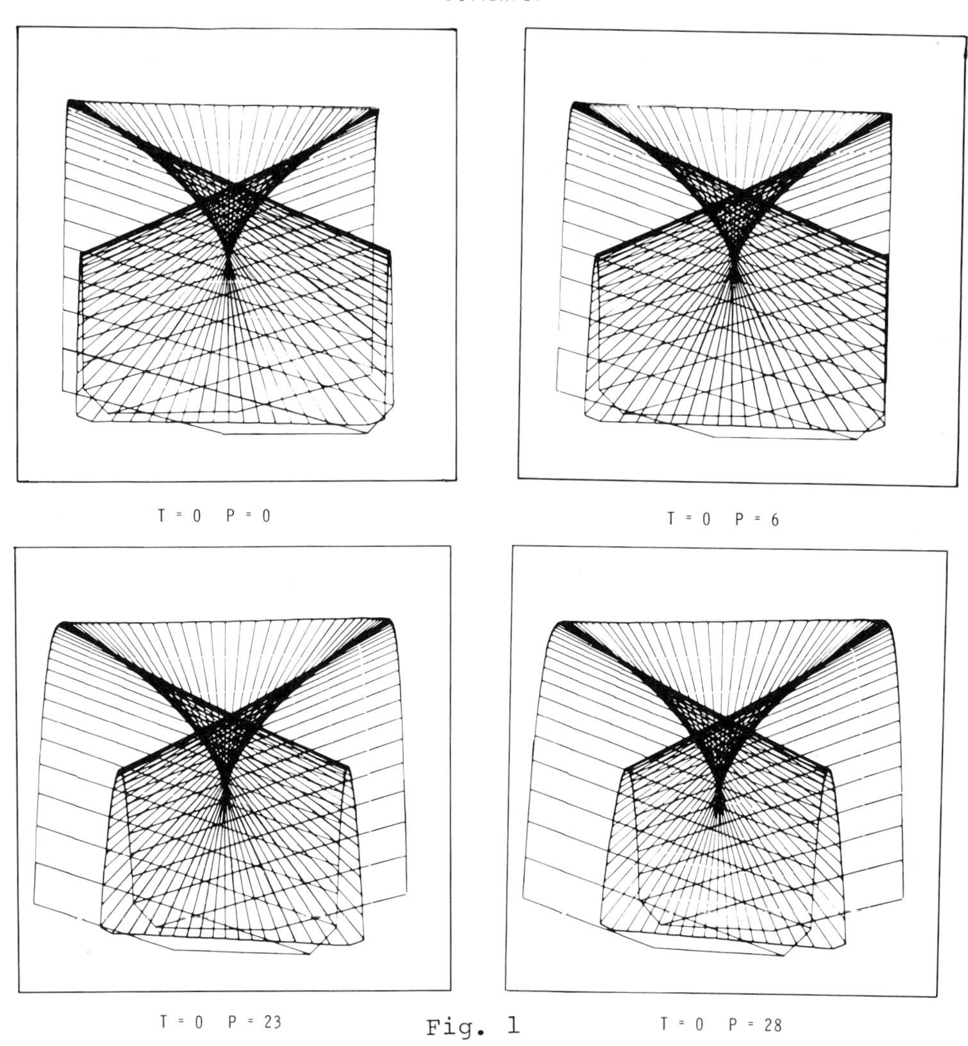

Fig. 1

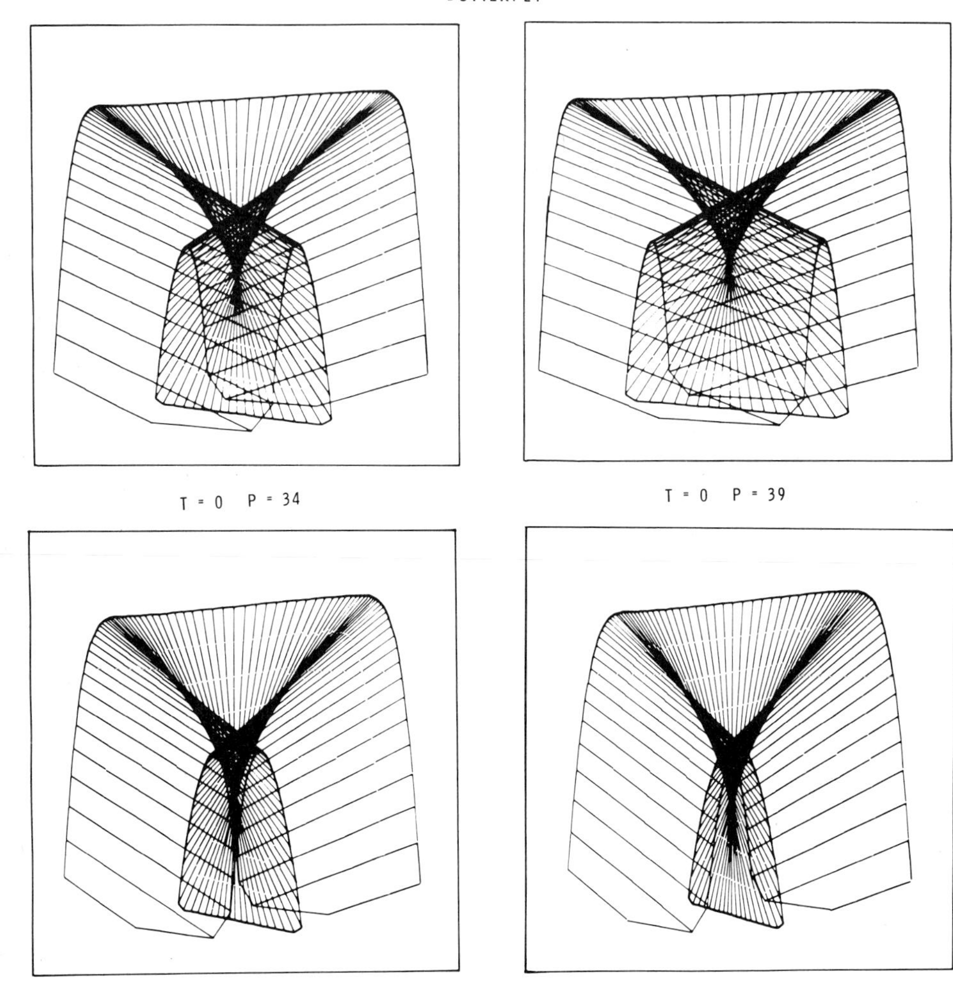

Fig. 2

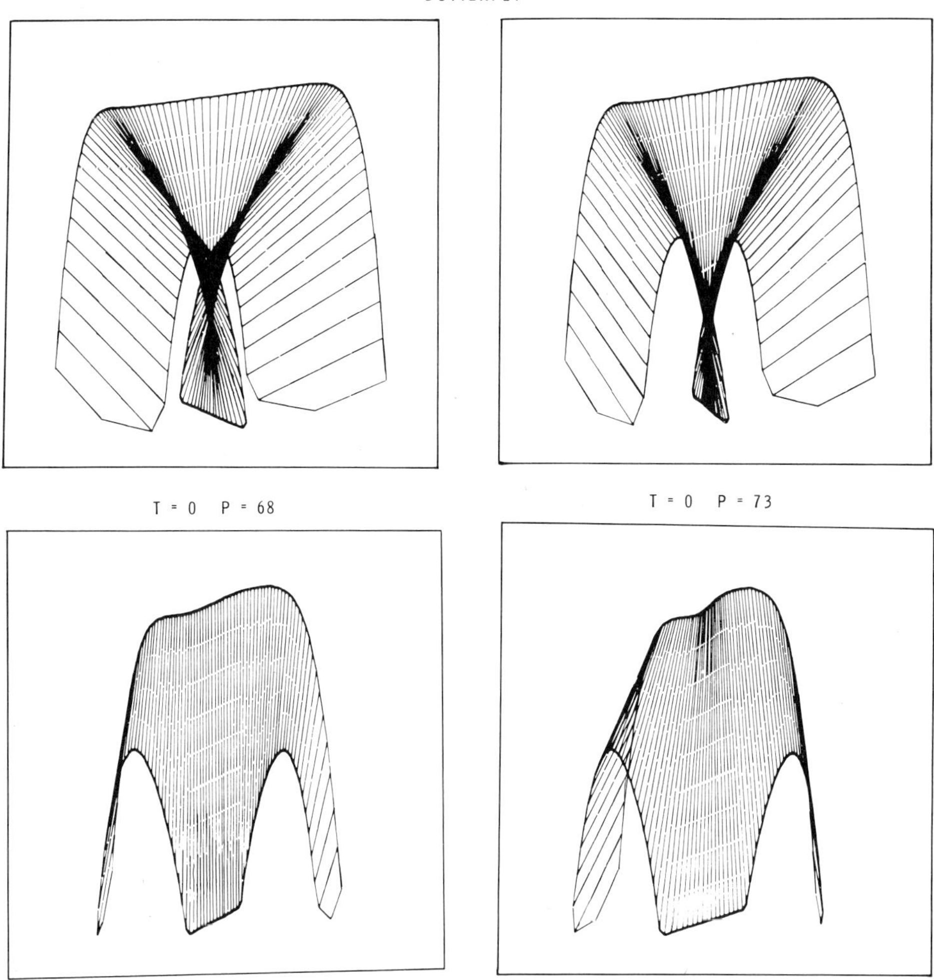

Fig. 3

BUTTERFLY

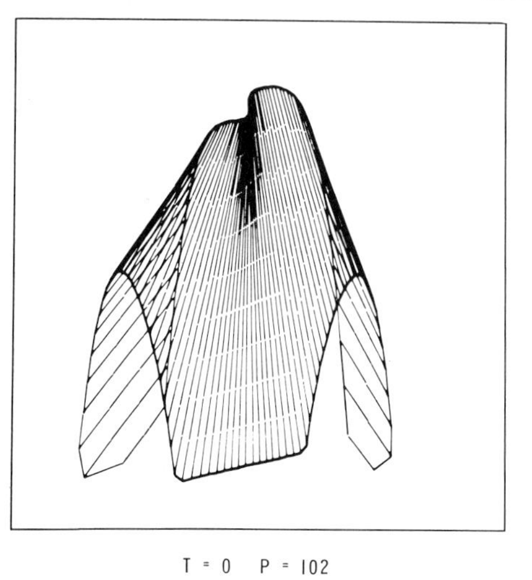

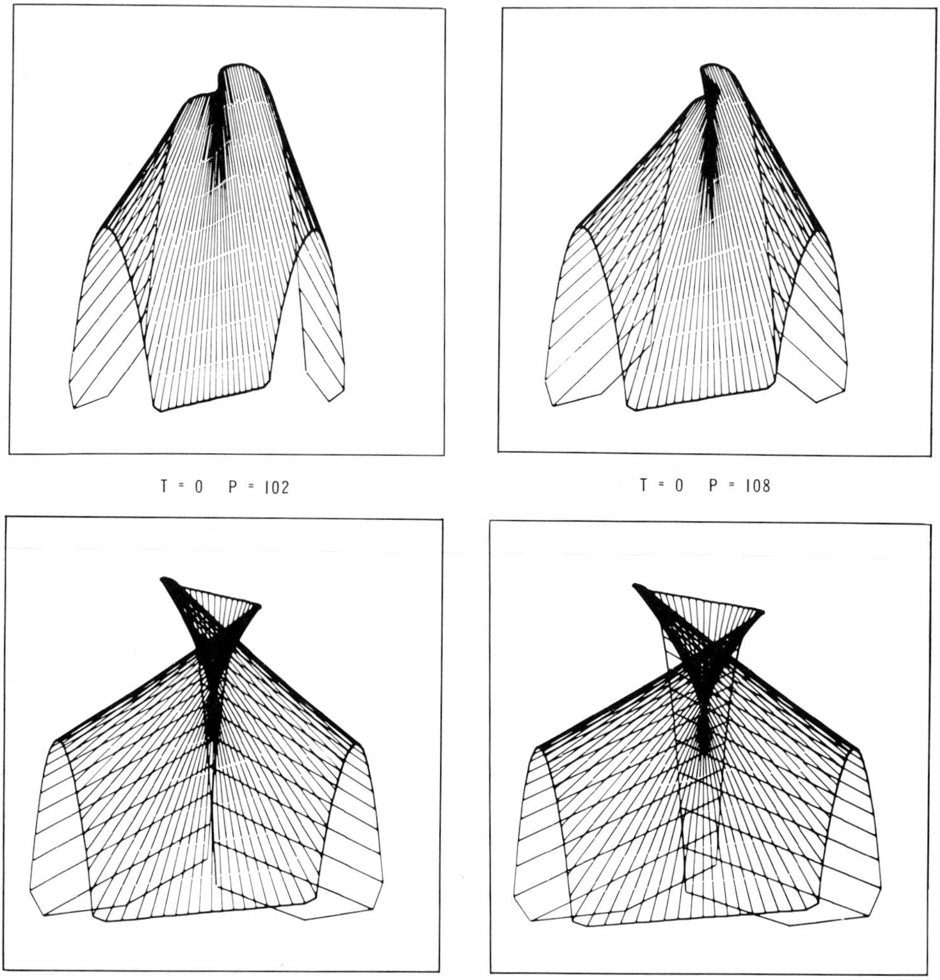

T = 0 P = 102

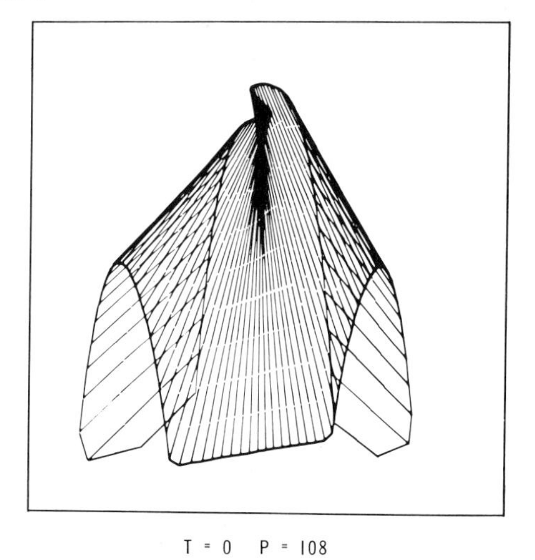

T = 0 P = 108

T = 0 P = 125

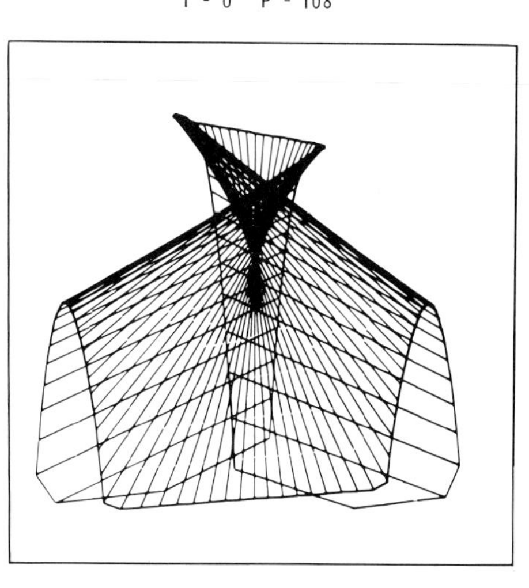

T = 0 P = 131

Fig. 4

Figs. 5, 6, 7 and 8 show projections from the Catastrophe Mfd of the Butterfly Catastrophe (x,A,B,C,D):

$$V = x^6 + A\frac{x^4}{4} + B\frac{x^3}{3} + C\frac{x^2}{2} + D\,x$$

P = 0 degrees throughout and T is varied in 6 degree increments from T = 0 degrees to T = 149 degrees.

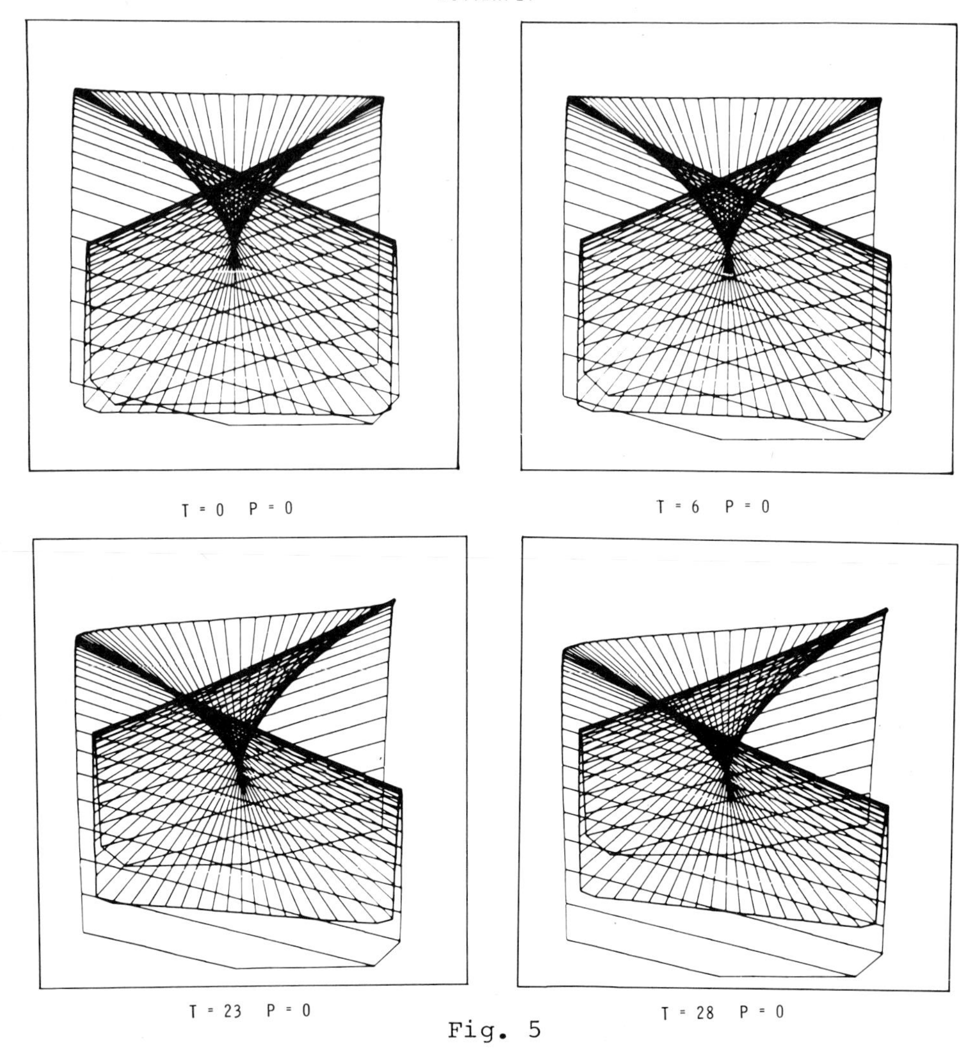

Fig. 5

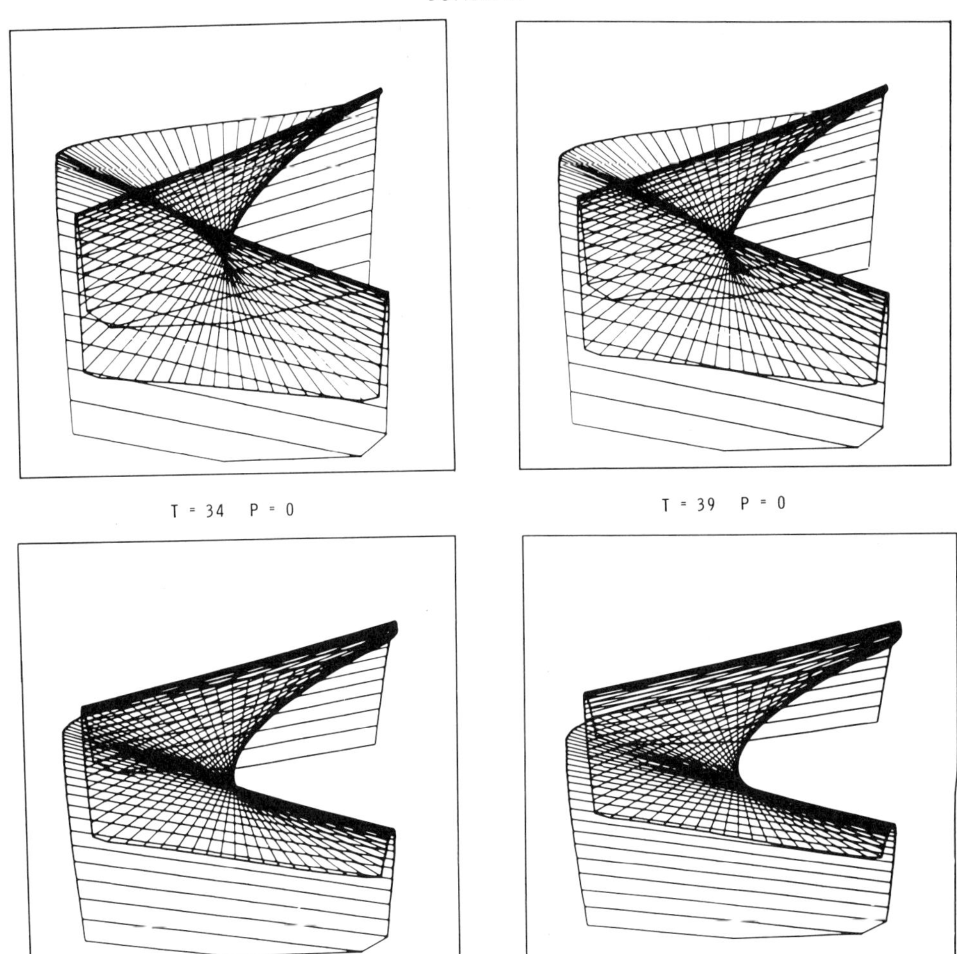

Fig. 6

BUTTERFLY

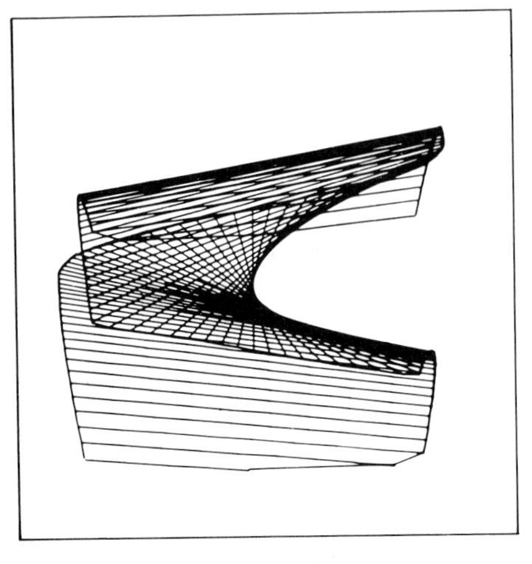

T = 79 P = 0

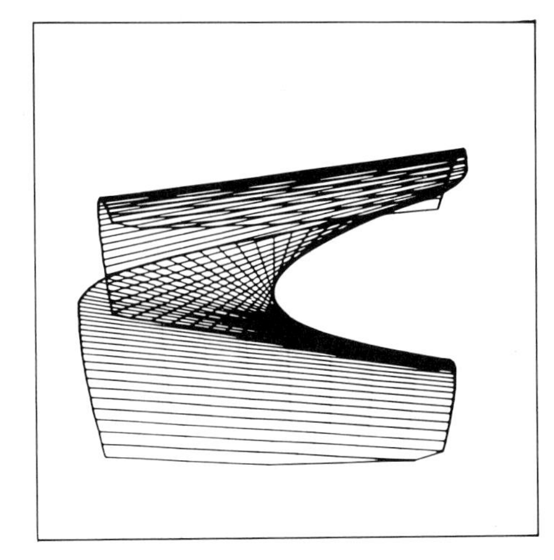

T = 84 P = 0

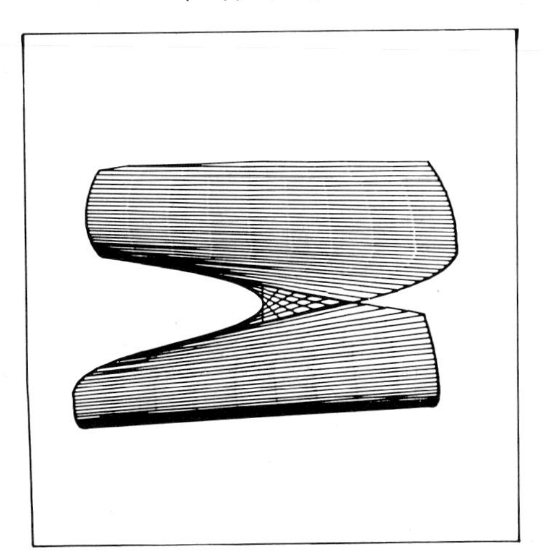

T = 108 P = 0

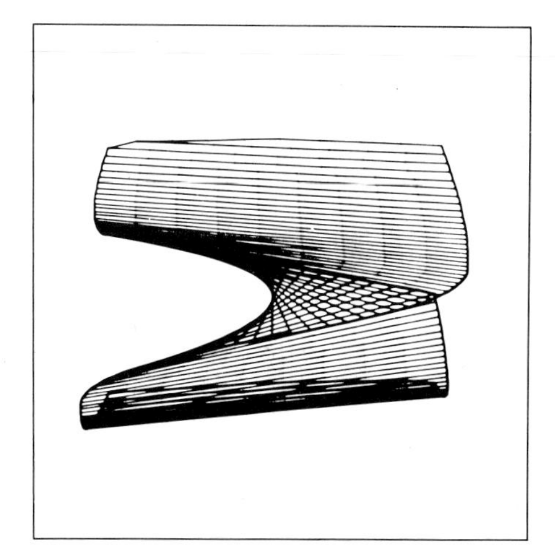

T = 113 P = 0

Fig. 7

BUTTERFLY

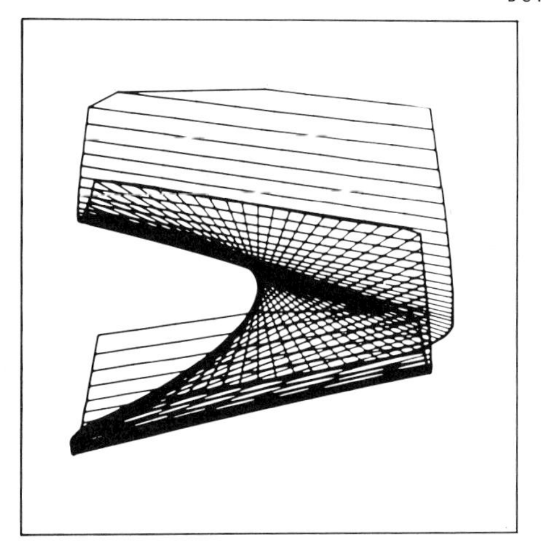

T = 119 P = 0

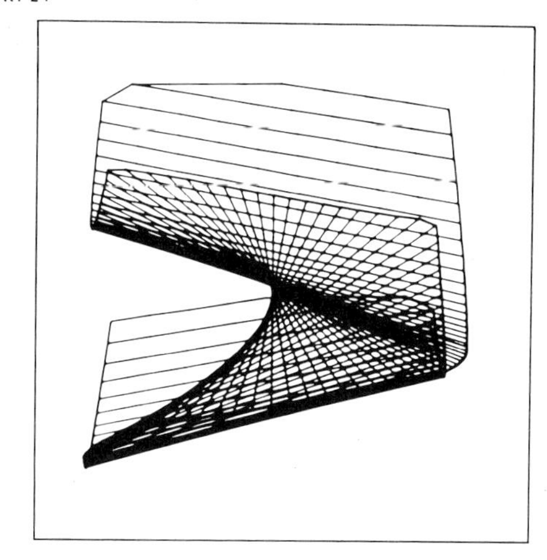

T = 125 P = 0

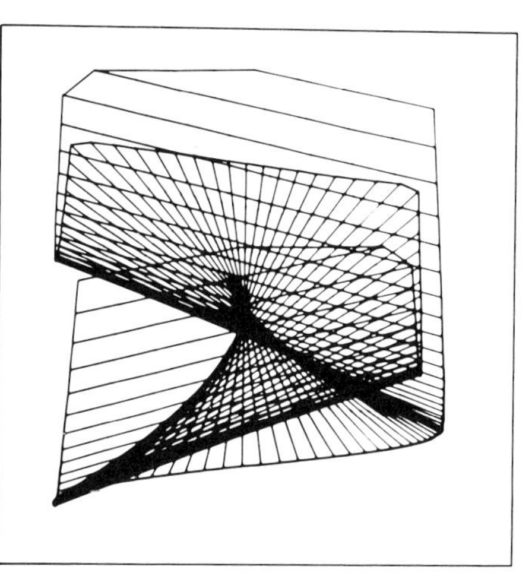

T = 143 P = 0

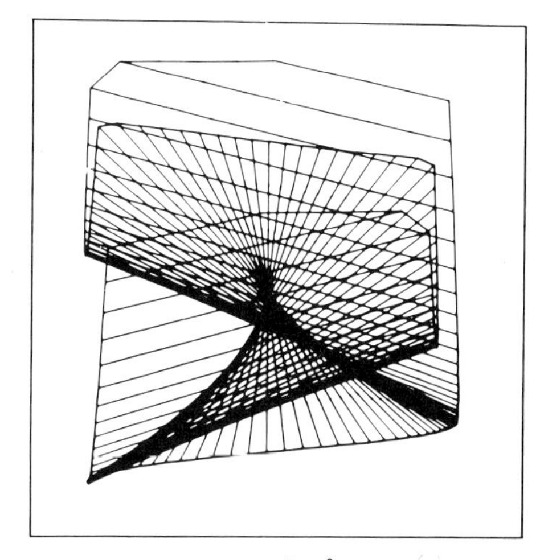

T = 149 P = 0

Fig. 8

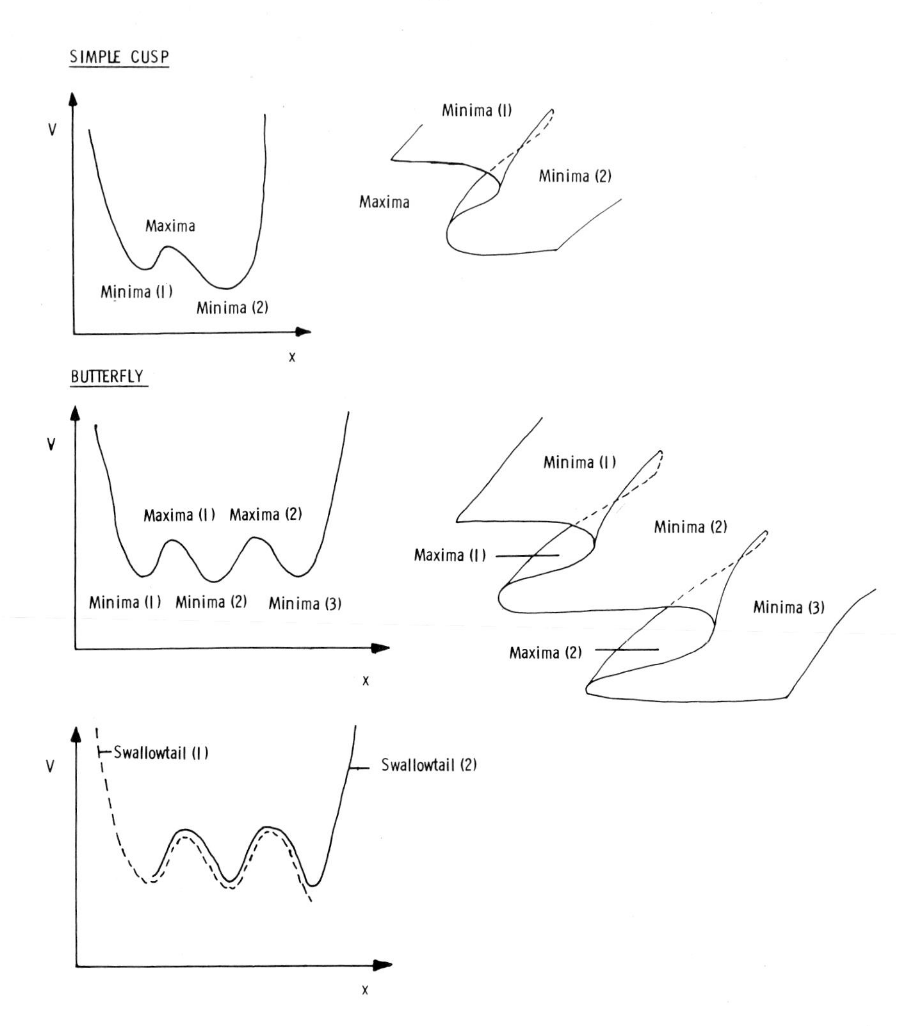
SIMPLE CUSP
V
Maxima
Minima (1)
Minima (2)
x
Minima (1)
Minima (2)
Maxima
BUTTERFLY
V
Maxima (1)
Maxima (2)
Minima (1)
Minima (2)
Minima (3)
x
Minima (1)
Minima (2)
Maxima (1)
Minima (3)
Maxima (2)
V
Swallowtail (1)
Swallowtail (2)
x

Fig. 9

Figs. 10, 11, 12 and 13. Reconstruction of the Catastrophe Mfd of the Butterfly Catastrophe. The figures are plotted in (x,B,C) space and are therefore equivalent to sections of the Swallowtail Catastrophe (2). Rotation of the three-dimensional surface is for fixed T (0 degrees) and increasing P. (P = 0 degrees to P = 131 degrees.)

BUTTERFLY(Swallowtail)

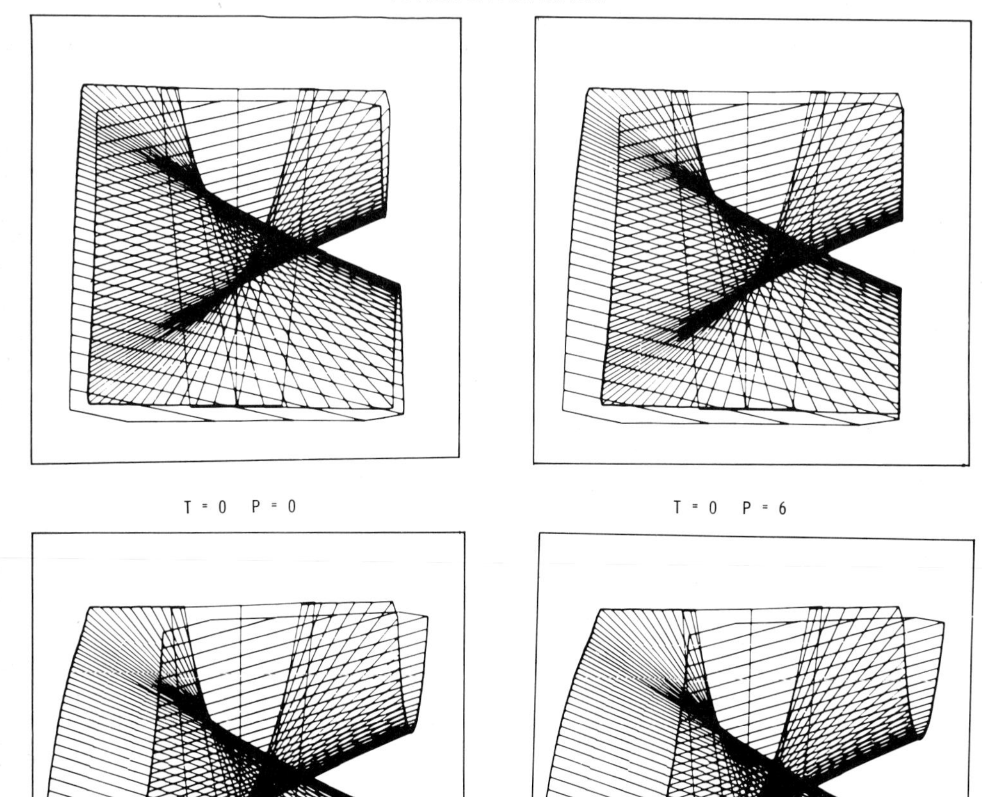

T = 0 P = 23

T = 0 P = 28

Fig. 10

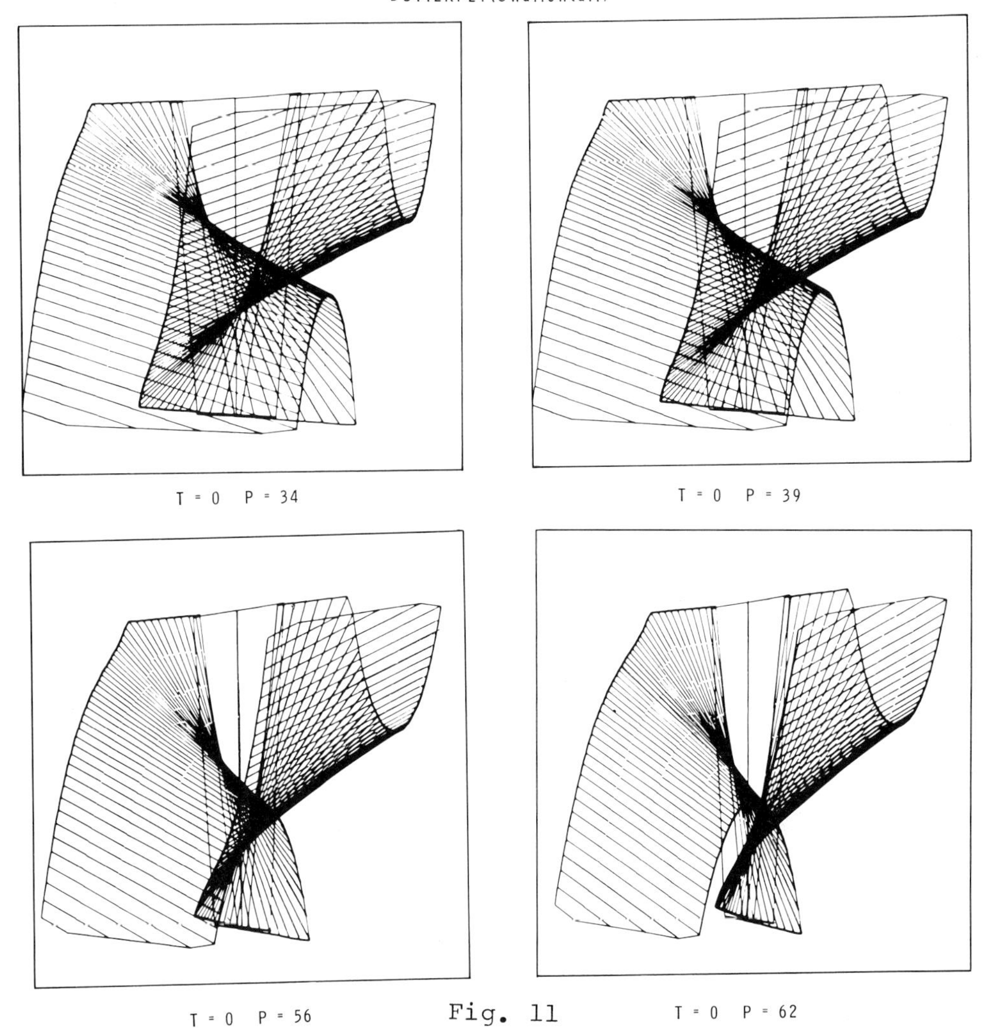

Fig. 11

BUTTERFLY(Swallowtail)

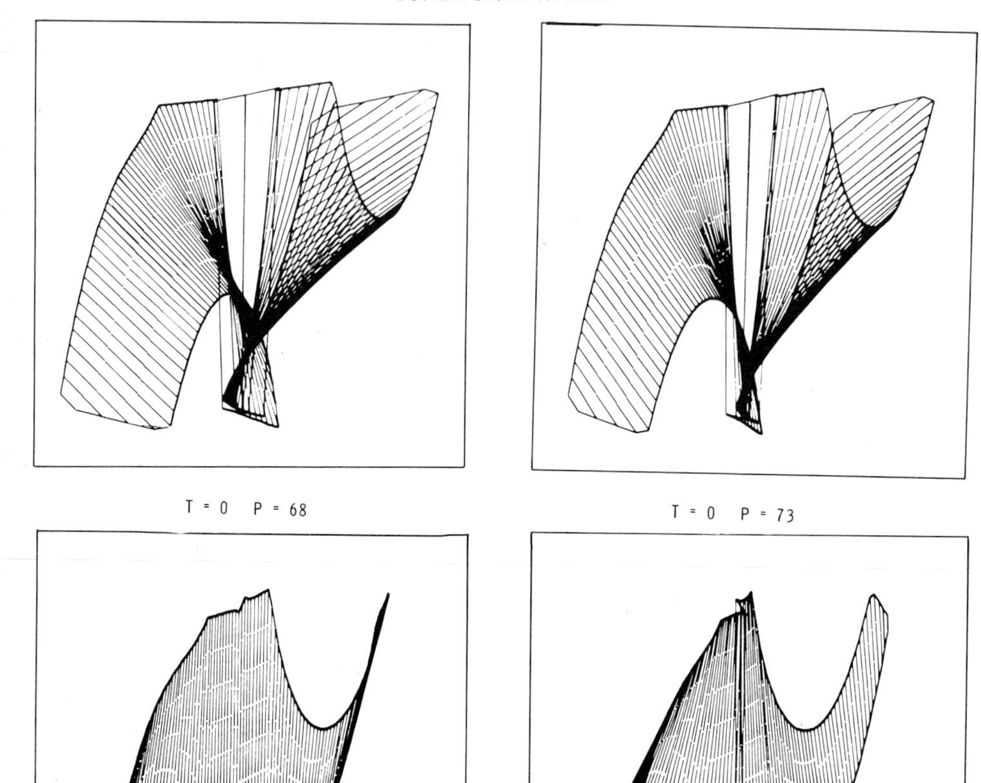

T = 0 P = 68

T = 0 P = 73

T = 0 P = 90

T = 0 P = 96

Fig. 12

BUTTERFLY(Swallowtail)

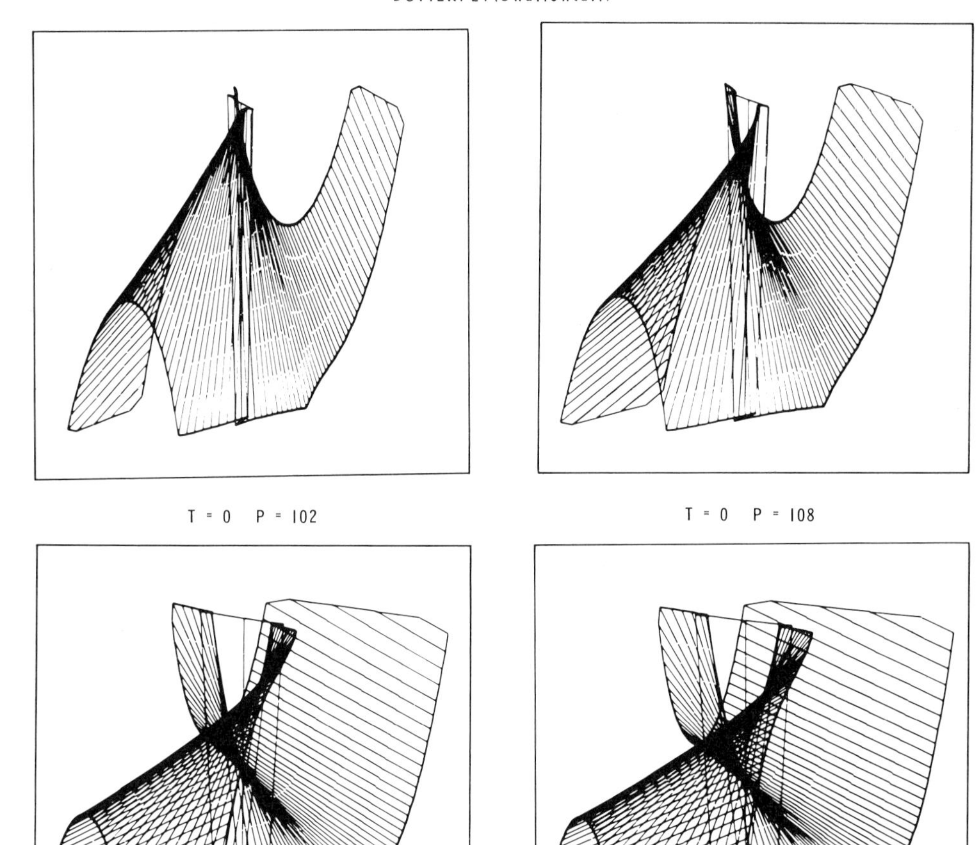

Fig. 13

Figs. 14, 15, 16, 17 and 18. Projections of Swallowtail-like sections (x,B,C) of the Butterfly Catastrophe Catastrophe Mfd (x,A,B,C,D). The value of P was fixed (0 degrees) and T increased in 6 degree increments from 0 degrees to 172 degrees.

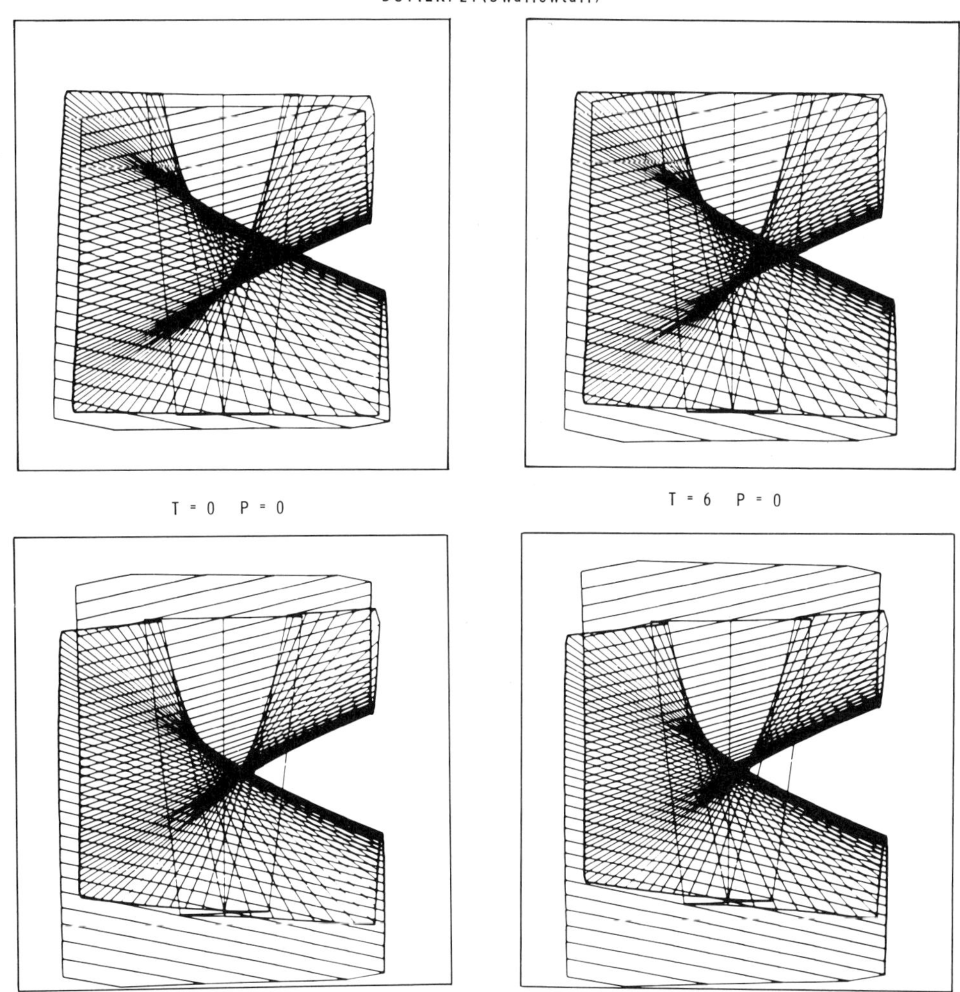
BUTTERFLY(Swallowtail)
T = 0 P = 0
T = 6 P = 0
T = 23 P = 0
T = 28 P = 0

Fig. 14

BUTTERFLY(Swallowtail)

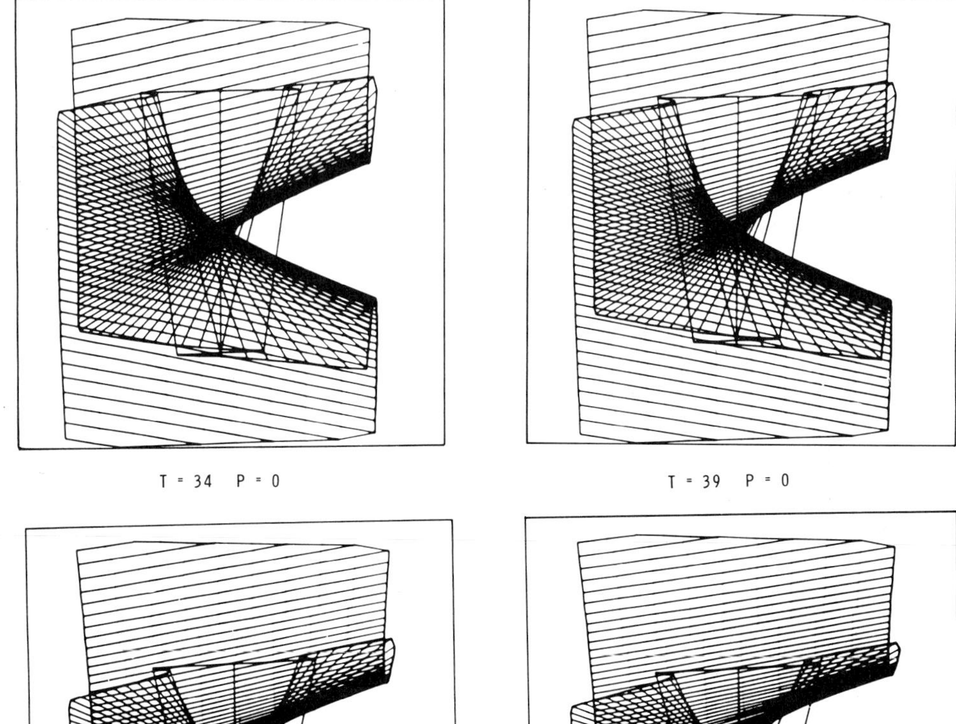

T = 34 P = 0

T = 39 P = 0

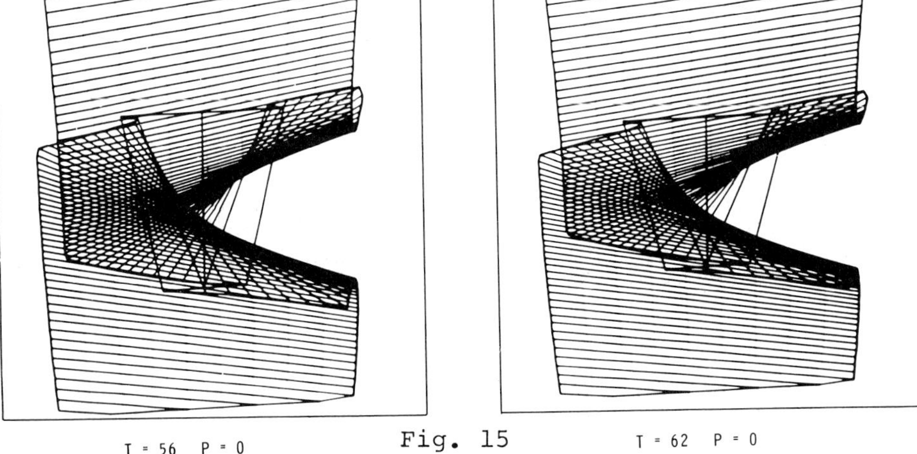

T = 56 P = 0

Fig. 15

T = 62 P = 0

BUTTERFLY(Swallowtail)

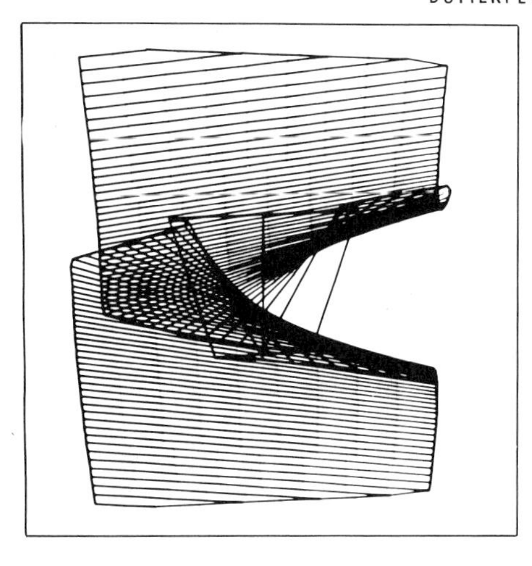

T = 68 P = 0

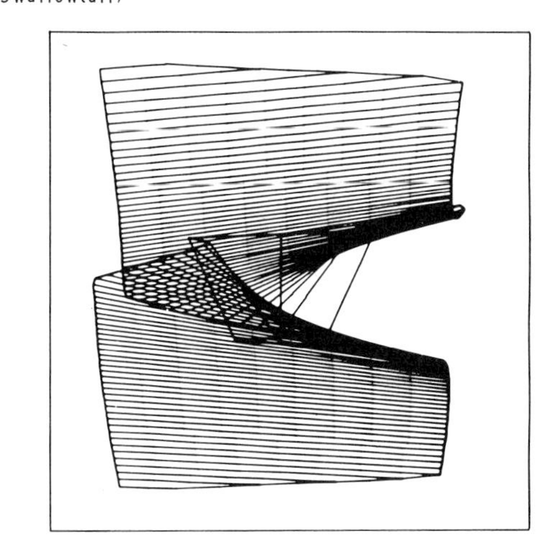

T = 73 P = 0

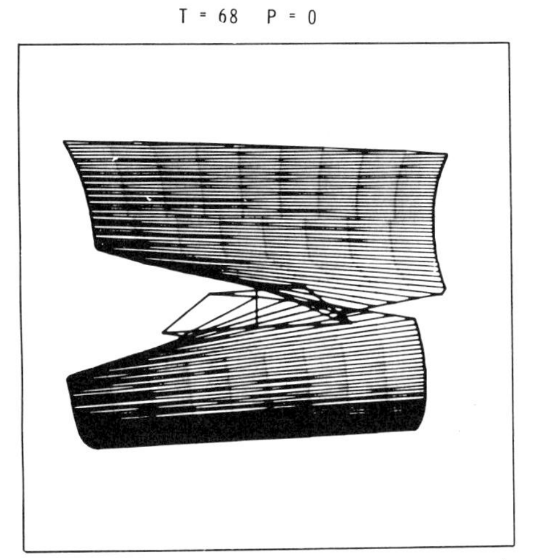

T = 96 P = 0

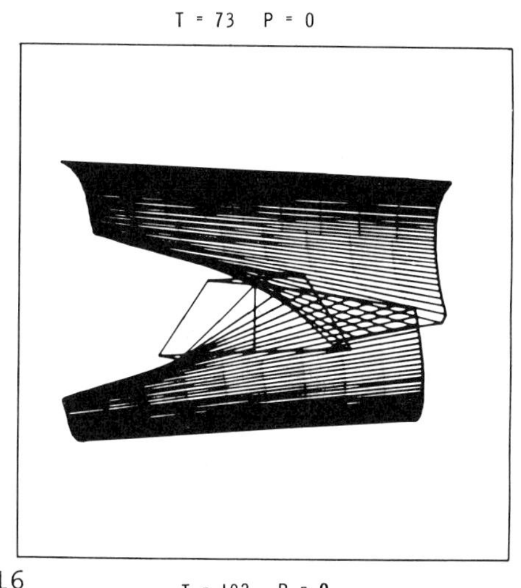

T = 102 P = 0

Fig. 16

BUTTERFLY(Swallowtail)

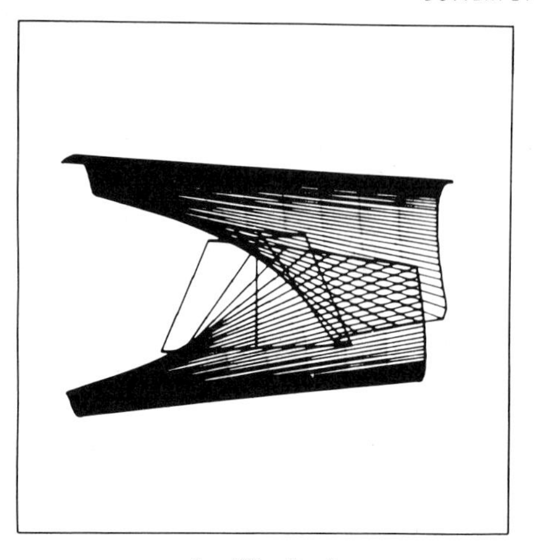

T = 108 P = 0

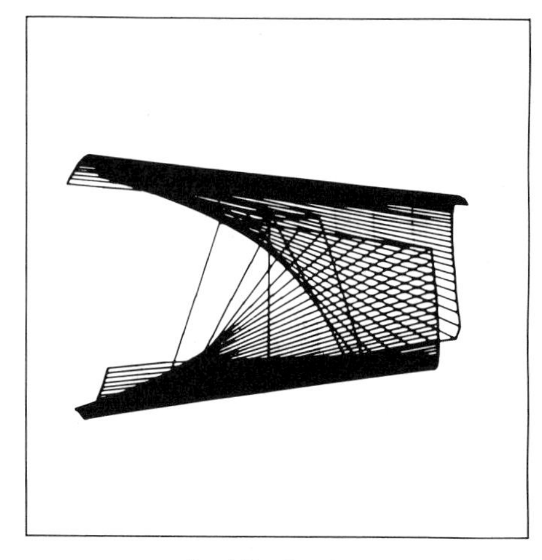

T = 113 P = 0

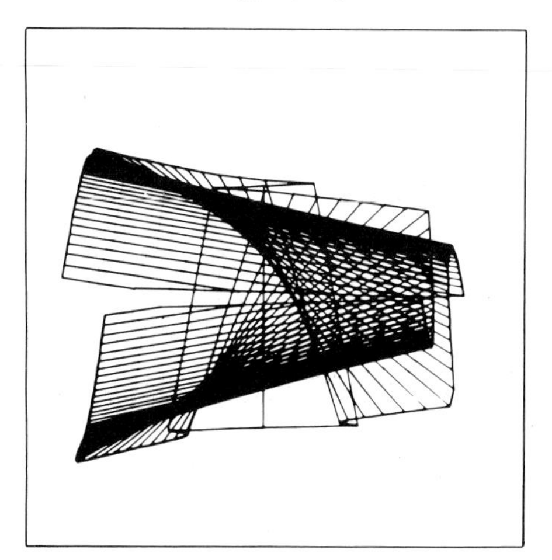

T = 131 P = 0

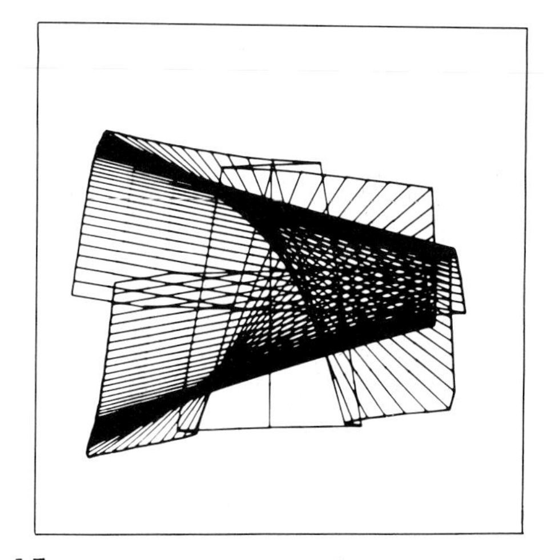

T = 137 P = 0

Fig. 17

BUTTERFLY(Swallowtail)

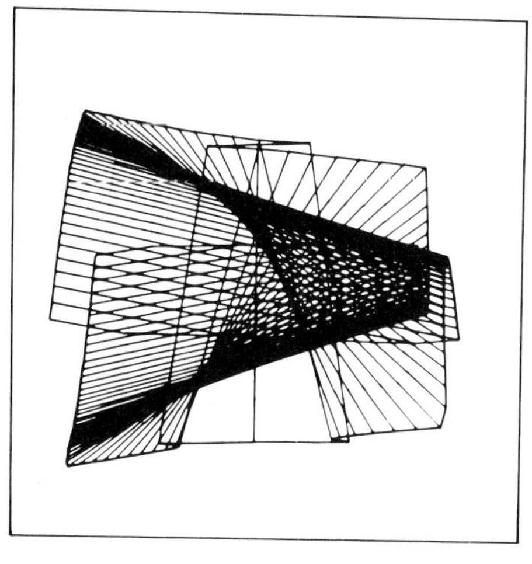

T = 143 P = 0

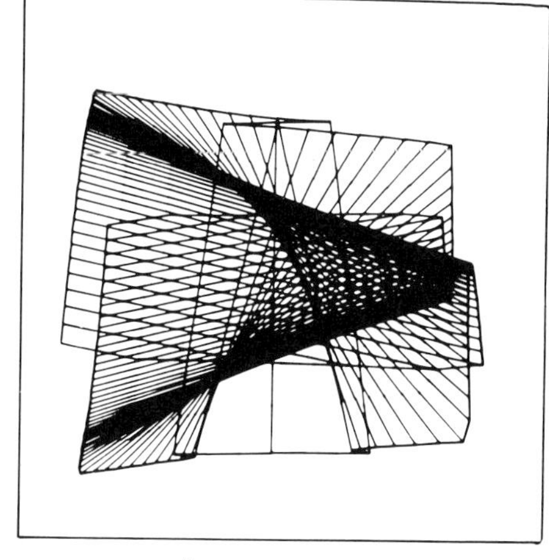

T = 149 P = 0

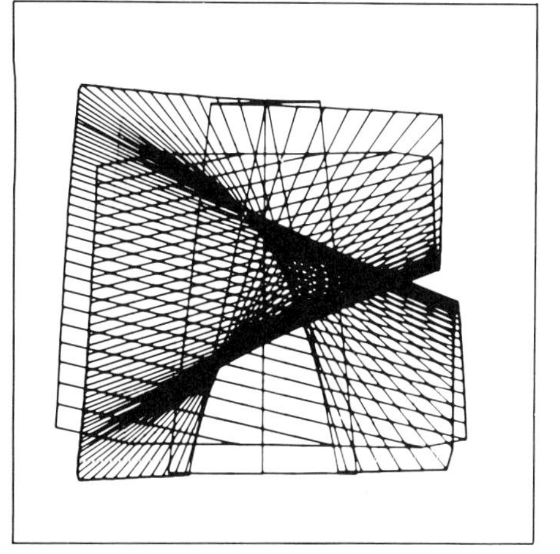

T = 166 P = 0

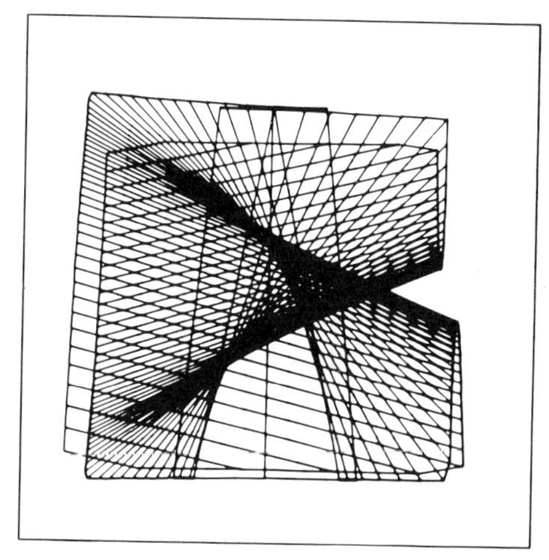

T = 172 P = 0

Fig. 18

Bibliography:

(1) Woodcock, A.E.R., Sterographic Reconstructions of the Catastrophe Mfds of the Cuspoids: (1) The Simple Cusp. (this volume.)

(2) *Ibid.*, (2) The Swallowtail. (this volume.)

(3) Woodcock, A.E.R. and Poston, T., The Geometrical Properties of the Elementary Catastrophes. (1) The Cuspoids. (this volume.)

STEREOGRAPHIC RECONSTRUCTIONS OF THE CATASTROPHE MANIFOLDS OF THE CUSPOIDS: (4) THE WIGWAM

Introduction:

The Wigwam Catastrophe arises from the following Generalized Potential Energy Function:

$$V = \frac{x^7}{7} + A\,\frac{x^5}{5} + B\,\frac{x^4}{4} + C\,\frac{x^3}{3} + D\,\frac{x^2}{2} + E\,x \qquad (1)$$

The Catastrophe Mfd (x,A,B,C,D,E) is five dimensional and since the equation

$$\frac{dV}{dx} = x^6 + A\,x^4 + B\,x^3 + C\,x^2 + D\,x + E = 0 \qquad (2)$$

is of order 6 then there are, at most 6 real roots of the equation and thus, at most 6 stationary (identifiable with maxima and minima) values of the function. Since there are an even number of roots of the equation (2) then the Wigwam is of the same family as the Fold and Swallowtail Catastrophes (2). The equivalent potential energy well will have no lowest minima, and will be as represented in Fig. (11).

The pictures were produced as previously described (2) and presented as stereographic pairs; in these pictures A = -8.0, B = 0.0 and C = +16.0.

Observations:

The Catastrophe Mfd of the Wigwam is rather similar to that of the Swallowtail and only differs, in general, by the introduction of one extra sheet of maxima and one of minima. (In the Catastrophe Mfd of the Wigwam, in regions where A is positive and C is negative, the morphology of the envelope is exactly that of the Swallowtail, Woodcock and Poston (1; Figs. 10 and 11). Comparison of the projection of the Catastrophe Mfd at T = 0 degrees, P = 73 degrees for the Wigwam (Fig. 1) and the Swallowtail (2; Fig. 3), for example, show that the gross shape of the surfaces are similar but that the Wigwam is more complicated in that an extra fold has been introduced into the area of the surface corresponding to low values of x. Figs. 1, 2 and 3 show projections for T = 0 degrees and increasing values of P (from P = 45 degrees to P = 143 degrees) and reveal the way in which the extra fold gives the surface the characteristic Wigwam appearance. Figs. 4, 5, 6 and 7 demonstrate two interesting phenomena. In Figs. 4 and 5 the envelope generated by the x - constant lines varies in a way that is closely similar to the changes observed in that envelope when the value of the parameter C was varied for negative values of A and C equal to zero; in this case, however, the Catastrophe Mfd was projected along the x-axis onto the (D,E) plane of the Control Space. (Woodcock and Poston, 1; Fig. 13, for example). Furthermore, as T is increased to 149 degrees (Figs. 5, 6 and 7) the surface behaves in a way similar to that observed in the Swallowtail Catastrophe (2), except that the requirement that the equation (2) has more stable values than the equivalent equation for the Swallowtail introduces an extra Swallowtail-like (folded) region of the surface; this extra folded part of the surface is seen, for example at T = 149 degrees (Fig. 7). Thus, the Wigwam may be considered either as two

Butterfly Catastrophes with a common limb, or as the conjunction of a Swallowtail Catastrophe and a Fold Catastrophe, Fig. (11), as degenerate sections of the two separate Butterfly Catastrophes.

Butterfly-like Sections of the Wigwam Catastrophe

Observations:

The apposition of two Swallowtail Catastrophes to generate the Butterfly Catastrophe is clearly shown in the Figs. 8, 9 and 10. In these pictures, the projection was from the Catastrophe Mfd of the Wigwam Catastrophe (x,A,B,C,D,E) onto the (C,D) control plane of the (A,B,C,D,E) Control Space. Choice of the non-zero value for E (in the present case, A = -8.0, B = 0.0, D = +0.5) means that the surface is not identically that of the Butterfly Catastrophe. At P = 73 degrees, T = 0 degrees, for example, the two regions of the surface (one for positive values of x, another for negative values of x) each separately resemble the surface of the Swallowtail Catastrophe at a similar orientation, except that one of the Swallowtails grows out of the edge of the other. (2, Fig. 3); this is particularly striking when the pictures are viewed stereographically.

Figs. 1, 2 and 3. Projections of the Catastrophe Mfd (x,A,B,C,D,E) of the Wigwam Catastrophe:

$$V = \frac{x^7}{7} + A\frac{x^5}{5} + B\frac{x^4}{4} + C\frac{x^3}{3} + D\frac{x^2}{2} + E\,x$$

for T = 0 degrees, P is increased from P = 45 degrees to P = 143 degrees in increments of 6 degrees.

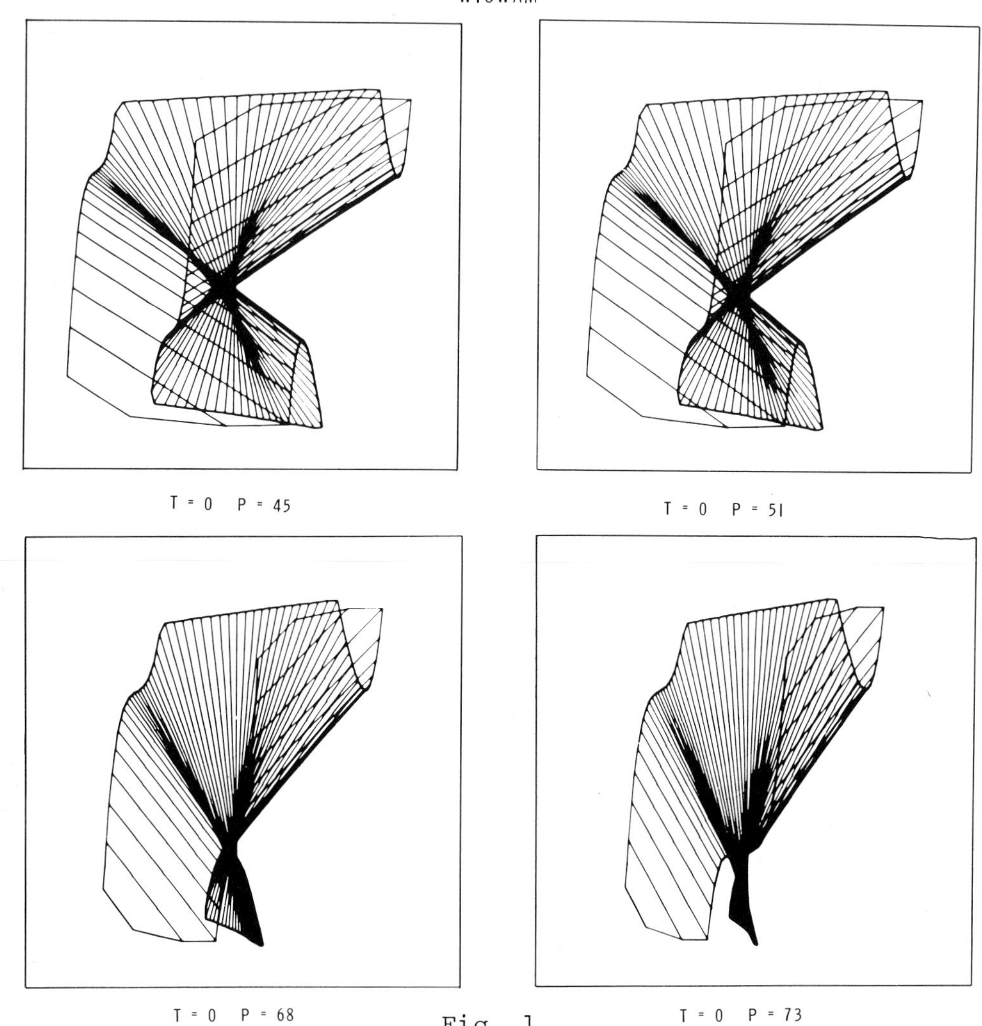

Fig. 1

WIGWAM

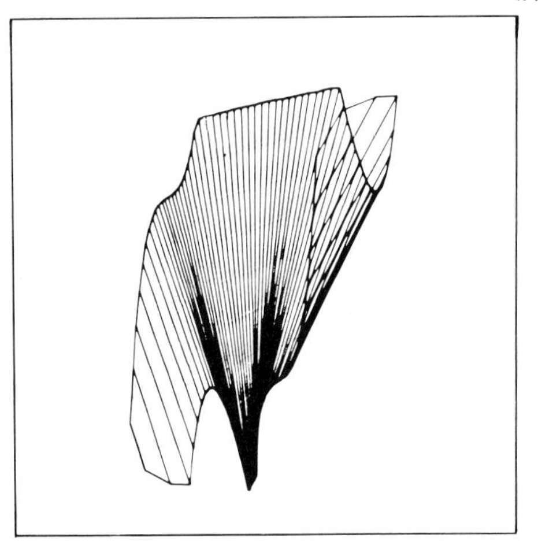

T = 0 P = 79

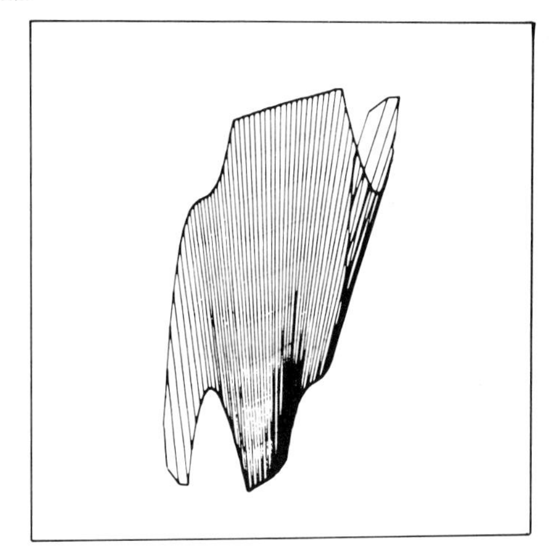

T = 0 P = 84

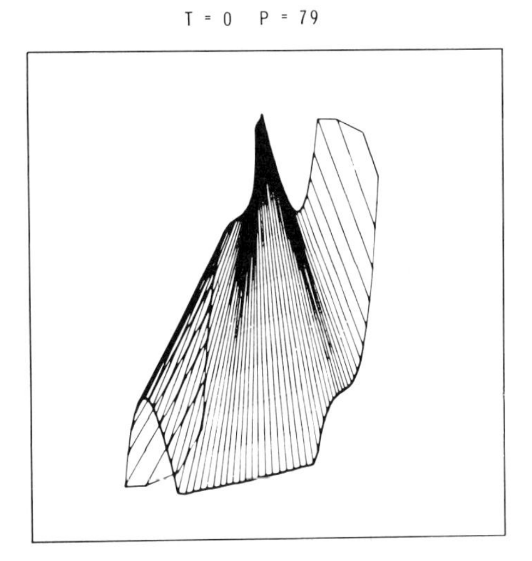

T = 0 P = 102

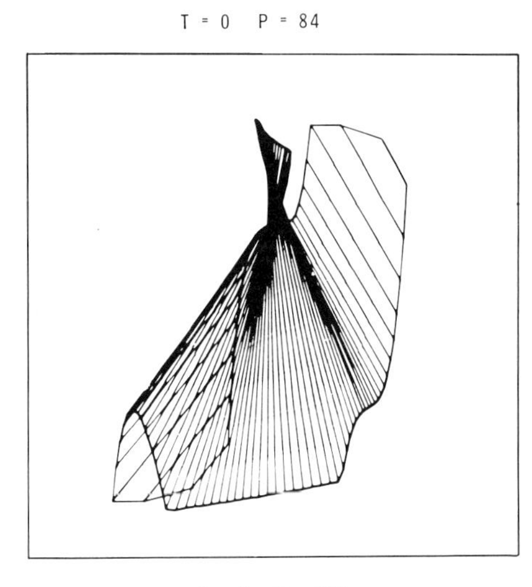

T = 0 P = 108

Fig. 2

WIGWAM

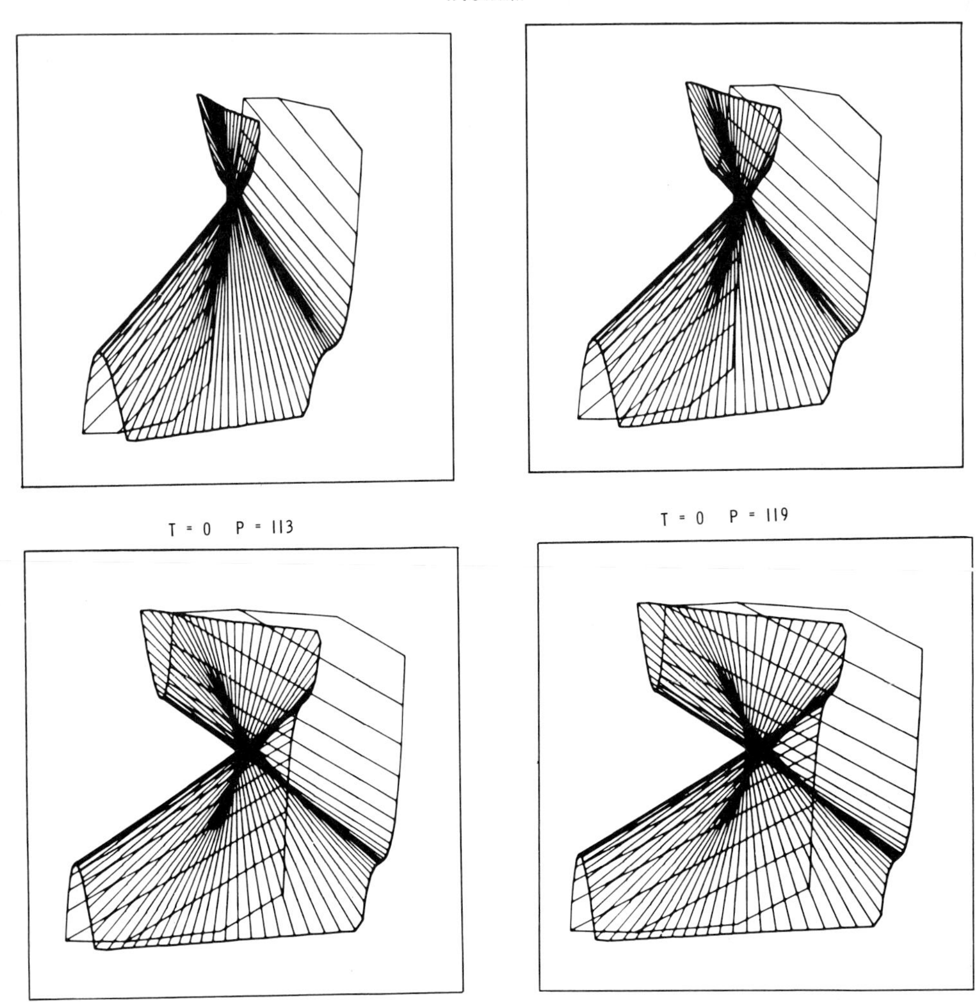

Fig. 3

Figs. 4, 5, 6 and 7. Projections of the Catastrophe Mfd (x,A,B,C,D,E) of the Wigwam Catastrophe:

$$V = \frac{x^7}{7} + A\frac{x^5}{5} + B\frac{x^4}{4} + C\frac{x^3}{3} + D\frac{x^2}{2} + E\,x$$

for P = 0 degrees, T is increased from T = 34 degrees to T = 149 degrees in increments of 6 degrees.

WIGWAM

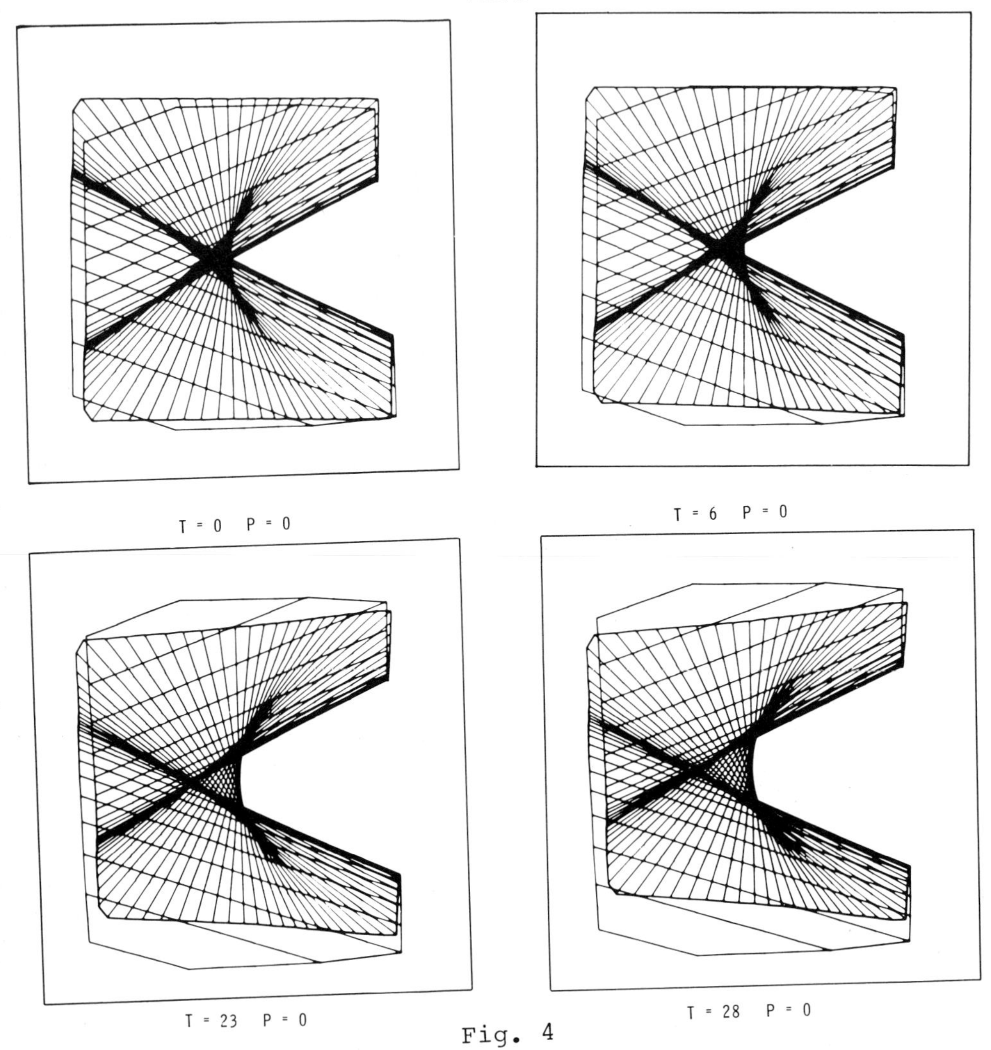

Fig. 4

WIGWAM

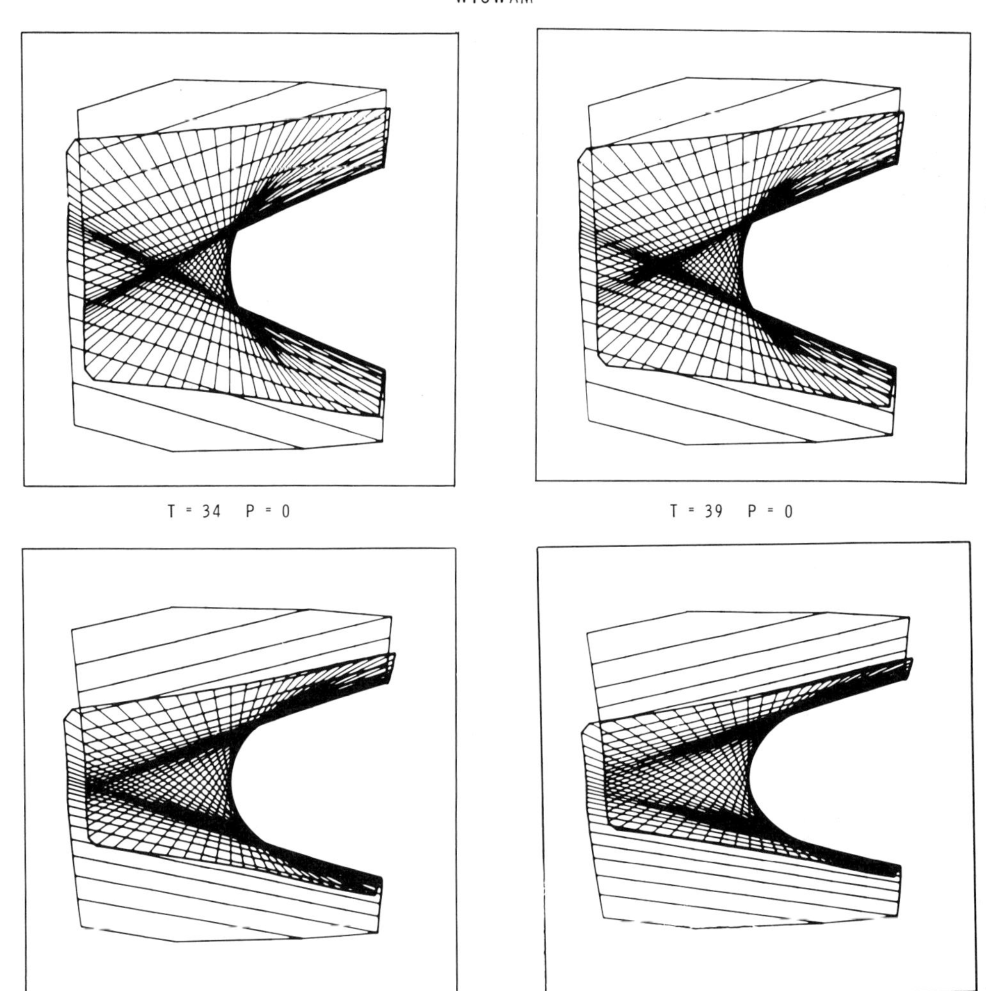
T = 34 P = 0
T = 39 P = 0
T = 56 P = 0
T = 62 P = 0

Fig. 5

WIGWAM

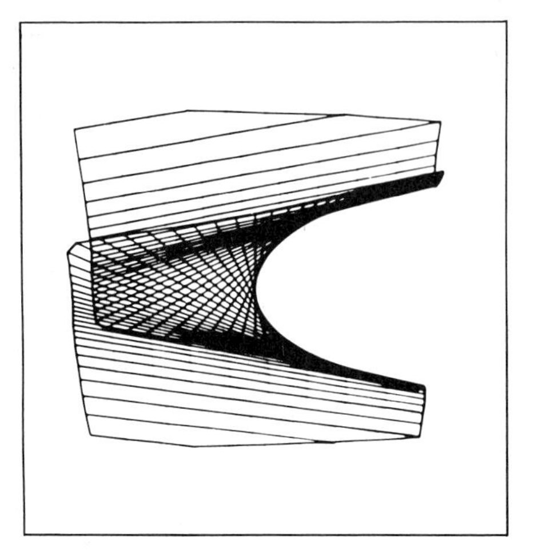

T = 68 P = 0

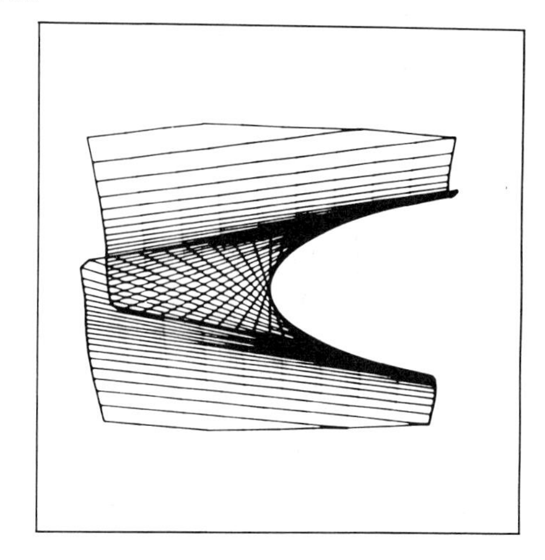

T = 73 P = 0

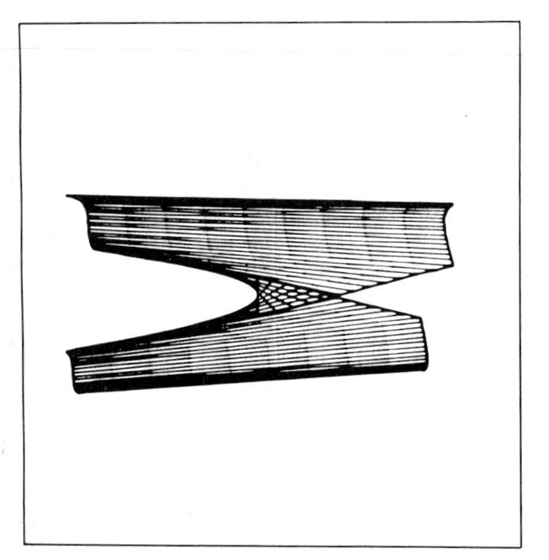

T = 96 P = 0

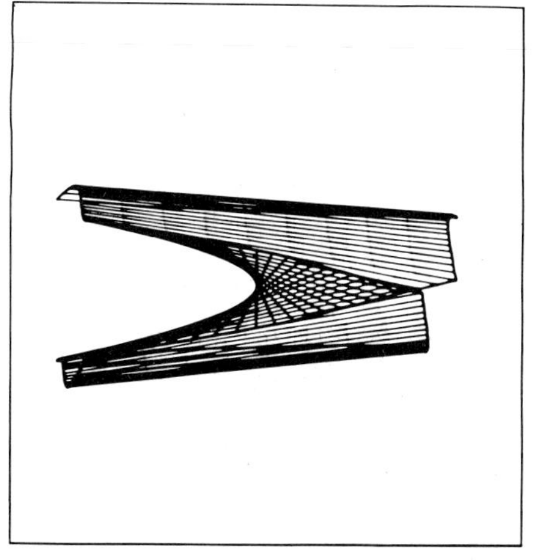

T = 102 P = 0

Fig. 6

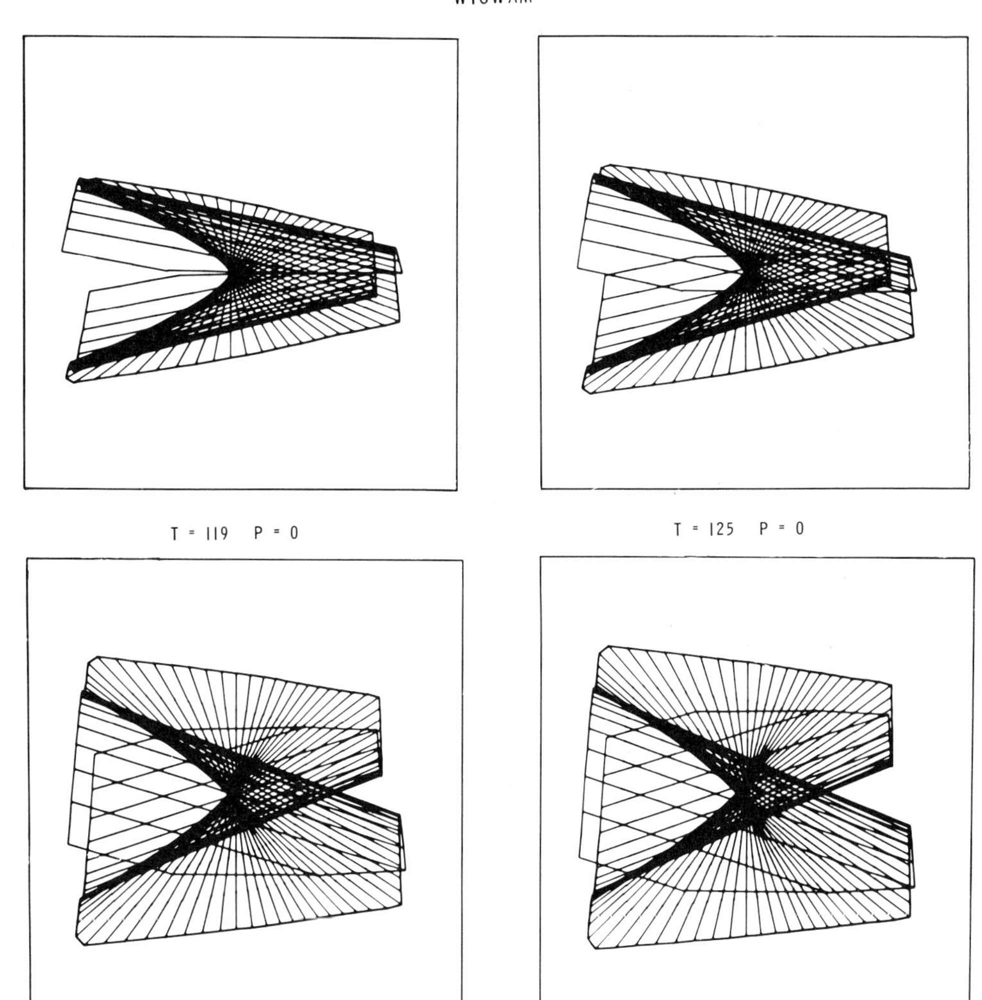

Fig. 7

Figs. 8, 9 and 10. Projections of the Wigwam Catastrophe:

$$V = \frac{x^7}{7} + A\,\frac{x^5}{5} + B\,\frac{x^4}{4} + C\,\frac{x^3}{3} + D\,\frac{x^2}{2} + E\,x$$

onto the (C,D) plane of the Control Space (A,B,C,D,E) for T = 0 degrees, the projected figure is rotated in 6 degree increments from P = 34 degrees to P = 131 degrees. The (C,D) plane in the Control Space of the Butterfly Catastrophe contains a true Butterfly as a projection from the (X,A,B,C,D) Catastrophe Mfd of that Catastrophe.

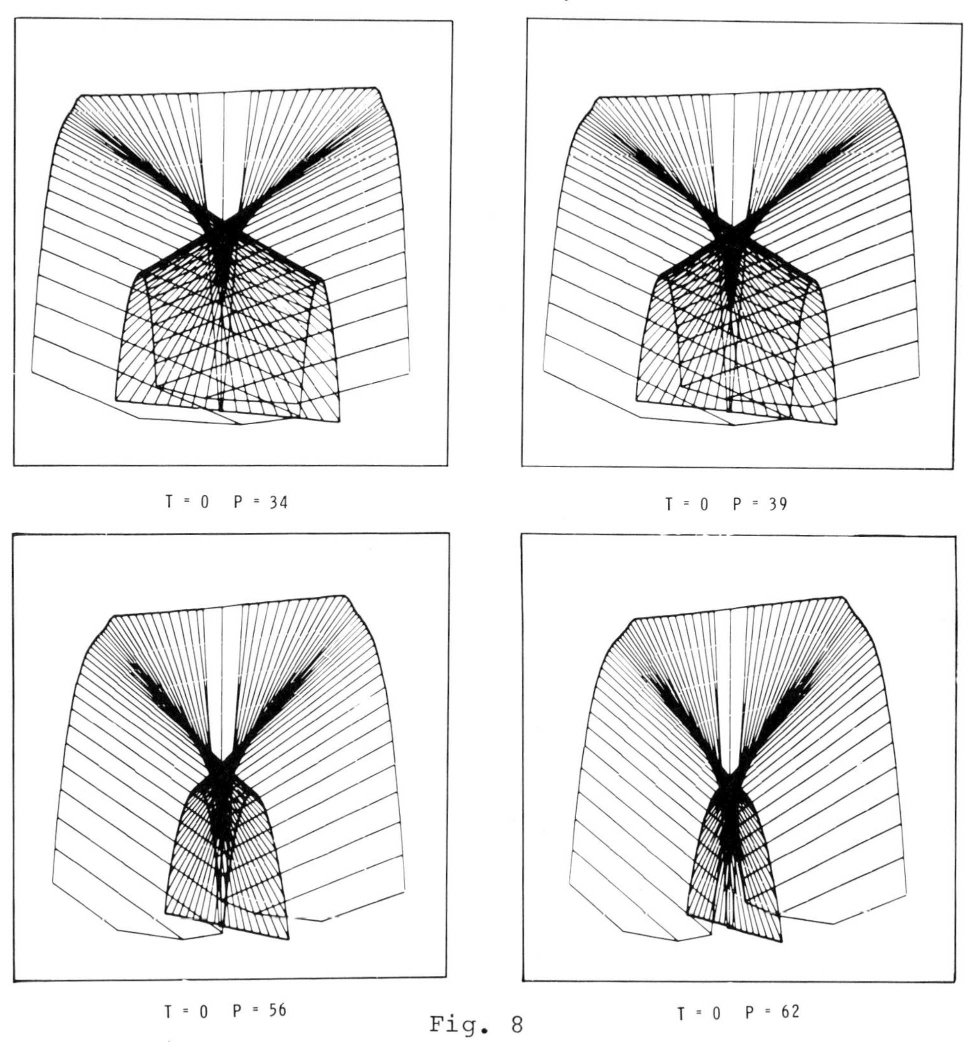

Fig. 8

WIGWAM(Butterfly)

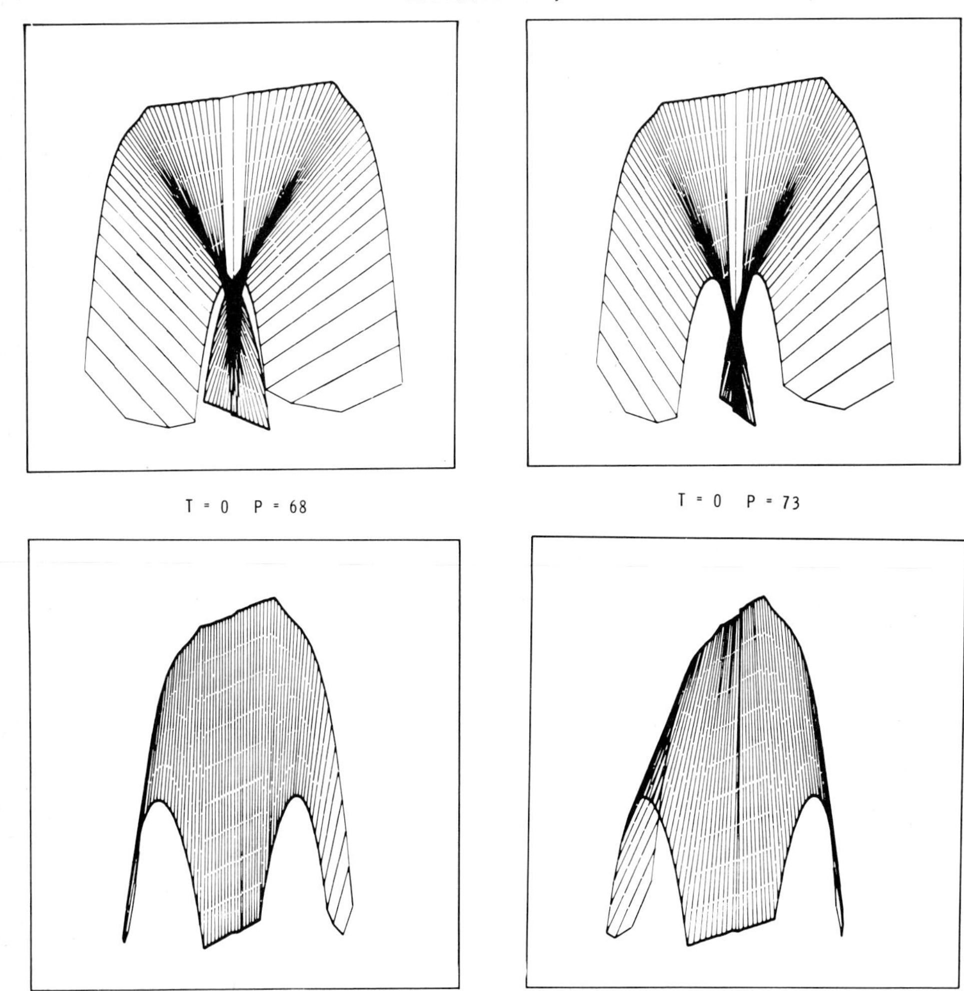

Fig. 9

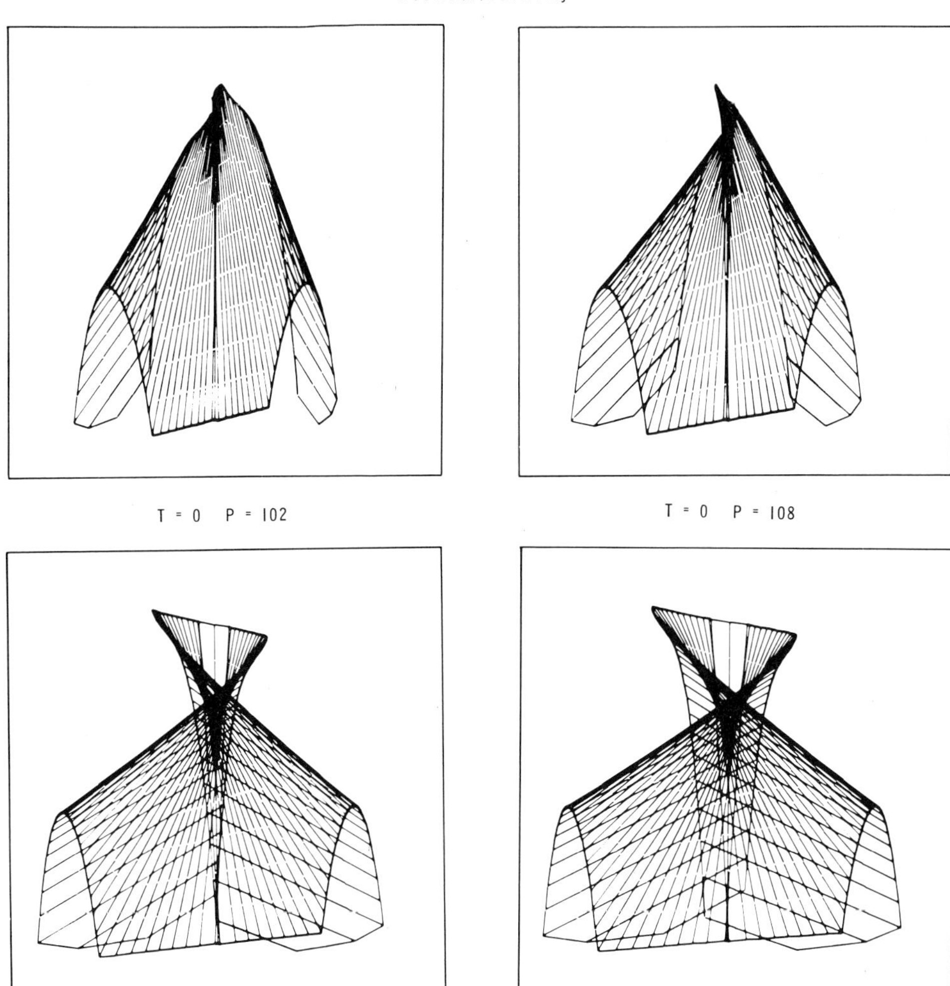
WIGWAM(Butterfly)
T = 0 P = 102
T = 0 P = 108
T = 0 P = 125
T = 0 P = 131

Fig. 10

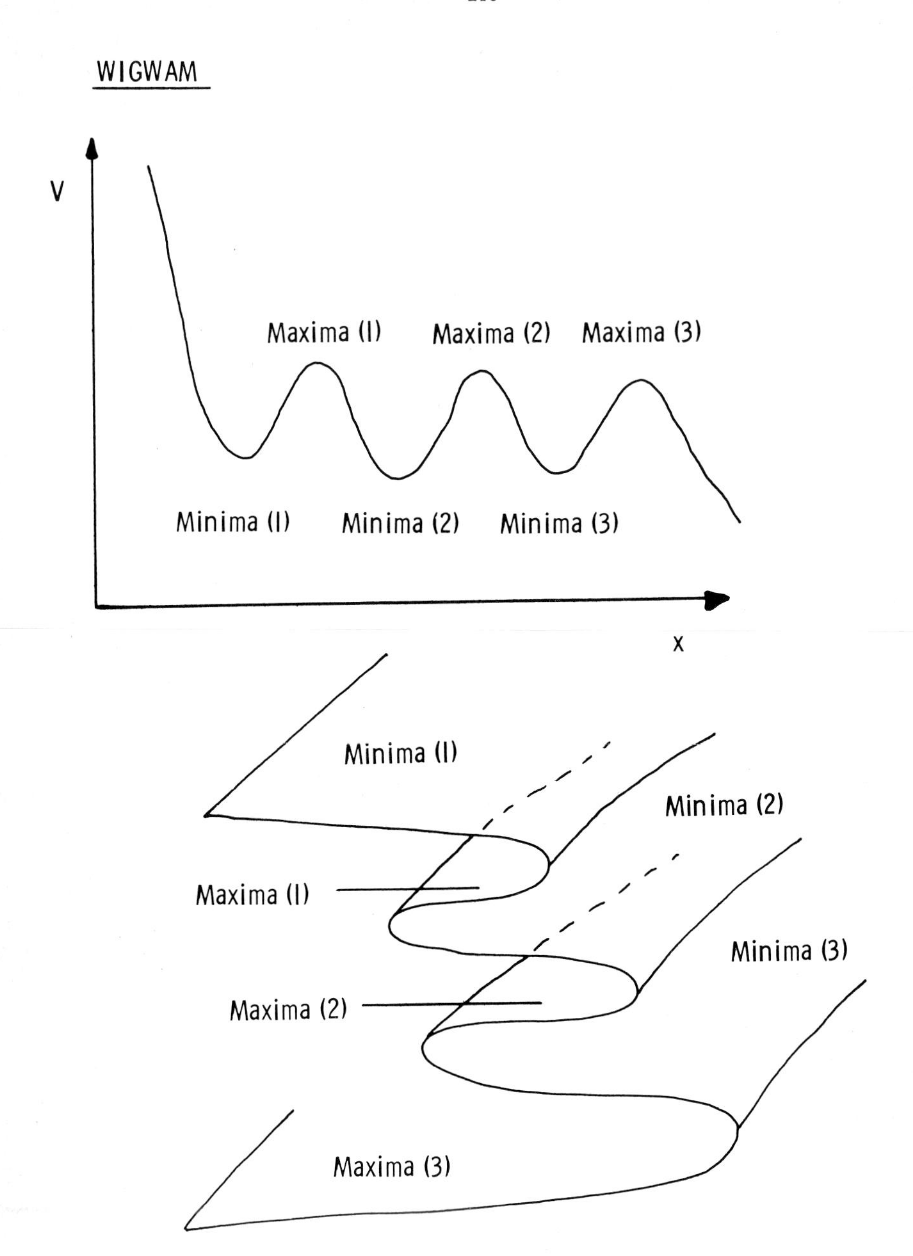
WIGWAM
V
Maxima (1)
Maxima (2)
Maxima (3)
Minima (1)
Minima (2)
Minima (3)
x
Minima (1)
Minima (2)
Maxima (1)
Minima (3)
Maxima (2)
Maxima (3)

Fig. 11

Bibliography:

(1) Woodcock, A.E.R. and Poston, T., The Geometrical Properties of the Elementary Catastrophes. (1) The Cuspoids. (this volume.)

(2) Woodcock, A.E.R., Stereographic Reconstructions of the Catastrophe Mfd of the Cuspoids: (2) The Swallowtail. (this volume.)

STEREOGRAPHIC RECONSTRUCTIONS OF THE CATASTROPHE MANIFOLDS OF THE CUSPOIDS: (5) THE STAR.

Introduction:

The Star Catastrophe:

$$V = \frac{x^8}{8} + A\,\frac{x^6}{6} + B\,\frac{x^5}{5} + C\,\frac{x^4}{4} + D\,\frac{x^3}{3} + E\,\frac{x^2}{2} + F\,x \qquad (1)$$

belongs to the same generic sub-family as the Simple Cusp (1) and the Butterfly (2) Catastrophes. Differentiation of equation (1) with respect to x gives:

$$\frac{dV}{dx} = x^7 + A\,x^5 + B\,x^4 + C\,x^3 + D\,x^2 + E\,x + F \qquad (2)$$

The stable (or stationary) values of the potential energy function (1) occur when equation (2) is zero. Under these circumstances, the function can have at most 7 real roots. As this is an odd number, the Catastrophe is not self-dual. There will be either 4 maxima and 3 minima or 4 minima and 3 maxima. This behavior is shown in Fig. (12) which shows both the (generalized) convolutions of the Catastrophe Mfd (x,A,B,C,D,E,F) as well as the nature of the equivalent potential energy well.

The figures were prepared and presented as described previously (1). In these figures A = -8.0, B = 0.0, C = +16.0 and D = 0.0.

Observation:

The essential similarity between the Butterfly (2; Figs. 1 and 2) and the Star (Figs. 1, 2 and 3; P = 34 degrees to P = 131 degrees, T = 0 degrees) is revealed by a study of the pictures, particularly when they are viewed stereographically. At P = 73 degrees, for example, the warped surface resembles that of the Butterfly very closely (2; Fig. 3) except that an extra fold has been introduced into the Butterfly to accommodate the requirement for two extra solutions of the equation (2) compared to the number of solutions of the similar equation for the Butterfly Catastrophe (2). The relative motion of the layers of the Catastrophe Mfd of the Star Catastrophe, revealed by increasing the value of T from T = 0 degrees to T = 184 degrees; P = 0 degrees (Figs. 4, 5, 6, 7 and 8), is very similar to that induced by changing the value of either the B or the D parameter (A negative, C positive) of the Star Catastrophe, and in which the projection is simply along the x-axis (Woodcock and Poston, 3). In Fig. 6, at T = 84 degrees, for example, the surface is essentially similar to projection to that of the Butterfly (2; Fig. 7) except that the two separate regions of the twisted surface are each composed of S-shaped regions, resembling somewhat the S-shaped appearance of the Simple Cusp (1, Fig. 4, P = 84 degrees, for example) and the overall nature of the surface resembles somewhat that of the Swallowtail Catastrophe under a similar angular projection (4; Fig. 7, T = 84 degrees, P = 0 degrees). Thus, the Star Catastrophe could be considered as being constructed by the apposition of two Simple Cusps and a Swallowtail, each sharing common limbs (Fig. 12), as well as from two copies of the Wigwam (5).

Wigwam-like Sections of the Star Catastrophe

Observations:

Sections of the Catastrophe Mfd of the Star Catastrophe (x,A,B,C,D,E,F), when projected onto the (D,E) plane of the Control Space, resemble very closely the projections of the (x,A,B,C,D,E) Wigwam Catastrophe Mfd onto the (D,E) plane. Using this technique, it is easily shown that this particular projection of the Wigwam Catastrophe is generated as the composition of the Swallowtail and a Simple Cusp. Here however, both the Swallowtail and the Simple Cusp arise as specific projections of two copies of the Butterfly Catastrophe, since both are contained in this higher dimensioned catastrophe (3; Figs. 8A and 8B). This is described in Fig. 13 and becomes very evident from study of Figs. 9, 10 and 11 (in which the projection is made for increasing value of P from P = 34 degrees to P = 131 degrees; T = 0 degrees). In these figures, A = -8.0, B = 0.0, C = +16.0 and F = + 0.5 throughout.

Figs. 1, 2 and 3 show the projection of the (x,A,B,C,D,E,F) Catastrophe Mfd of the Star Catastrophe onto the (E,F) plane of the Control Space. For T = 0 degrees, the (x,E,F) cube is rotated from P = 34 degrees to P = 131 degrees in 6 degree increments.

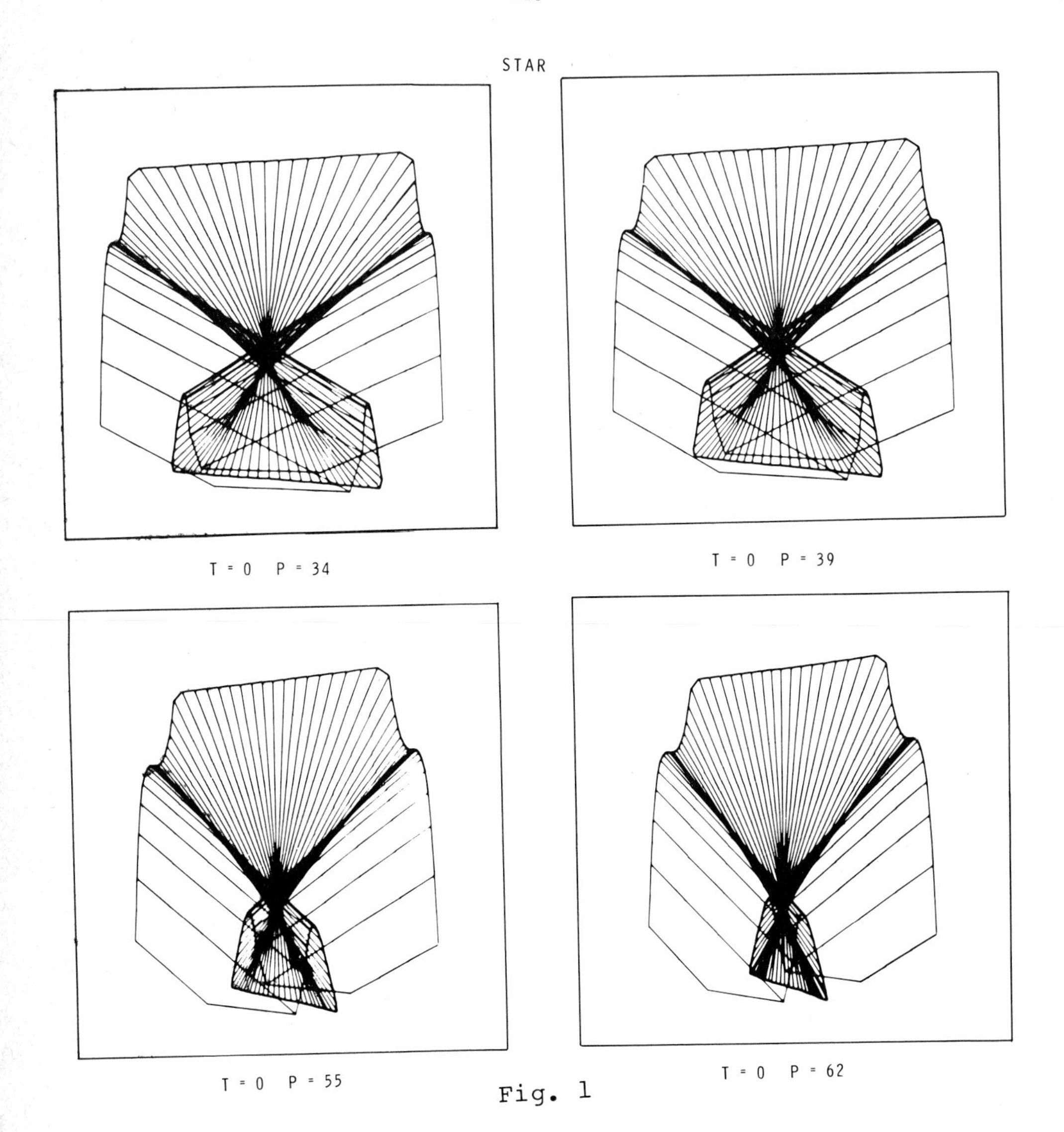

Fig. 1

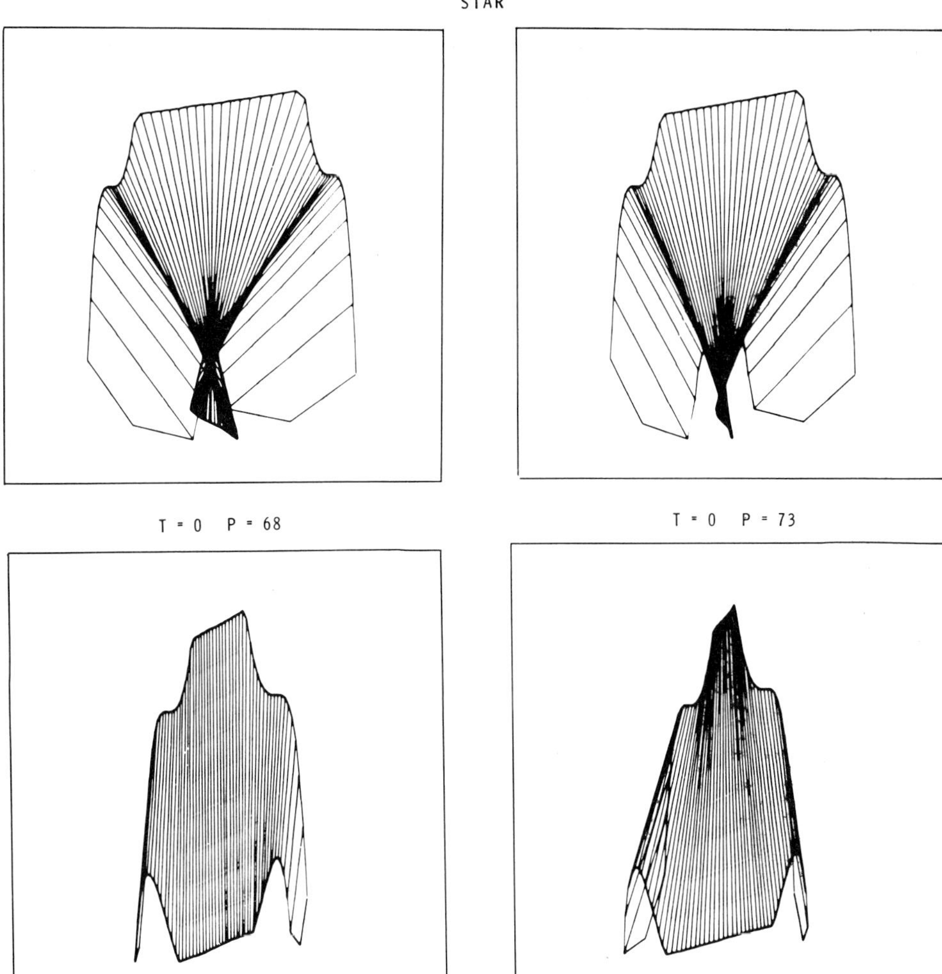

Fig. 2

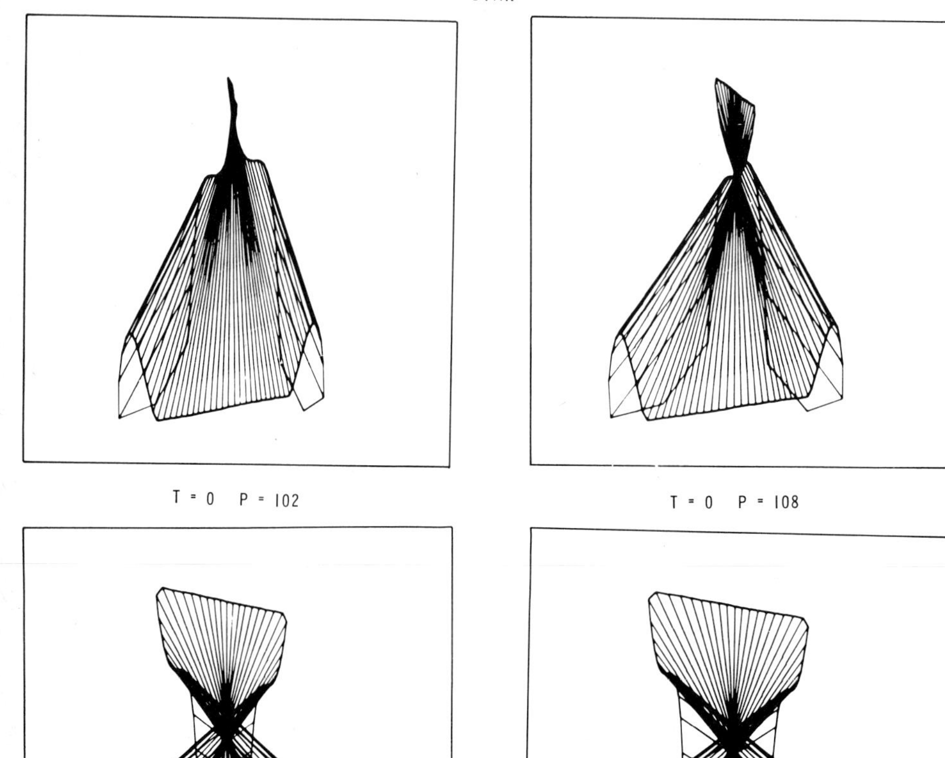

Fig. 3

Figs. 4, 5, 6, 7 and 8 show the projection of the (x,A,B,C,D,E,F) Catastrophe Mfd of the Star Catastrophe onto the (E,F) plane of the Control Space. For P = 0 degrees, the (x,E,F) cube is rotated from T = 0 degrees to T = 184 degrees in 6 degree increments.

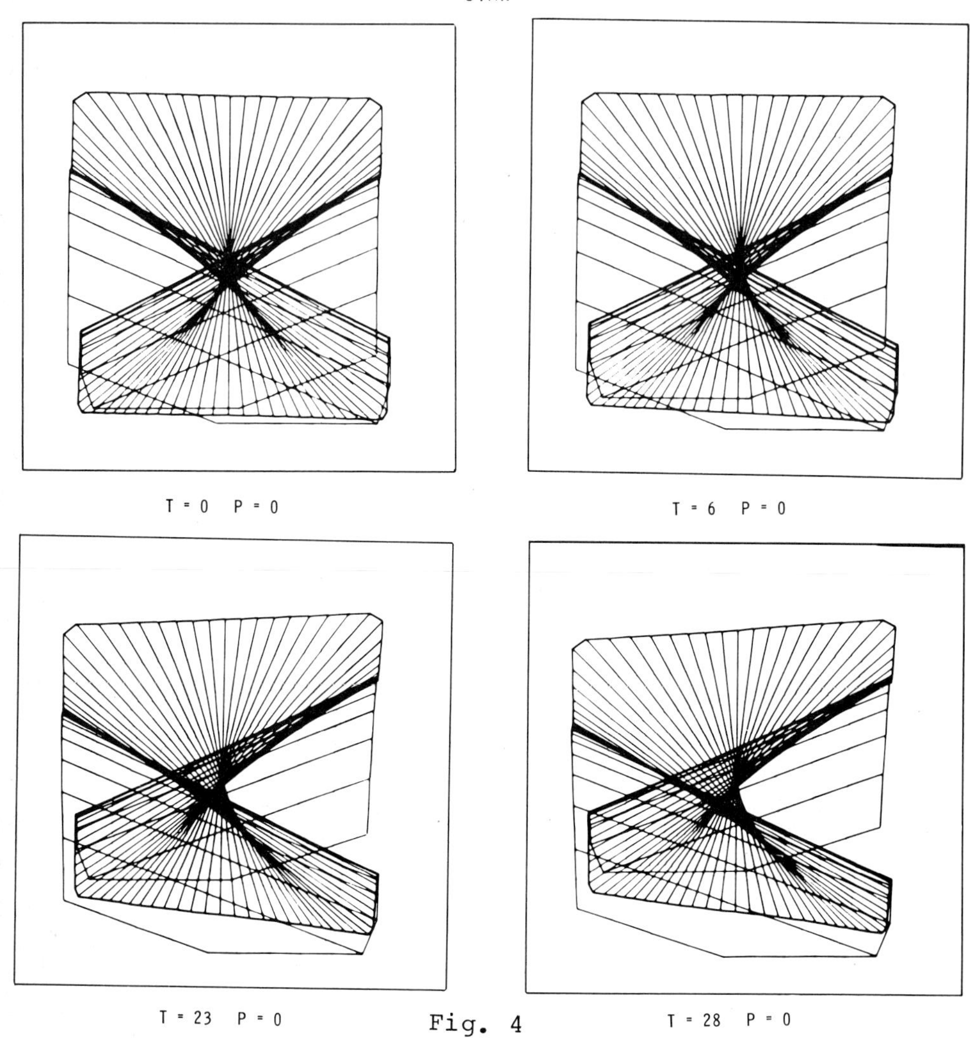

Fig. 4

STAR

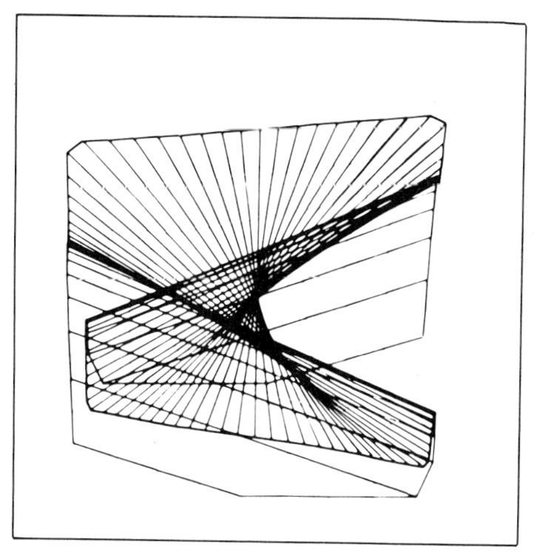

T = 34 P = 0

T = 39 P = 0

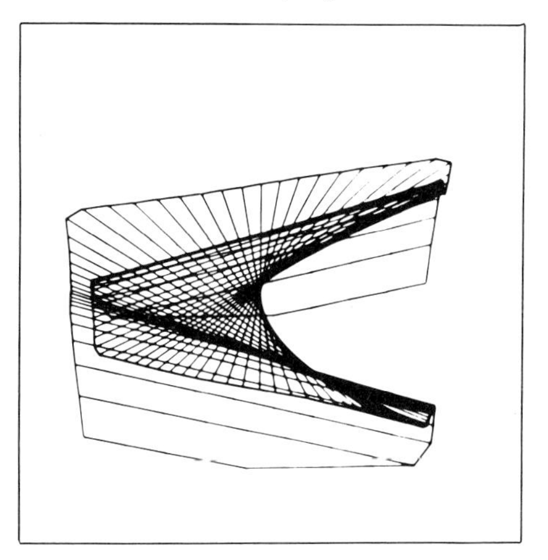

T = 56 P = 0

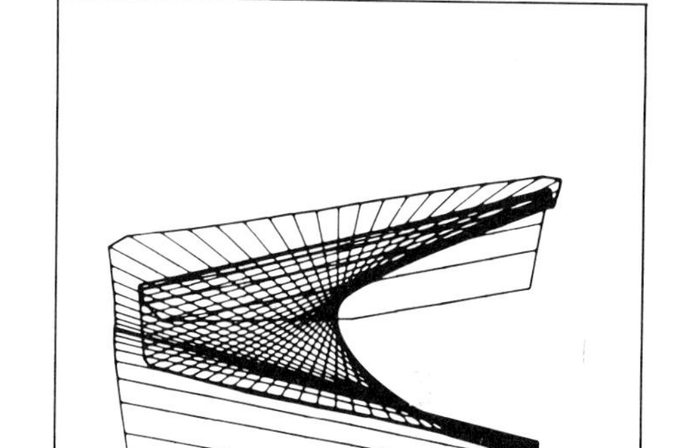

T = 62 P = 0

Fig. 5

STĀR

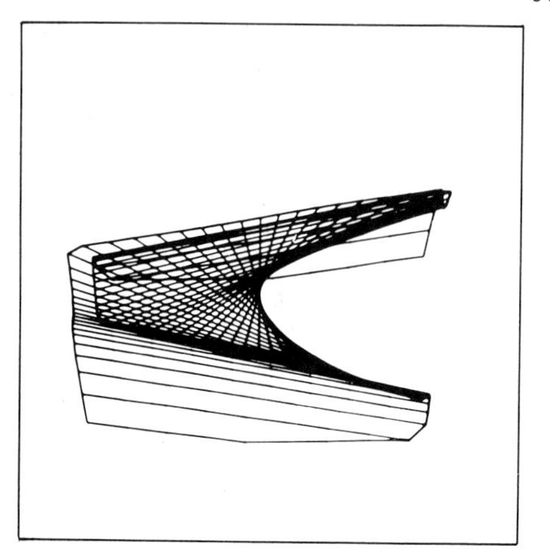

T = 68   P = 0

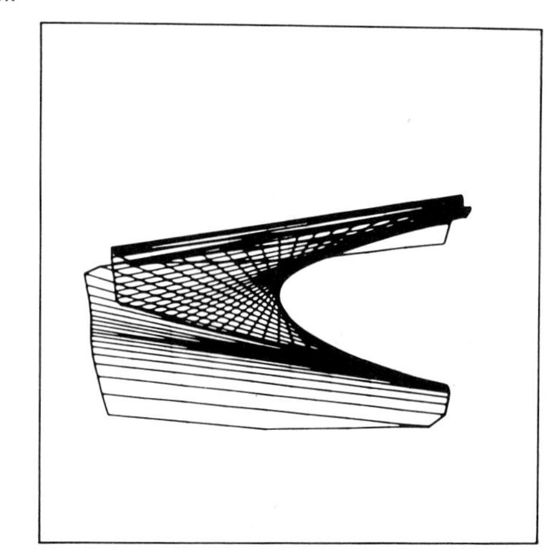

T = 73   P = 0

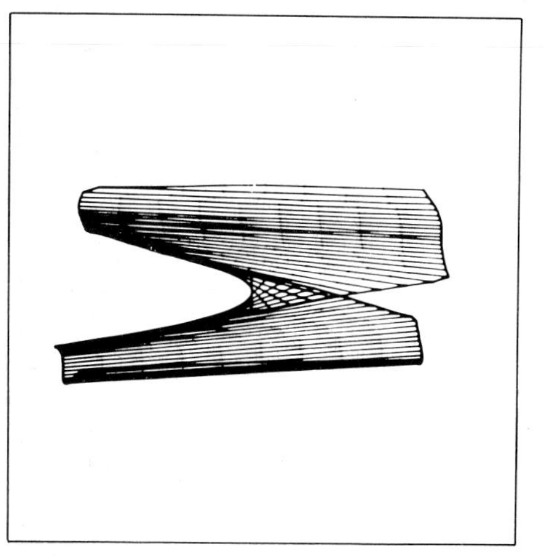

T = 96   P = 0

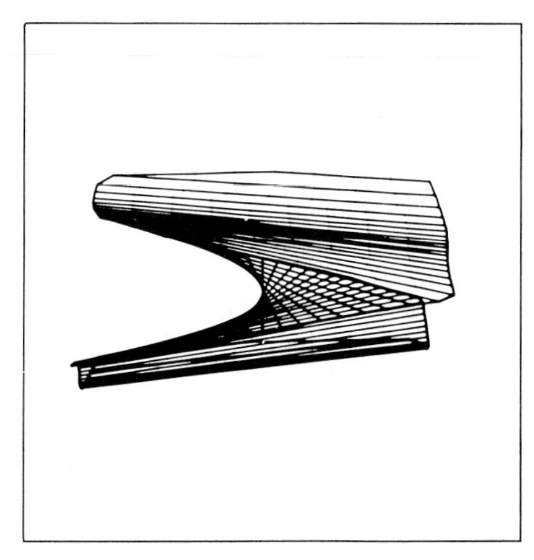

T = 102   P = 0

Fig. 6

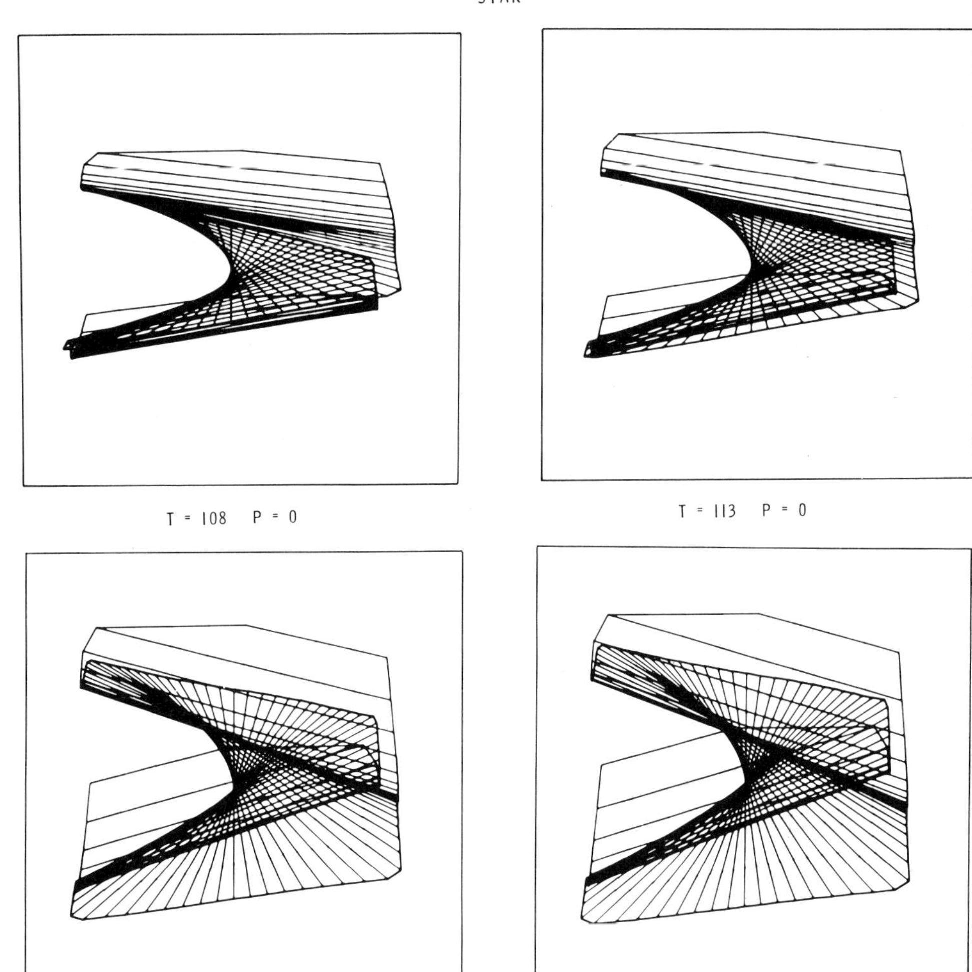

Fig. 7

STAR

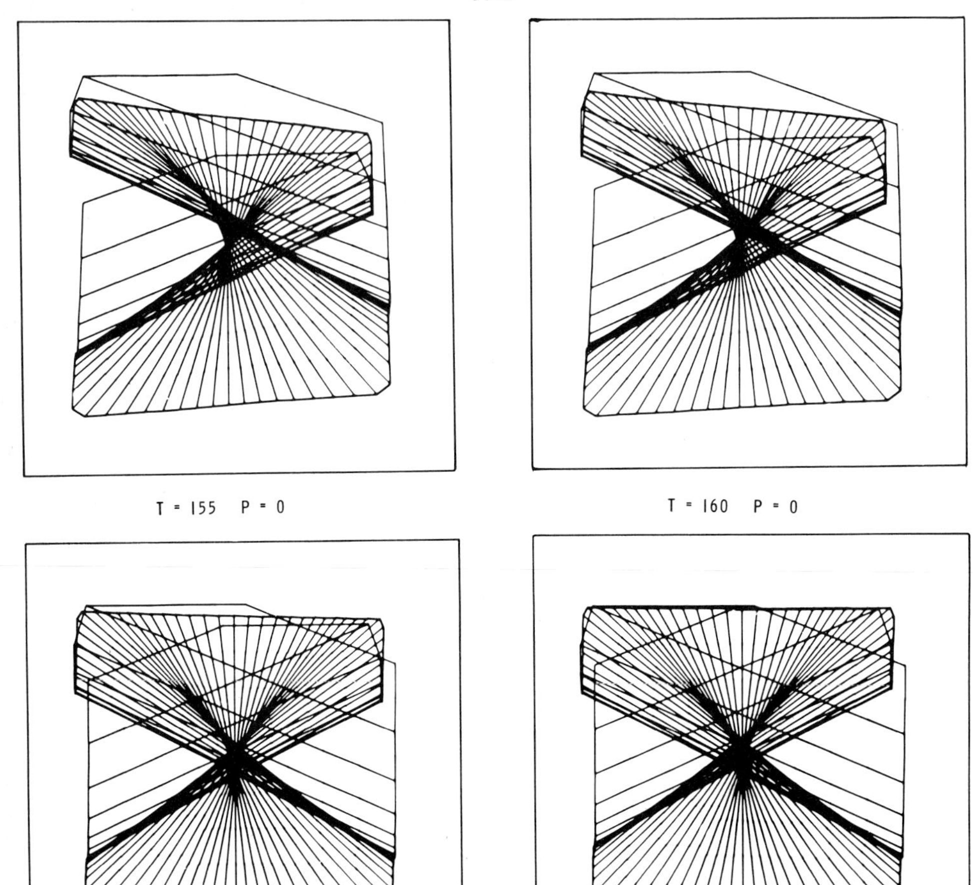

Fig. 8

Figs. 9, 10 and 11 show the projection of the (x,A,B,C,D,E,F) Catastrophe Mfd of the Star Catastrophe onto the (D,E) plane of the Control Space. This is the plane in the (A,B,C,D,E) Control Space of the Wigwam Catastrophe onto which, projection from the (x,A,B,C,D,E) Catastrophe Mfd generates typical Wigwam-like morphologies. The (x,D,E) cube is rotated from P = 34 degrees to P = 131 degrees (T being fixed at T = 0 degrees).

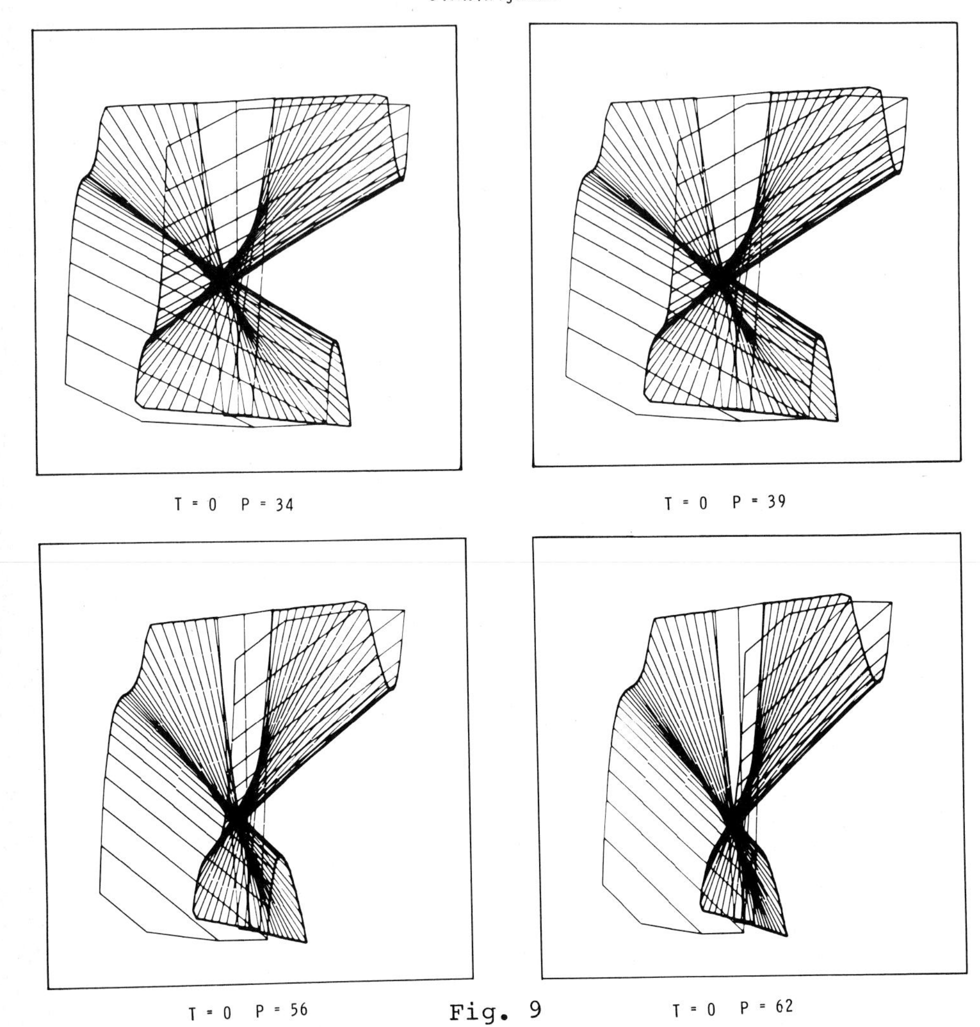

Fig. 9

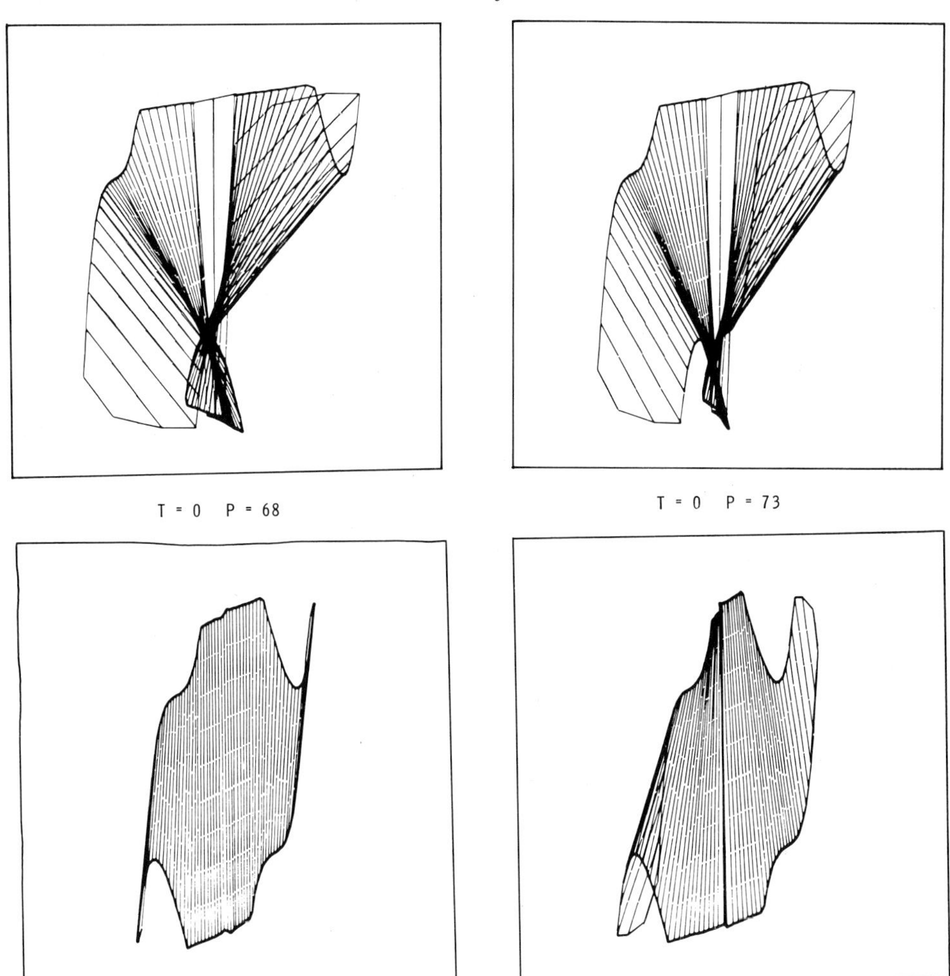
STAR(Wigwam)
T = 0 P = 68
T = 0 P = 73
T = 0 P = 90
T = 0 P = 96

Fig. 10

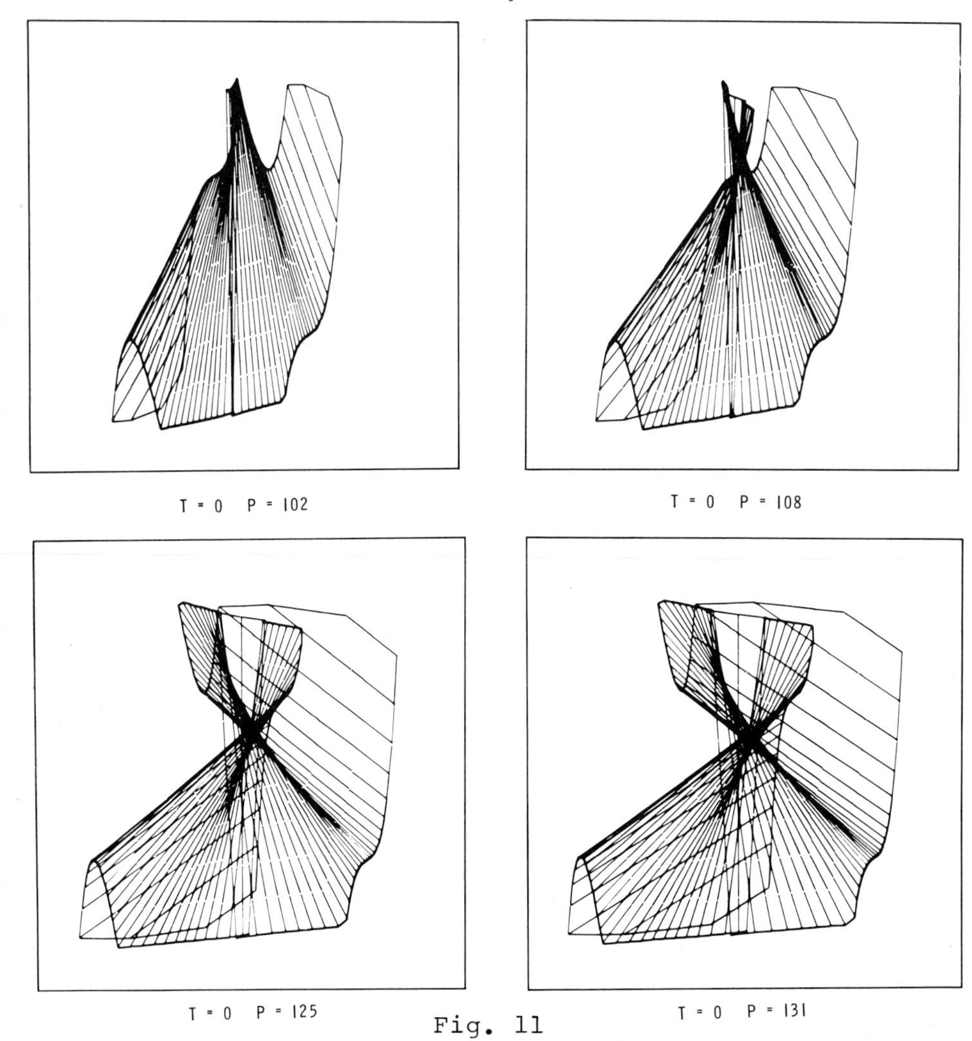
STAR(Wigwam)
T = 0 P = 102
T = 0 P = 108
T = 0 P = 125
T = 0 P = 131

Fig. 11

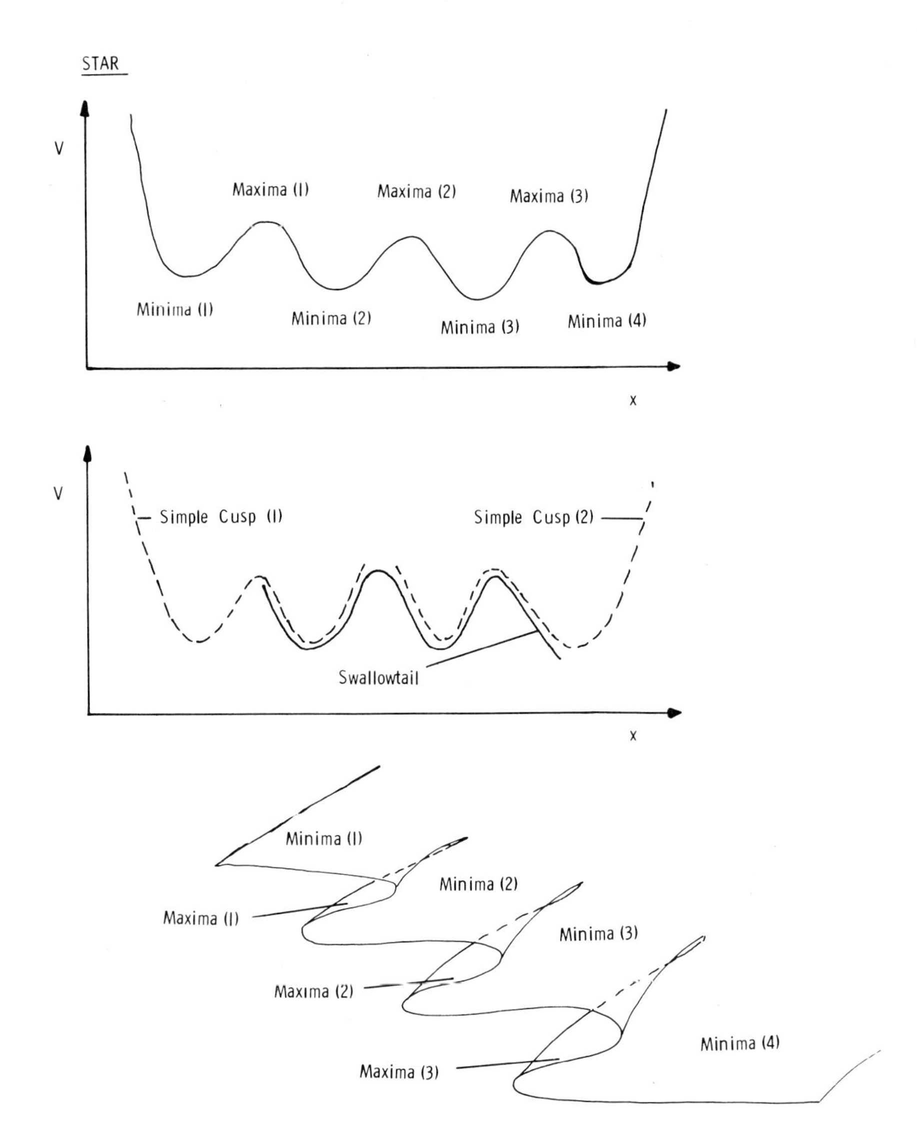
STAR
V
Maxima (1)
Maxima (2)
Maxima (3)
Minima (1)
Minima (2)
Minima (3)
Minima (4)
x
V
Simple Cusp (1)
Simple Cusp (2)
Swallowtail
x
Minima (1)
Minima (2)
Maxima (1)
Minima (3)
Maxima (2)
Minima (4)
Maxima (3)

Fig. 12

WIGWAM sections of the STAR Catastrophe.

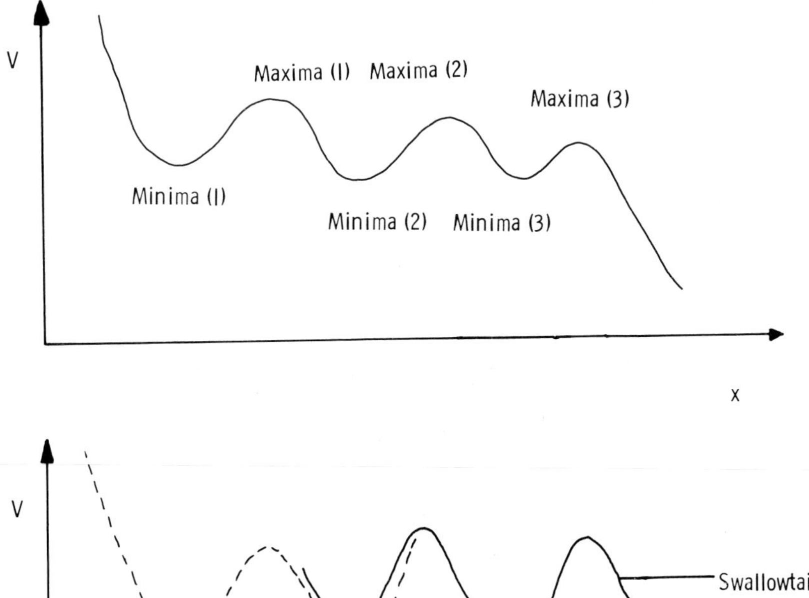

Fig. 13

Bibliography:

(1) Woodcock, A.E.R., Stereographic Reconstruction of the Catastrophe Mfd of the Cuspoids: (1) The Simple Cusp. (this volume.)

(2) Ibid., (3) The Butterfly. (this volume.)

(3) Woodcock, A.E.R. and Poston, T., The Geometrical Properties of the Elementary Catastrophes: (1) The Cuspoids. (this volume.)

(4) Woodcock, A.E.R., Stereographic Reconstruction of the Catastrophe Mfd of the Cuspoids: (2) The Swallowtail. (this volume.)

(5) Woodcock, A.E.R., Ibid., (4) The Wigwam. (this volume.)

THE GEOMETRICAL PROPERTIES OF THE REDUCED DOUBLE CUSP.

by

A. E. R. Woodcock
IBM Thomas J. Watson Research Center
Yorktown Heights, N.Y. 10598, U.S.A.

and

T. Poston
Instituto de Mathemática Pura e Aplicada
Rio-de-Janeiro, Brazil

ABSTRACT: The Bifurcation Sets of the Reduced Double Cusp $V = A(x^4 + y^4 - 6x^2y^2) + B(x^3 - 3xy^2) + C(x^2+y^2) - ux - vy$ were drawn to investigate the possibility of the existence of three-, four- and five-fold symmetry in the projected envelope. While three-and four-fold symmetry could be demonstrated, it was impossible to demonstrate five-fold symmetry. However, modifying one of the parametic equations in a heuristic way did permit the production of a figure with five-fold symmetry.

This work was begun when both authors were at the Institute of Mathematics, University of Warwick, Coventry CV4 7AL, Warwickshire, England.

INTRODUCTION

René Thom (1), in discussing the properties of the so-called "Double Cusp" states that the unfolding of the Double Cusp Singularity has certain symmetry properties, which he enumerates. It is one purpose of this paper to investigate the validity of those statements using the graphical techniques that were developed earlier (Woodcock and Poston, 2, 3, 4). The version of the Double Cusp described by Thom is given by the following equation:

$$V = A(x^4 + y^4 - 6x^2y^2) + B(x^3 - 3\,x\,y^2) + C(x^2 + y^2) - ux - vy \qquad (1).$$

(we will term the singularity given by equation (1), the Reduced Double Cusp) We will investigate the three regions of interest that Thom has considered, namely:

1. When A is zero and B and C are non-zero, Thom says that "the curve of critical values in the (u,v) plane is a hypocycloid with three cusps, H_3, and hence that it has tertiary symmetry". In this case, the equation reduces to that of the Elliptic Umbilic, which, indeed does have tertiary symmetry (Woodcock and Poston, (3)).

2. When B is zero and A and C are non-zero, "the projection of the critical curve in the u,v plane is an hypocycloid with four cusps, or quaternary symmetry".

3. When C is zero and A and B are non-zero, "the corresponding curve in the (u,v) plane is a cycloidal curve which defines the stellated pentagon (the pentagram): it is invariant under rotation of angle $2k\Pi/5$."

We will draw the appropriate sections of the Reduced Double Cusp Catastrophe control surface in order to study the validity of these statements. Furthermore, we will show sections other than those described above to demonstrate how one type of critical curve may be transformed into another.

OBSERVATIONS

The critical curve for A = 0, B $\neq$ 0 and C $\neq$ 0 of the Reduced Double Cusp is shown in Fig. 1. With B and C positive and relatively large, the critical curve does, indeed, possess tertiary symmetry; this symmetry is maintained as C is reduced and then changes sign (Fig. 1). At C = 0, the tertiary symmetry disappears.

The critical curve for B = 0 and A $\neq$ 0, C $\neq$ 0 is shown in Fig. 2. For A = -60, B = + 60, the critical curve has quaternary symmetry, again in agreement with Thom's original assertions. As the value of C is changed from +60 to -60, the curve of critical values rotates by $\Pi/2$, the size of the critical curve being dependent upon the magnitude of C for fixed value of A.

For C = 0 and A $\neq$ 0 and B $\neq$ 0, Fig. 3., shows that the curve of critical values of the Reduced Double Cusp does not have pentagonal symmetry. For A and B large, the 'curve' is simple two regions of apparently singular points. As B is reduced, these separate points appear to coalesce (see, for example, A = + 60, B = + 20, Fig. 3). This is a major difference in behavior from that predicted by René Thom, we will show, later, how it is possible to derive the property of pentagonal symmetry using the full unfolding of the Double Cusp (see Woodcock and Poston (5) and Woodcock, Poston and Stewart (6)). It is possible, however, to obtain a pentagonally symmetric critical curve in the Reduced version of the Double Cusp in a heuristic way (see below).

Fig. 4. shows the genesis of the tertiary symmetry (that is, of an Elliptic Umbilic section) of the Reduced Double Cusp. As A, for instance, is reduced to zero, and C subsequently increased, the two singular points coalesce, as previously described, and the resultant singular point serves as a 'growth point' for the Elliptic Umbilic-like section, which is shown fully developed at A = 0, B = C = + 60, Fig. 4).

The development of the quarternally symmetric critical curve from the separated singular points (at A = B = + 60, C = 0) is shown in Fig. 5. In this series of diagrams, A is maintained at + 60 and B is reduced to zero and C, originally zero, in increased to + 60.

Transformation of the tertiary into quaternary symmetry is shown in Fig. 6. As B is reduced, from +60, the structure with tertiary symmetry opens up and the y=constant lines become almost parallel (at B = +5). With B zero and increasing values of A the critical curve adopts quaternary symmetry, which is displayed completely at A = C = +60, B = 0.

Further transformations between types of symmetries of the critical curves are shown in Figs. 7 and 8. These figures show pictures of the critical curves for A, B and C all non-zero. In particular, Fig. 7 shows the transition from a curve with approximately quaternary symmetry (A = + 60, B = C = -60) through a series of pictures with tertiary symmetry (A =+5, B = C = -60, for example) into a region where the critical curve appears as an 'oblique' section of the quaternary symmetry curve. Fig. 8. shows how this oblique section may be transformed into the full quarternary symmetrical figure.

Thus, by drawing the appropriate sections of the control space of the Reduced Double Cusp, we have been able to demonstrate that two of the assertions made by Thom (mentioned above) are correct. However, we could not confirm the validity of the third assertion, namely, that pentagonal symmetry exists for certain values of the A,B, C (Control) parameters.

It is possible, however, to adjust the parametric equations that were used to draw the critical curves shown earlier in this paper to permit the drawing of a critical curve with pentagonal symmetry. Differentiation of the equation (1) with respect to x and y produces expressions for the u and v coordinates of the points in the control space. These equations are:

$$u = A(4x^3 - 12\,x\,y^2) + 3B(x^2 - y^2) + 2\,C\,x. \tag{2}$$

$$v = A(4y^3 - 12\,x^2y) - 6\,B\,x\,y + 2\,C\,y \tag{3}$$

Use of these equation failed to generate any pictures that demonstrated the pentagonal symmetry predicted by Thom. However, we were able to obtain critical curves that showed pentagonal symmetry, simply by changing one of the signs in equation (3), thus the equation (3) was ammended to:

$$v = A(4\,y^3 - 12\,x^2y) + 6\,B\,x\,y + 2\,C\,y. \tag{3,a}$$

Use of the equations (2) and (3,a) with C zero produced critical curves with pentagonal symmetry. (The remainder of the figures in the paper will all be drawn using this <u>Modified</u> version of the Reduced Double Cusp). Note, however, that these equations do <u>not</u> arise from the differentiation of a potential, so that their relationship with Catastrophe Theory is uncertain.

<u>The Reduced Double Cusp (Modified Version).</u>

The development of pentagonal symmetry is well demonstrated in Fig. 9. Initially, with A zero, the y = constant lines are superimposed. However, when A is positive (+5 or +20, for example), C is zero and B = +60 (B remains fixed in value throughout this series of pictures), the first indications of pentagonal symmetry occur. The pentagonal symmetry is well developed at A = + 40 and complete at A = + 60.

The transformation from quasi-quaternary symmetry (B non-zero but relatively small) into pentagonal symmetry is shown in Fig. (11). Increasing the magnitude of B reduces the size of one of the four component cusps of the cricital curve (B = -40, then B = -60). Subsequent reduction in the magnitude of C results in a splitting of this reduced cusp into two component cusps, which themselves increase in size as C is reduced to zero and give rise to the pentagonally symmetric critical curve. Starting from A = B = C = -60 (Fig. 12) and reducing A instead of C results in the generation of a singular point (see, for example A = +5). A further increase in the value of A (to, for example A = +30) causes the now inner ends of the y=constant lines to curve sharply inwards, opening a fissure in the outline of the control space. Further variants of this type of behavior are shown in Fig. 13., which shows transitions in which both quasi-tertiary and quasi-quaternary symmetric critical curves are generated.

The transition from quaternary to pentagonal symmetry in the modified, Reduced Double Cusp is shown in Fig. 14. The curve has quaternary symmetry at A = C = +60, B = 0; as these values are changed to A = B = +60, C = 0, pentagonal symmmetry develops. Fig. 15 shows the transitions that occur in the shape of the critical curve as the sign of the C parameter is changed from positive to negative (A = -60, B = +20), showing the apparant rotation by ($\Pi/2$) of the critical curve outline). The changes in the morphology of the critical curve as the sign of B is changed from negative to positive (A = -60, C = +60) is shown in Fig. 16.

Bibliography

(1) Thom, R., In: American Mathematical Society, Lectures on Mathematics in the Life Sciences, Ed. by J. Cowan, Providence, Rhode Island, 1972.

(2) Woodcock, A. E. R. and Poston, T., The Geometrical Properties of the Elementary Catastrophes, The Cuspoids, (this volume).

(3) Woodcock, A. E. R. and Poston, T., The Geometrical Properties of the Elementary Catastrophes (2) The Elliptic and Hyperbolic Umbilics (this volume).

(4) Woodcock, A. E. R. and Poston, T., The Geometrical Properties of the Elementary Catastrophes (3) The Parabolic Umbilic (this volume).

(5) Woodcock, A. E. R. and Poston, T., On the Geometry of Certain Potential Functions with Two-Fold Symmetry. (To appear)

(6) Woodcock, A. E. R., Poston, T. and Stewart, I.N., The Full Unfolding of the Double Cusp (to appear).

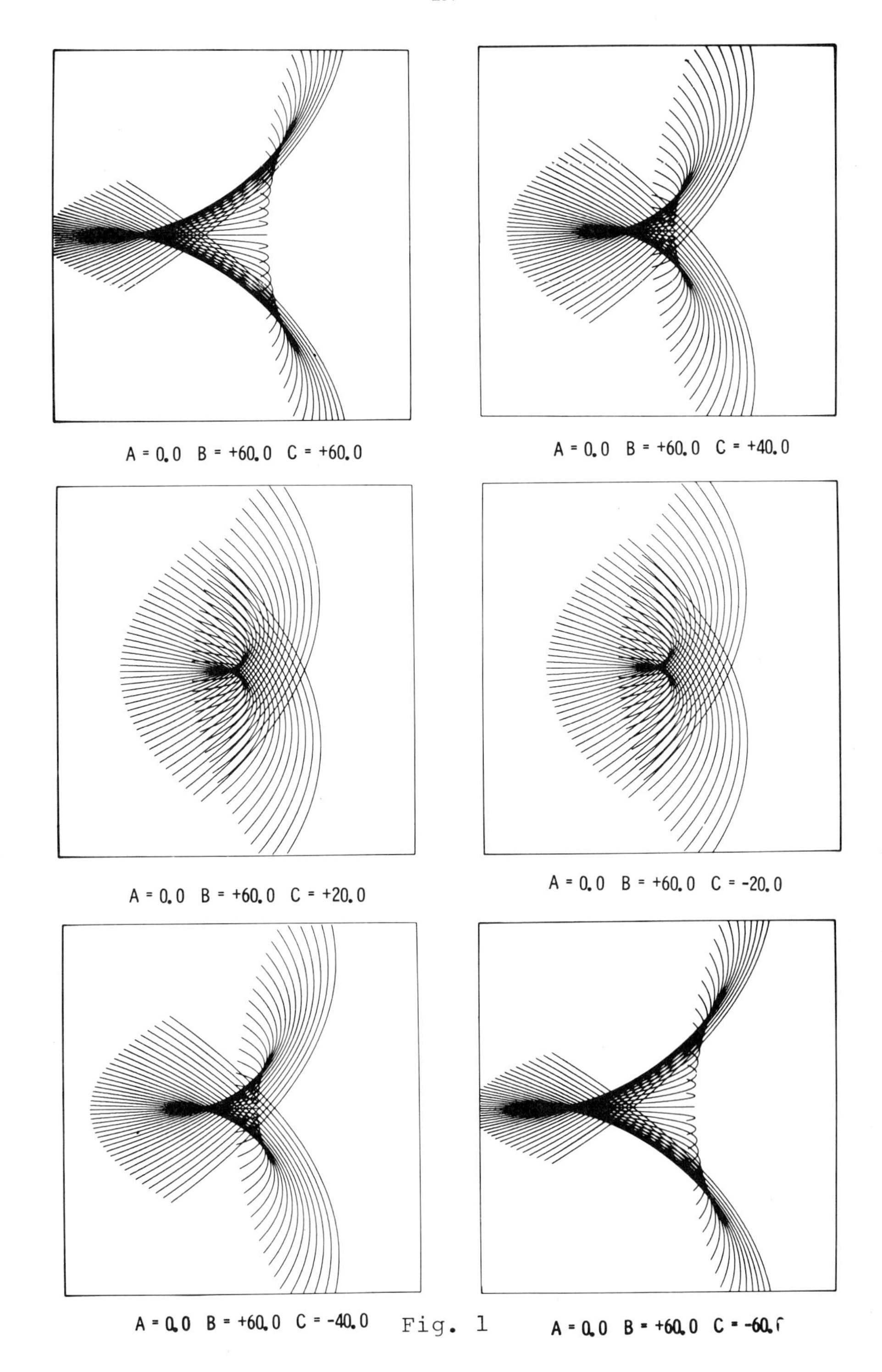
A = 0.0 B = +60.0 C = +60.0
A = 0.0 B = +60.0 C = +40.0
A = 0.0 B = +60.0 C = +20.0
A = 0.0 B = +60.0 C = -20.0
A = 0.0 B = +60.0 C = -40.0
A = 0.0 B = +60.0 C = -60.

Fig. 1

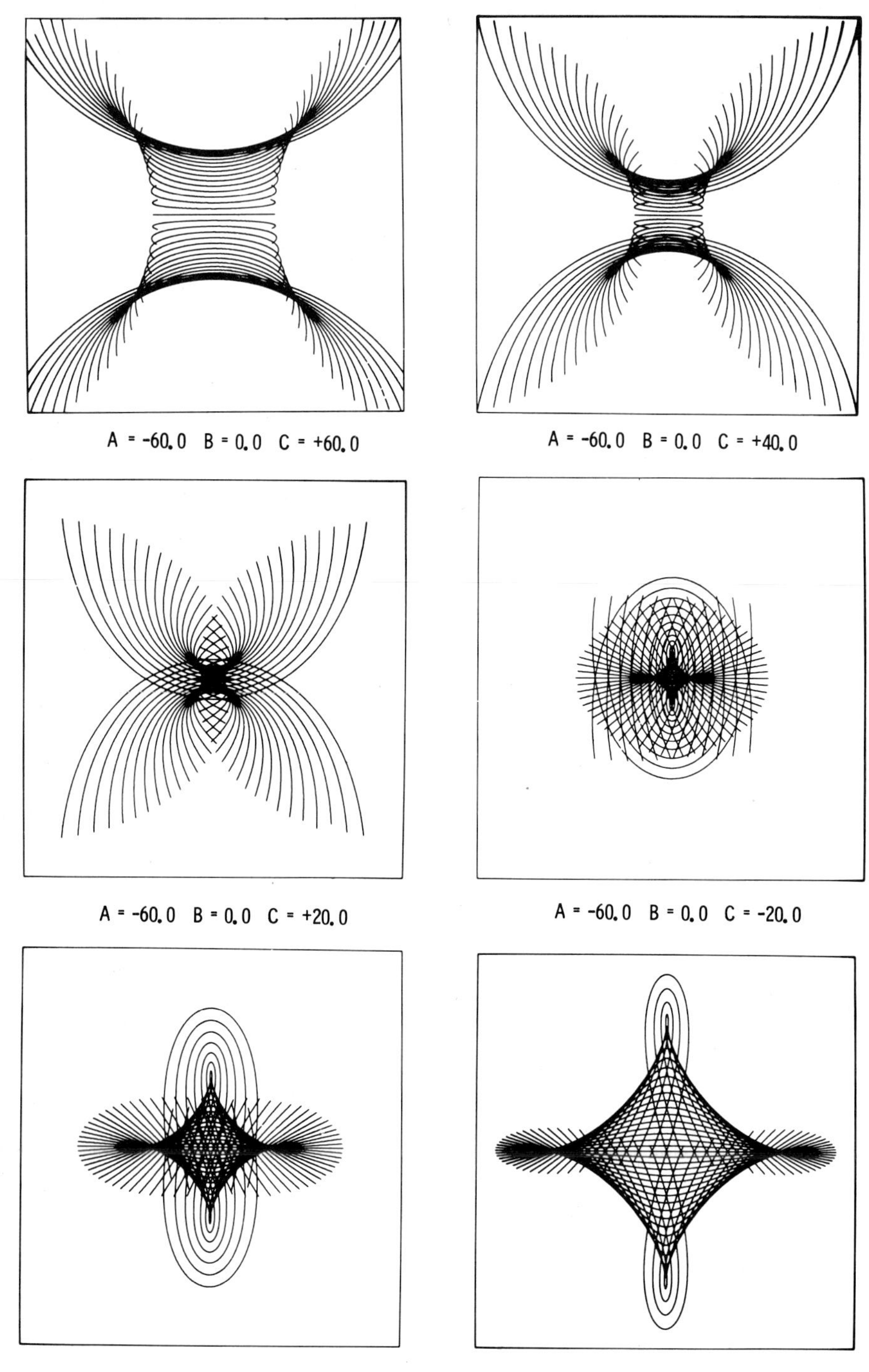
A = -60.0 B = 0.0 C = +60.0
A = -60.0 B = 0.0 C = +40.0
A = -60.0 B = 0.0 C = +20.0
A = -60.0 B = 0.0 C = -20.0
A = -60.0 B = 0.0 C = -40.0
A = -60.0 B = 0.0 C = -60.0

Fig. 2

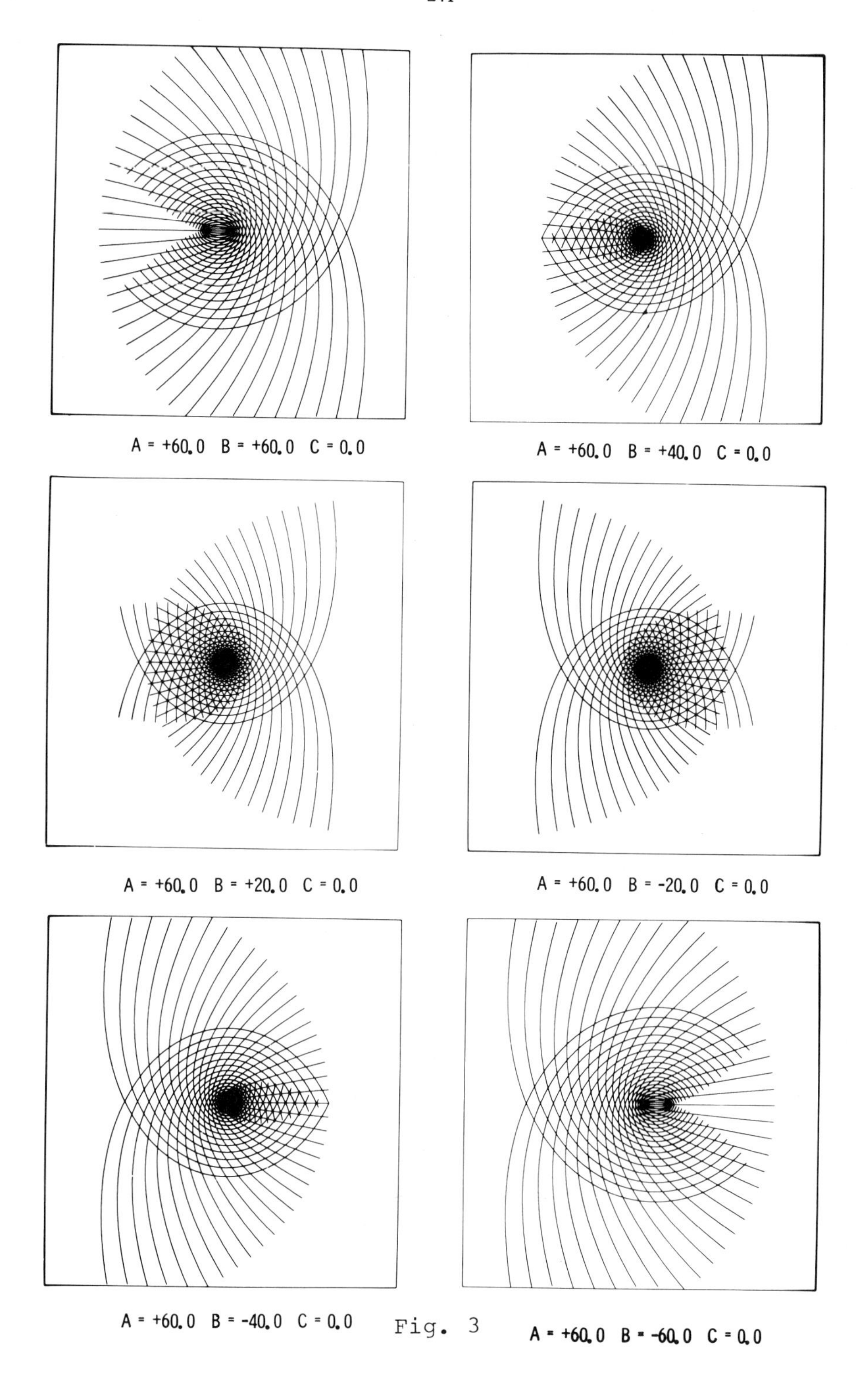
A = +60.0 B = +60.0 C = 0.0
A = +60.0 B = +40.0 C = 0.0
A = +60.0 B = +20.0 C = 0.0
A = +60.0 B = -20.0 C = 0.0
A = +60.0 B = -40.0 C = 0.0
A = +60.0 B = -60.0 C = 0.0

Fig. 3

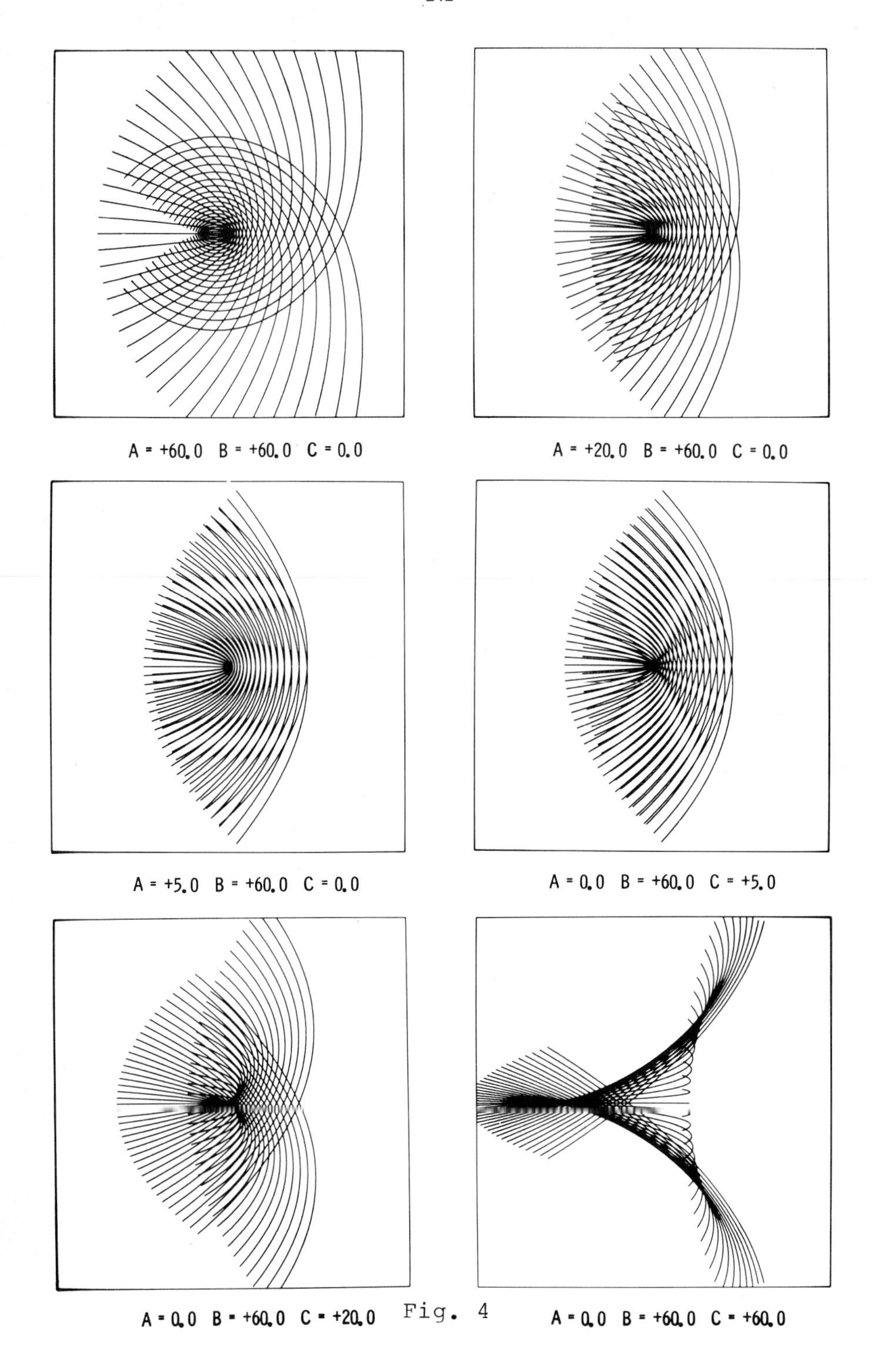
A = +60.0 B = +60.0 C = 0.0
A = +20.0 B = +60.0 C = 0.0
A = +5.0 B = +60.0 C = 0.0
A = 0.0 B = +60.0 C = +5.0
A = 0.0 B = +60.0 C = +20.0
A = 0.0 B = +60.0 C = +60.0

Fig. 4

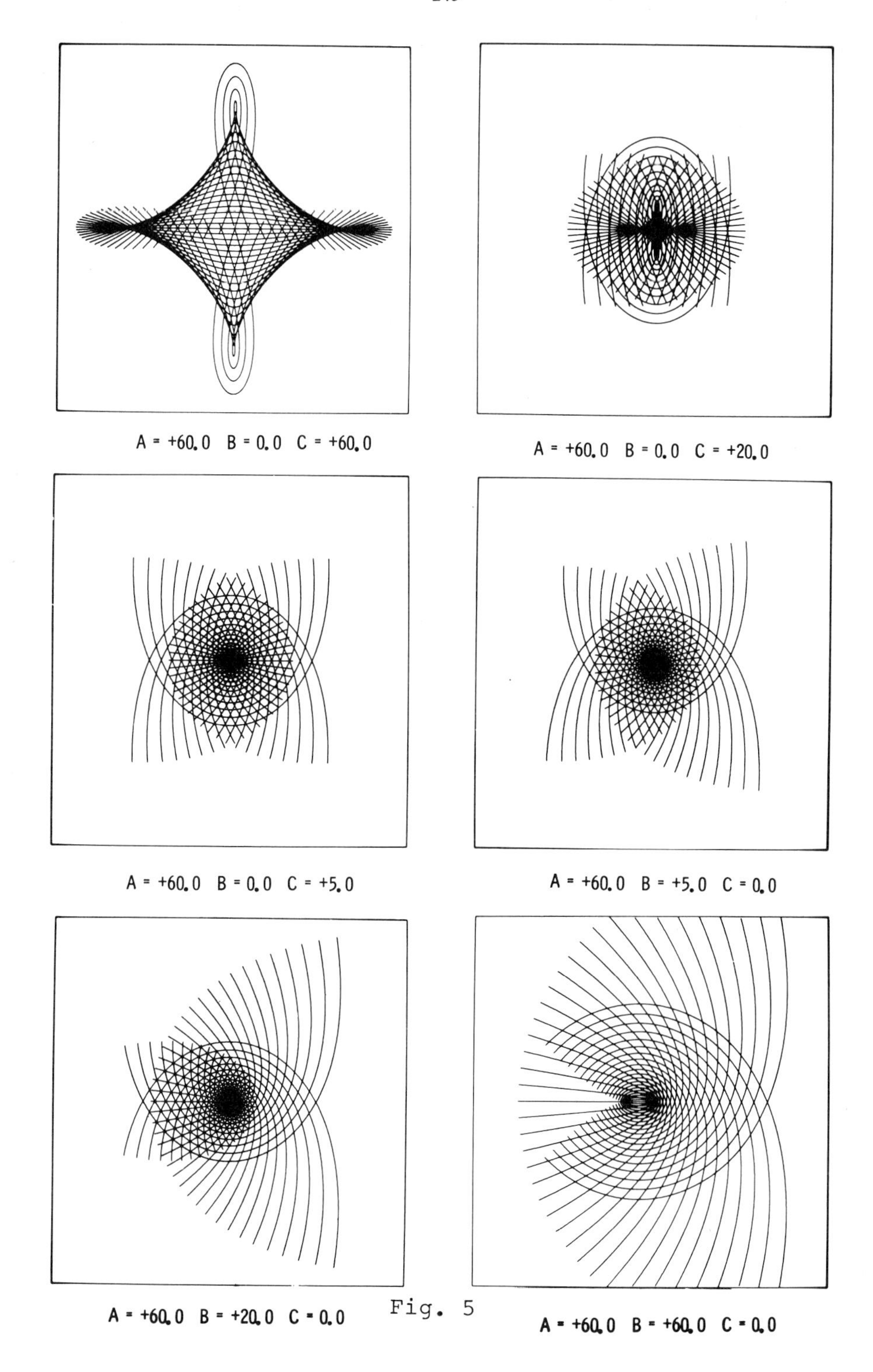
A = +60.0 B = 0.0 C = +60.0
A = +60.0 B = 0.0 C = +20.0
A = +60.0 B = 0.0 C = +5.0
A = +60.0 B = +5.0 C = 0.0
A = +60.0 B = +20.0 C = 0.0
A = +60.0 B = +60.0 C = 0.0

Fig. 5

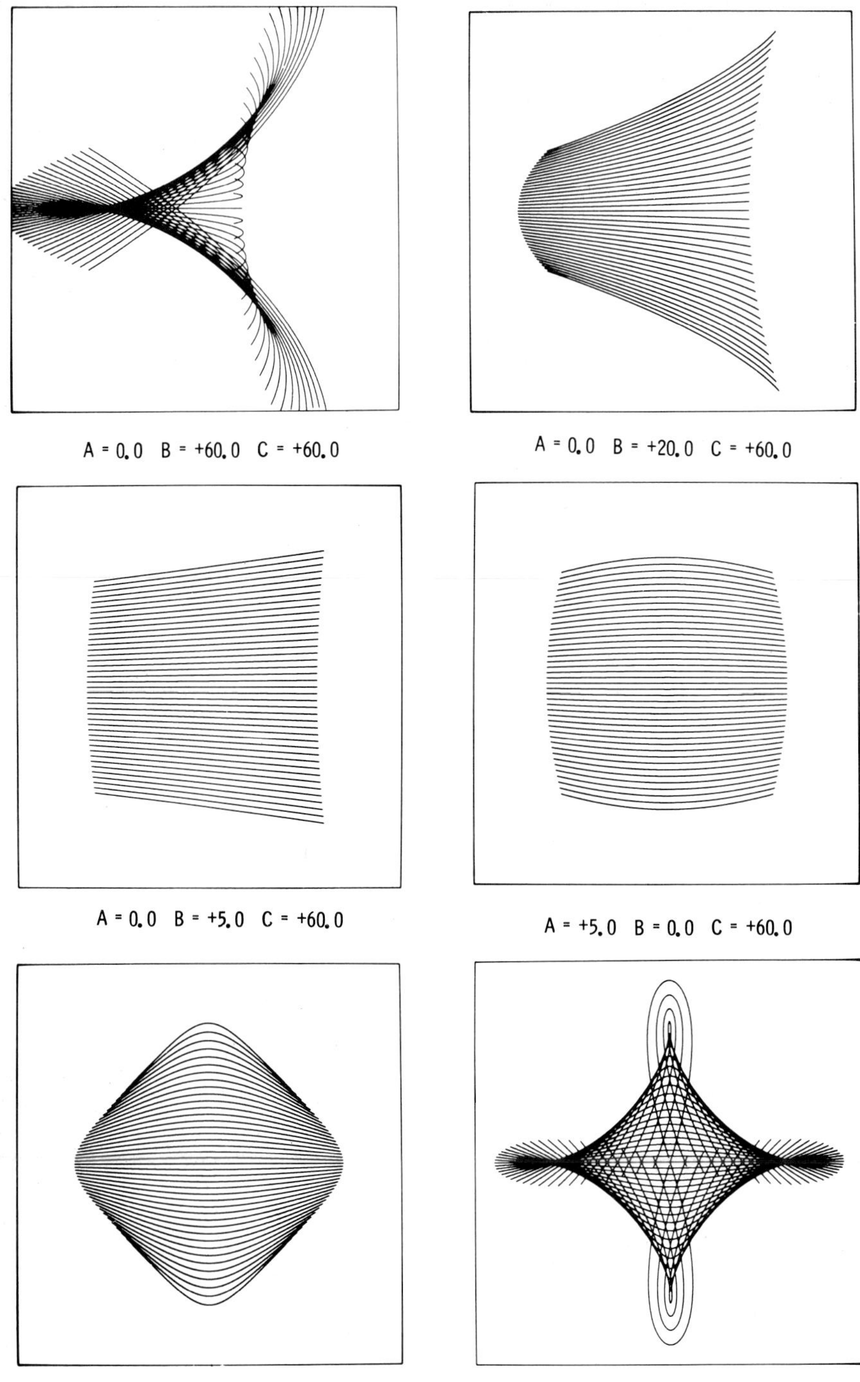

A = +20.0 B = 0.0 C = +60.0 Fig. 6 A = +60.0 B = 0.0 C = +60.0

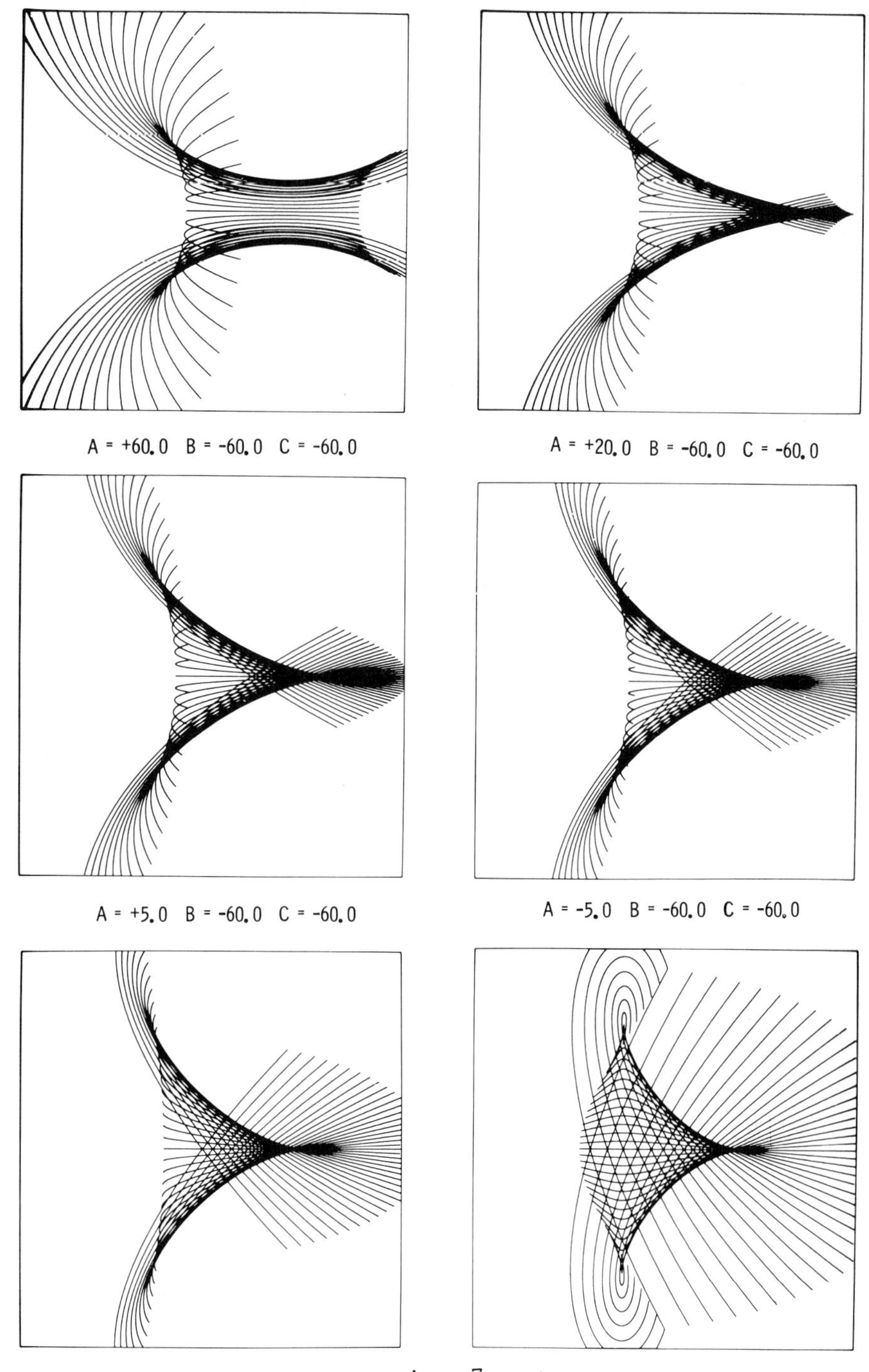
A = +60.0 B = -60.0 C = -60.0
A = +20.0 B = -60.0 C = -60.0
A = +5.0 B = -60.0 C = -60.0
A = -5.0 B = -60.0 C = -60.0
A = -20.0 B = -60.0 C = -60.0
A = -60.0 B = -60.0 C = -60.0

Fig. 7

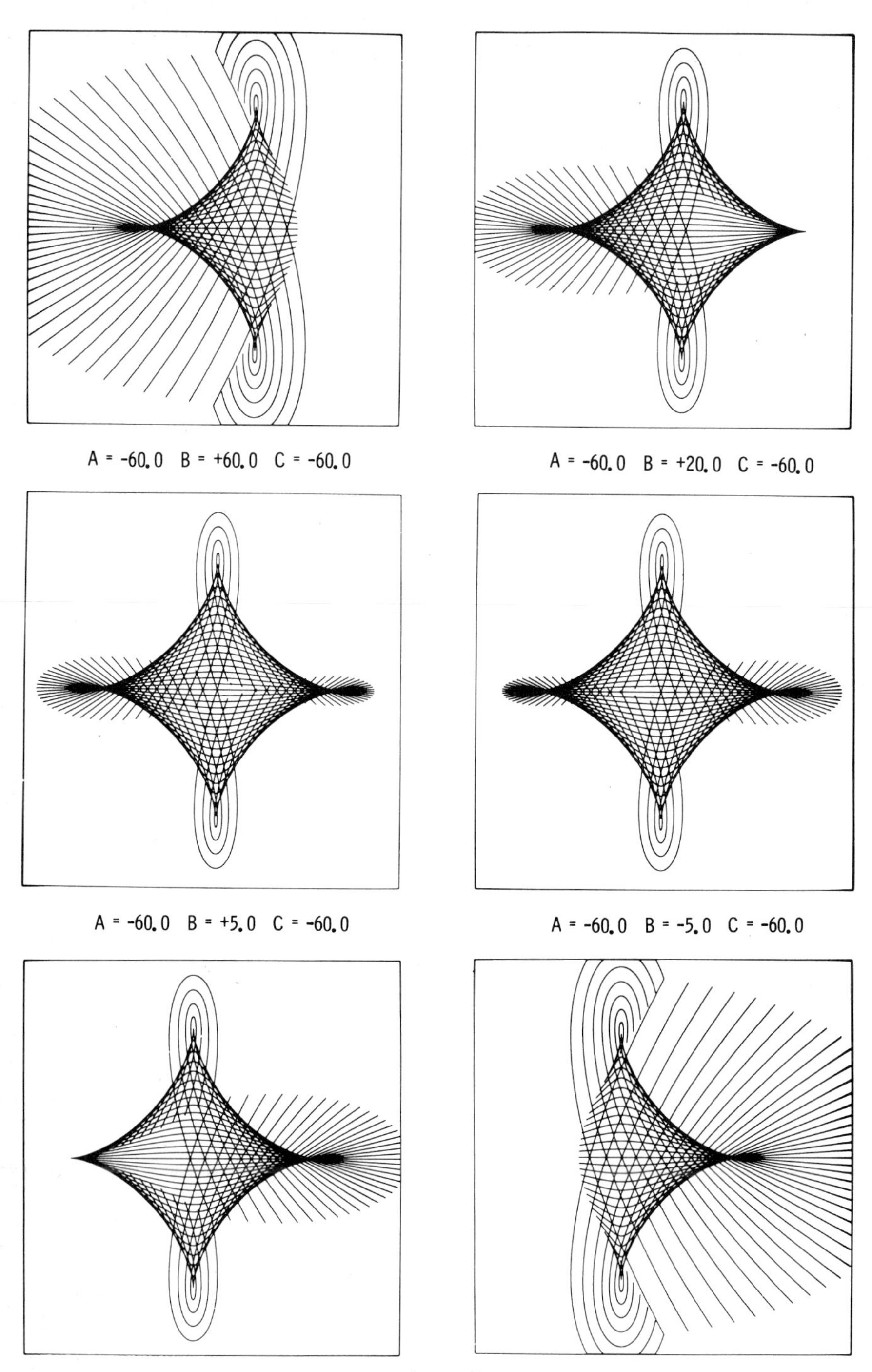
A = -60.0 B = +60.0 C = -60.0
A = -60.0 B = +20.0 C = -60.0
A = -60.0 B = +5.0 C = -60.0
A = -60.0 B = -5.0 C = -60.0
A = -60.0 B = -20.0 C = -60.0
A = -60.0 B = -60.0 C = -60.0

Fig. 8

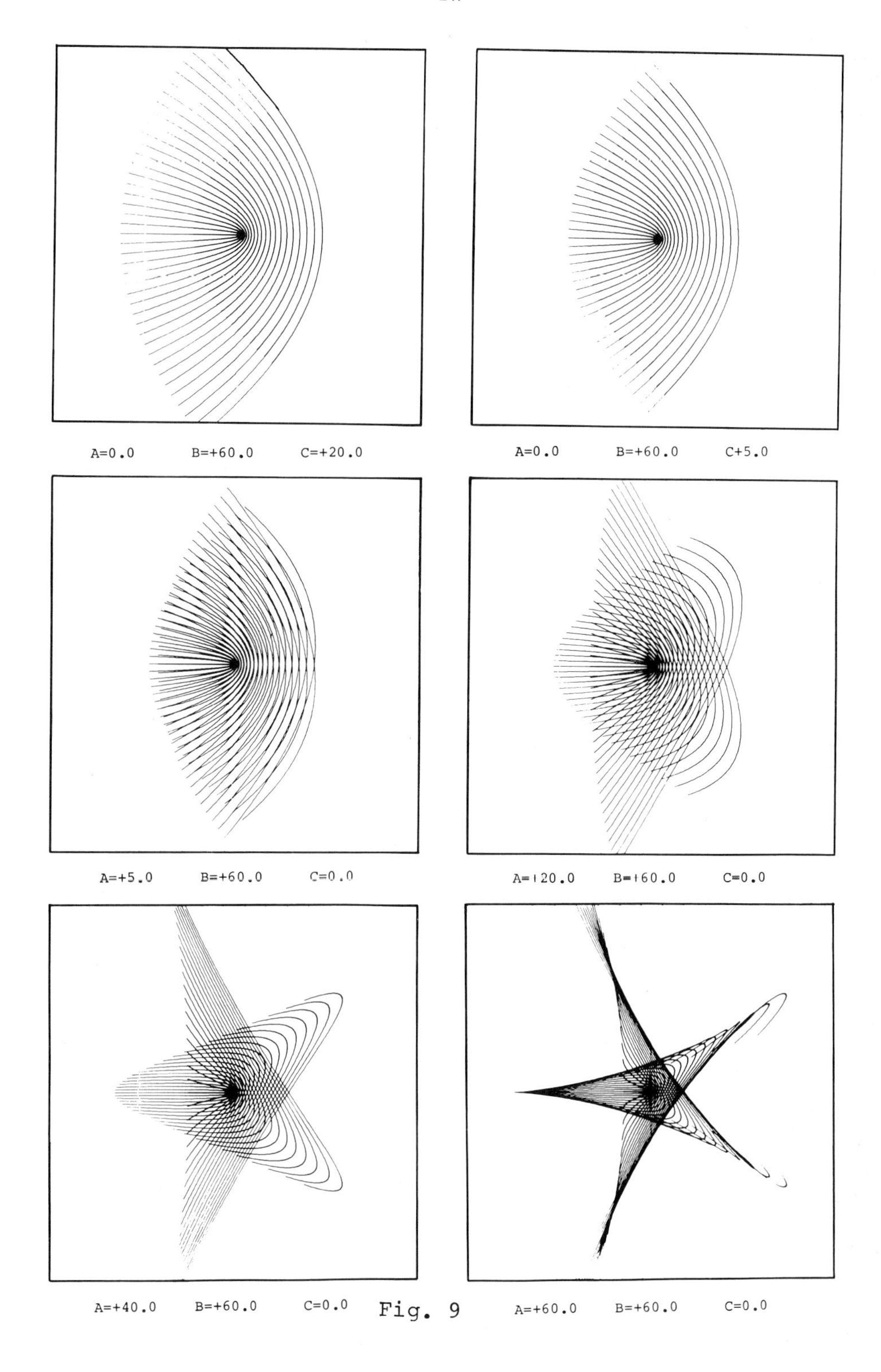
A=0.0 B=+60.0 C=+20.0
A=0.0 B=+60.0 C+5.0
A=+5.0 B=+60.0 C=0.0
A=+20.0 B=+60.0 C=0.0
A=+40.0 B=+60.0 C=0.0
A=+60.0 B=+60.0 C=0.0

Fig. 9

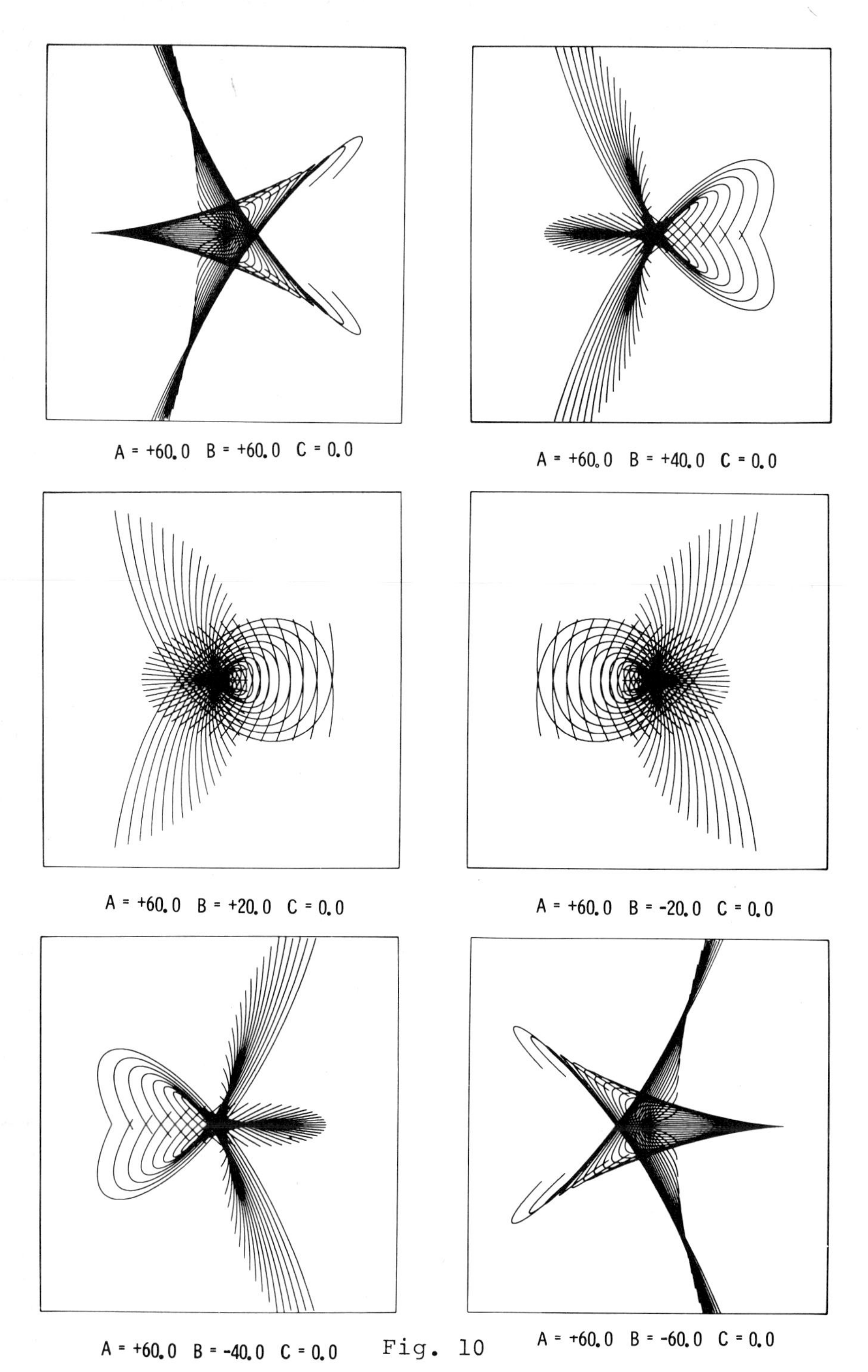
A = +60.0 B = +60.0 C = 0.0
A = +60.0 B = +40.0 C = 0.0
A = +60.0 B = +20.0 C = 0.0
A = +60.0 B = -20.0 C = 0.0
A = +60.0 B = -40.0 C = 0.0
A = +60.0 B = -60.0 C = 0.0

Fig. 10

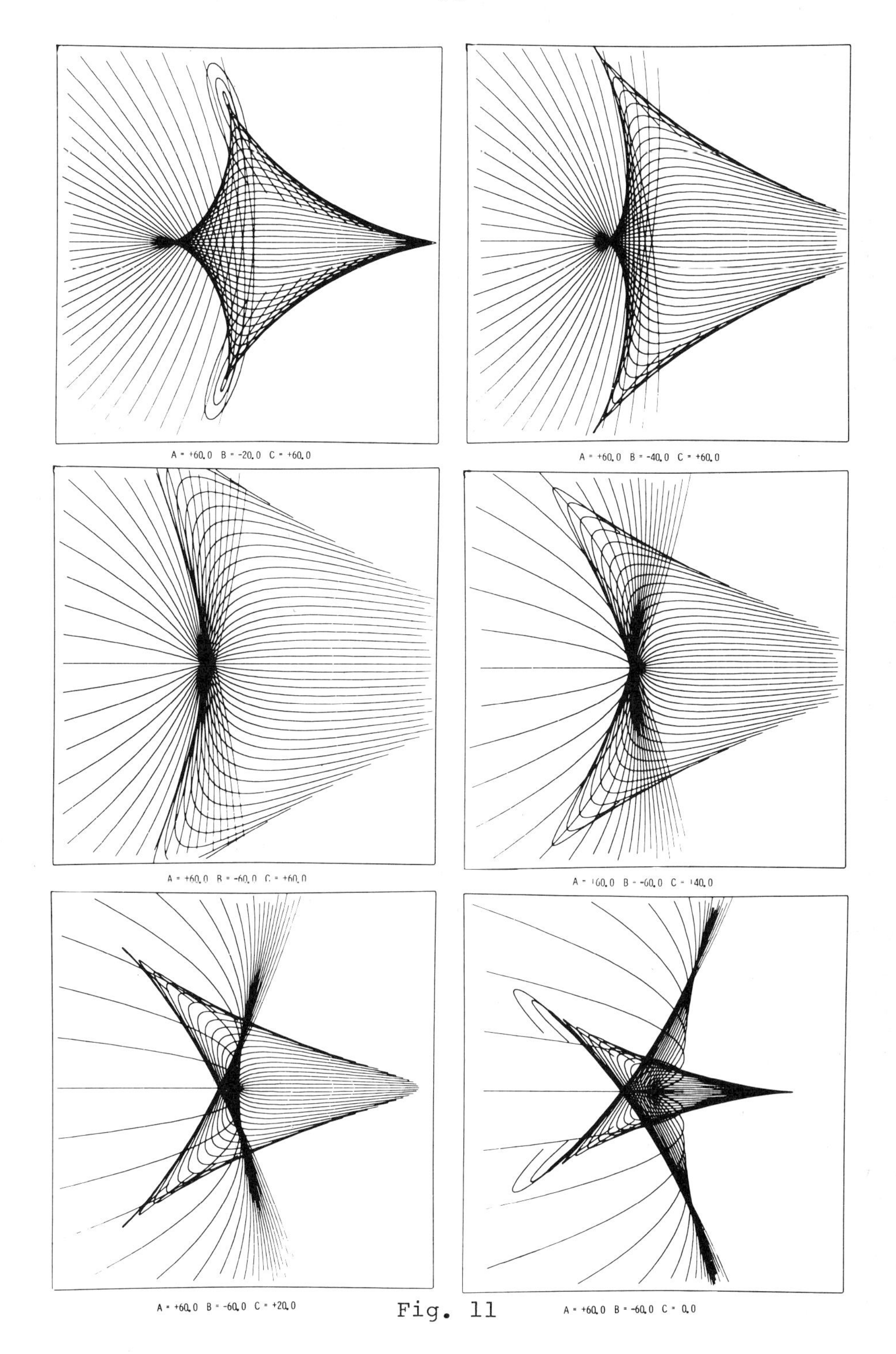
A = +60.0 B = -20.0 C = +60.0
A = +60.0 B = -40.0 C = +60.0
A = +60.0 B = -60.0 C = +60.0
A = +60.0 B = -60.0 C = +40.0
A = +60.0 B = -60.0 C = +20.0
A = +60.0 B = -60.0 C = 0.0

Fig. 11

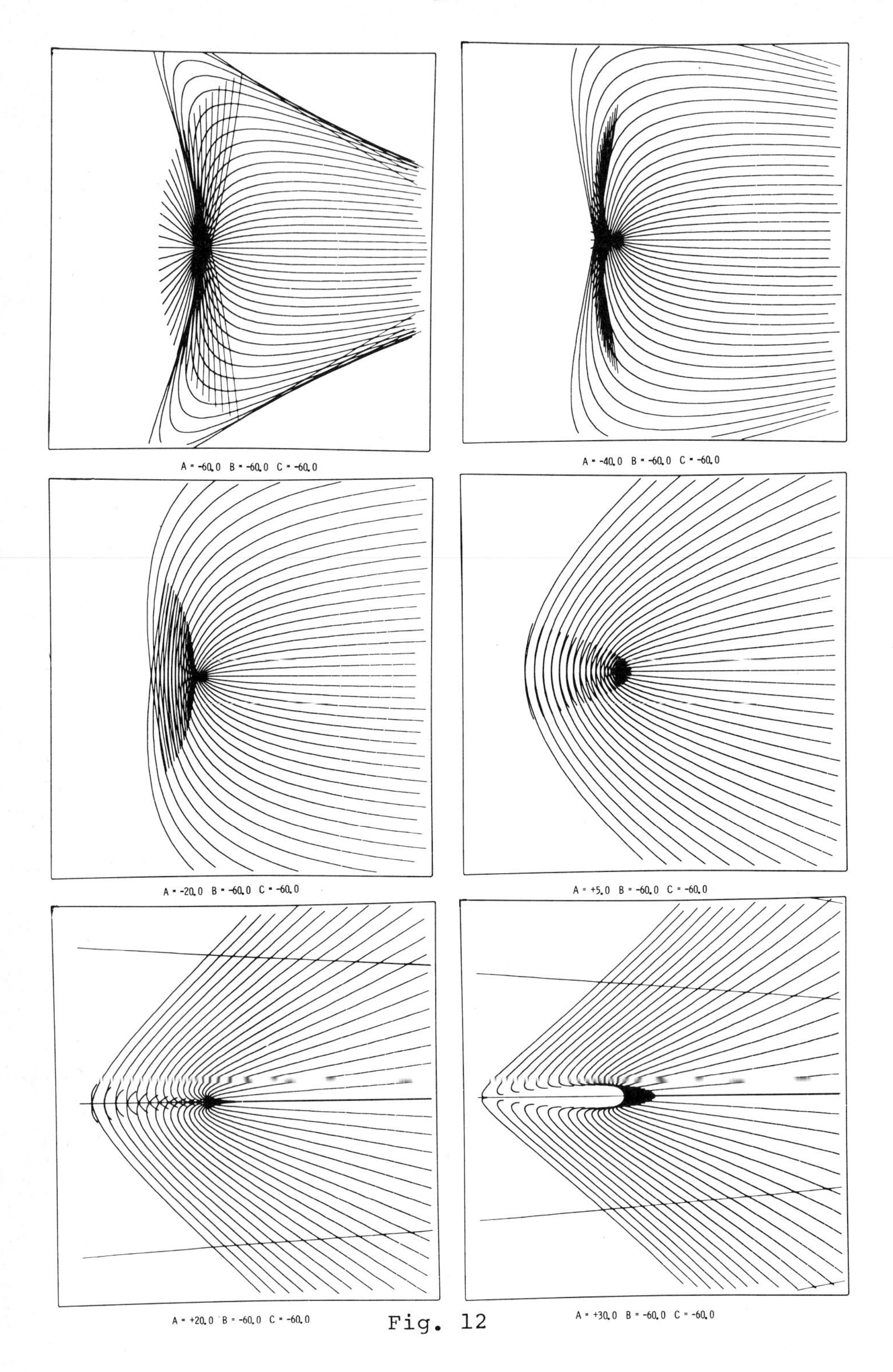
A = -60.0 B = -60.0 C = -60.0
A = -40.0 B = -60.0 C = -60.0
A = -20.0 B = -60.0 C = -60.0
A = +5.0 B = -60.0 C = -60.0
A = +20.0 B = -60.0 C = -60.0
A = +30.0 B = -60.0 C = -60.0

Fig. 12

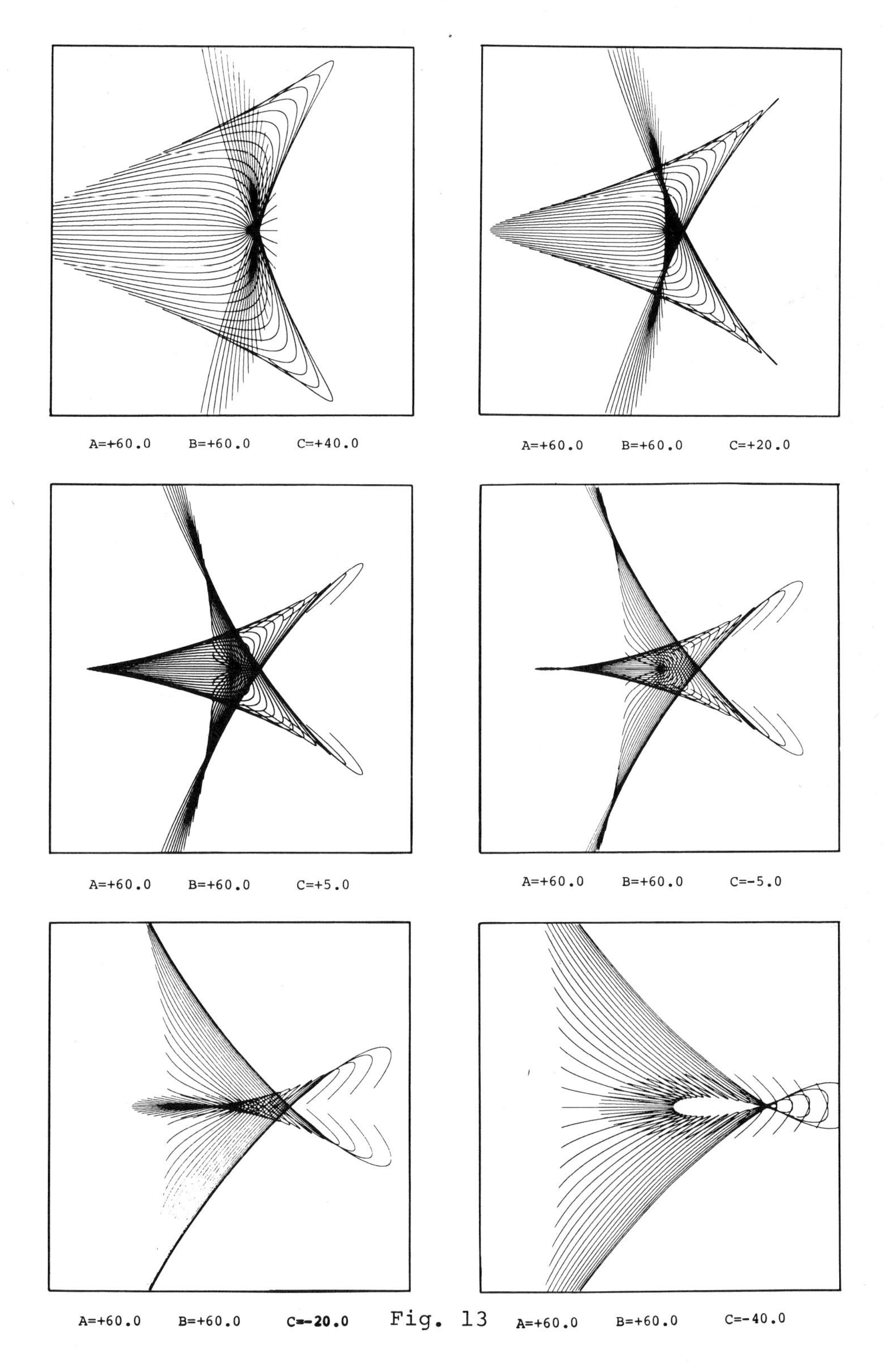
A=+60.0 B=+60.0 C=+40.0
A=+60.0 B=+60.0 C=+20.0
A=+60.0 B=+60.0 C=+5.0
A=+60.0 B=+60.0 C=-5.0
A=+60.0 B=+60.0 C=-20.0
A=+60.0 B=+60.0 C=-40.0

Fig. 13

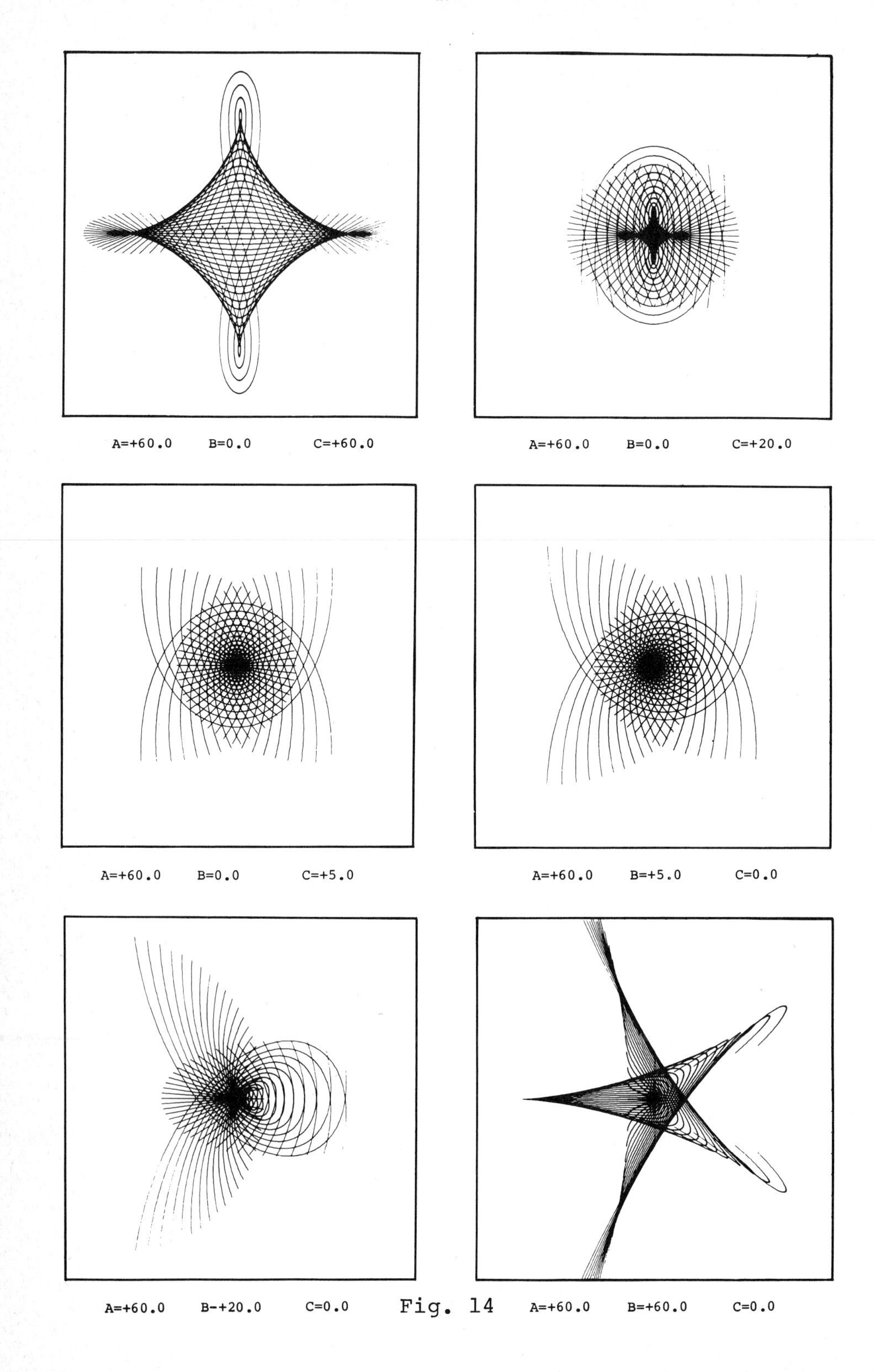
A=+60.0 B=0.0 C=+60.0
A=+60.0 B=0.0 C=+20.0
A=+60.0 B=0.0 C=+5.0
A=+60.0 B=+5.0 C=0.0
A=+60.0 B-+20.0 C=0.0
A=+60.0 B=+60.0 C=0.0

Fig. 14

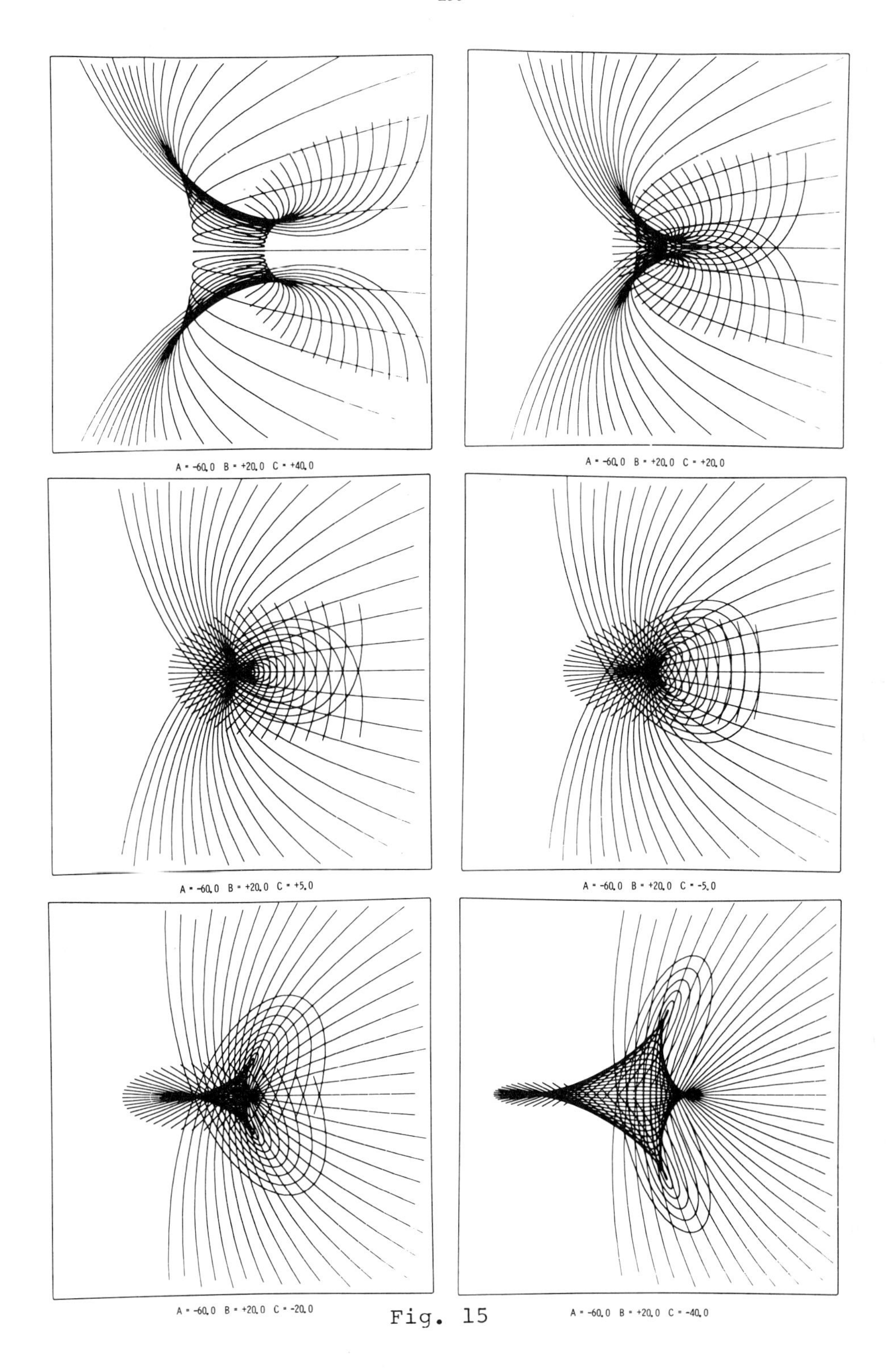
A = -60.0 B = +20.0 C = +40.0
A = -60.0 B = +20.0 C = +20.0
A = -60.0 B = +20.0 C = +5.0
A = -60.0 B = +20.0 C = -5.0
A = -60.0 B = +20.0 C = -20.0
A = -60.0 B = +20.0 C = -40.0

Fig. 15

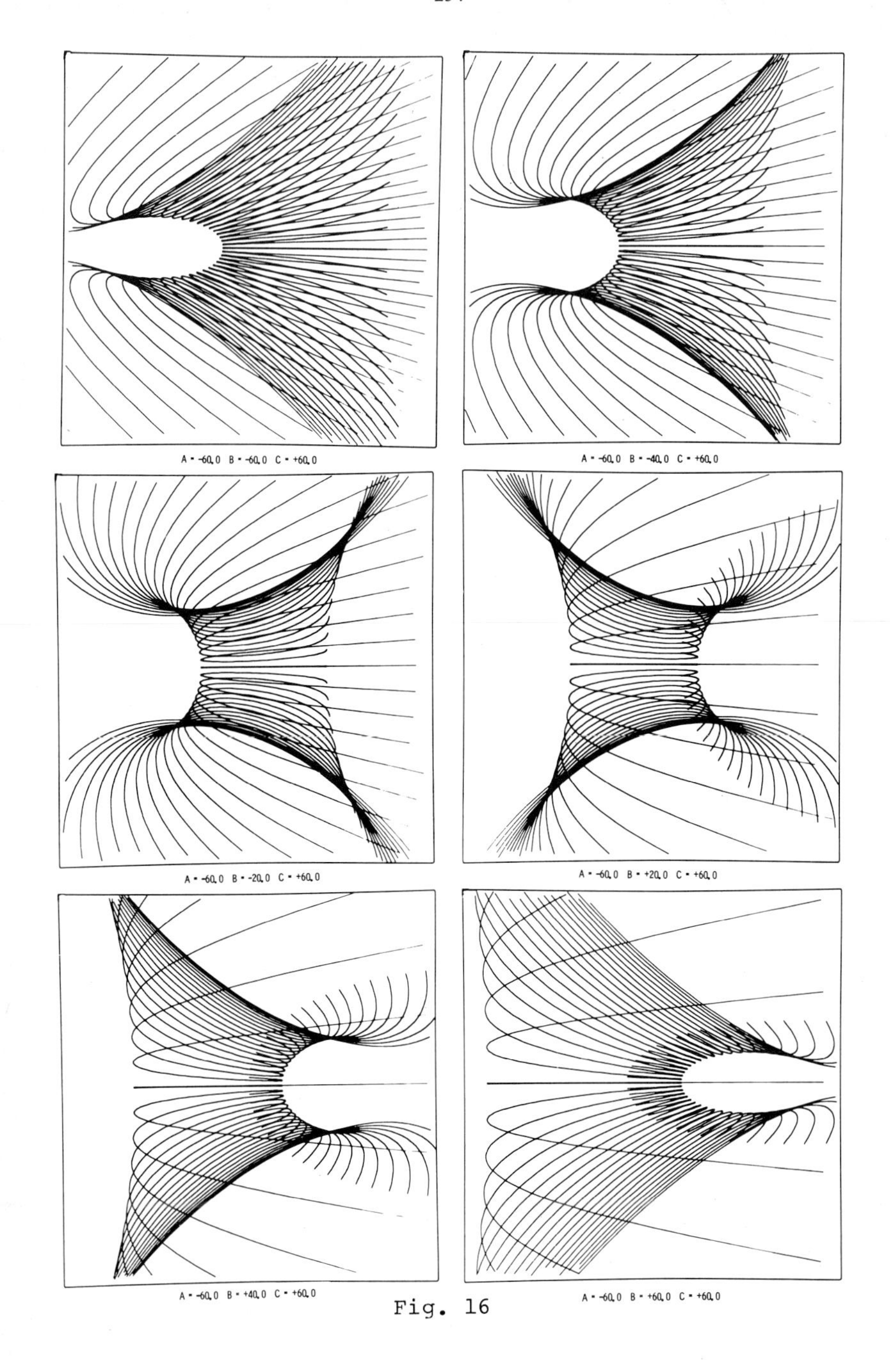
A = -60.0 B = -60.0 C = +60.0
A = -60.0 B = -40.0 C = +60.0
A = -60.0 B = -20.0 C = +60.0
A = -60.0 B = +20.0 C = +60.0
A = -60.0 B = +40.0 C = +60.0
A = -60.0 B = +60.0 C = +60.0

Fig. 16

Acknowledgments :

This work was performed at the University of Warwick, Coventry, Warwickshire, England; The T.J. Watson Research Center of the I.B.M. Corporation, Yorktown Heights, New York, U.S.A. and at the Instituto de Mathemática Pura e Aplicada, Rio-de-Janerio, Brazil. A.E.R.W. was in receipt of an I.B.M. Research Fellowship at Warwick, and an I.B.M. World Trade - I.B.M. (U.K.) Fellowship at Yorktown. T.P. was a Research Associate in Rio-de-Janerio. We would like to thank the Computing Center and Graphics Center Staffs at Warwick and Yorktown, and the Secretarial Staff at Yorktown for their help in the preparation of these papers.

Index:

(Note: Underlined numbers represent references to text illustrations.)

Vol. 215: P. Antonelli, D. Burghelea and P. J. Kahn, The Concordance-Homotopy Groups of Geometric Automorphism Groups. X, 140 pages. 1971. DM 16,–

Vol. 216: H. Maaß, Siegel's Modular Forms and Dirichlet Series. VII, 328 pages. 1971. DM 20,–

Vol. 217: T. J. Jech, Lectures in Set Theory with Particular Emphasis on the Method of Forcing. V, 137 pages. 1971. DM 16,–

Vol. 218: C. P. Schnorr, Zufälligkeit und Wahrscheinlichkeit. IV, 212 Seiten. 1971. DM 20,–

Vol. 219: N. L. Alling and N. Greenleaf, Foundations of the Theory of Klein Surfaces. IX, 117 pages. 1971. DM 16,–

Vol. 220: W. A. Coppel, Disconjugacy. V, 148 pages. 1971. DM 16,–

Vol. 221: P. Gabriel und F. Ulmer, Lokal präsentierbare Kategorien. V, 200 Seiten. 1971. DM 18,–

Vol. 222: C. Meghea, Compactification des Espaces Harmoniques. III, 108 pages. 1971. DM 16,–

Vol. 223: U. Felgner, Models of ZF-Set Theory. VI, 173 pages. 1971. DM 16,–

Vol. 224: Revètements Etales et Groupe Fondamental. (SGA 1). Dirigé par A. Grothendieck XXII, 447 pages. 1971. DM 30,–

Vol. 225: Théorie des Intersections et Théorème de Riemann-Roch. (SGA 6). Dirigé par P. Berthelot, A. Grothendieck et L. Illusie. XII, 700 pages. 1971. DM 40,–

Vol. 226: Seminar on Potential Theory, II. Edited by H. Bauer. IV, 170 pages. 1971. DM 18,–

Vol. 227: H. L. Montgomery, Topics in Multiplicative Number Theory. IX, 178 pages. 1971. DM 18,–

Vol. 228: Conference on Applications of Numerical Analysis. Edited by J. Ll. Morris. X, 358 pages. 1971. DM 26,–

Vol. 229: J. Väisälä, Lectures on n-Dimensional Quasiconformal Mappings. XIV, 144 pages. 1971. DM 16,–

Vol. 230: L. Waelbroeck, Topological Vector Spaces and Algebras. VII, 158 pages. 1971. DM 16,–

Vol. 231: H. Reiter, L^1-Algebras and Segal Algebras. XI, 113 pages. 1971. DM 16,–

Vol. 232: T. H. Ganelius, Tauberian Remainder Theorems. VI, 75 pages. 1971. DM 16,–

Vol. 233: C. P. Tsokos and W. J. Padgett. Random Integral Equations with Applications to stochastic Systems. VII, 174 pages. 1971. DM 18,–

Vol. 234: A. Andreotti and W. Stoll. Analytic and Algebraic Dependence of Meromorphic Functions. III, 390 pages. 1971. DM 26,–

Vol. 235: Global Differentiable Dynamics. Edited by O. Hájek, A. J. Lohwater, and R. McCann. X, 140 pages. 1971. DM 16,–

Vol. 236: M. Barr, P. A. Grillet, and D. H. van Osdol. Exact Categories and Categories of Sheaves. VII, 239 pages. 1971. DM 20,–

Vol. 237: B. Stenström, Rings and Modules of Quotients. VII, 136 pages. 1971. DM 16,–

Vol. 238: Der kanonische Modul eines Cohen-Macaulay-Rings. Herausgegeben von Jürgen Herzog und Ernst Kunz. VI, 103 Seiten. 1971. DM 16,–

Vol. 239: L. Illusie, Complexe Cotangent et Déformations I. XV, 355 pages. 1971. DM 26,–

Vol. 240: A. Kerber, Representations of Permutation Groups I. VII, 192 pages. 1971. DM 18,–

Vol. 241: S. Kaneyuki, Homogeneous Bounded Domains and Siegel Domains. V, 89 pages. 1971. DM 16,–

Vol. 242: R. R. Coifman et G. Weiss, Analyse Harmonique Non-Commutative sur Certains Espaces. V, 160 pages. 1971. DM 16,–

Vol. 243: Japan-United States Seminar on Ordinary Differential and Functional Equations. Edited by M. Urabe. VIII, 332 pages. 1971. DM 26,–

Vol. 244: Séminaire Bourbaki – vol. 1970/71. Exposés 382–399. IV, 356 pages. 1971. DM 26,–

Vol. 245: D. E. Cohen, Groups of Cohomological Dimension One. V, 99 pages. 1972. DM 16,–

Vol. 246: Lectures on Rings and Modules. Tulane University Ring and Operator Theory Year, 1970–1971. Volume I. X, 661 pages. 1972. DM 40,–

Vol. 247: Lectures on Operator Algebras. Tulane University Ring and Operator Theory Year, 1970–1971. Volume II. XI, 786 pages. 1972. DM 40,–

Vol. 248: Lectures on the Applications of Sheaves to Ring Theory. Tulane University Ring and Operator Theory Year, 1970–1971. Volume III. VIII, 315 pages. 1971. DM 26,–

Vol. 249: Symposium on Algebraic Topology. Edited by P. J. Hilton. VII, 111 pages. 1971. DM 16,–

Vol. 250: B. Jónsson, Topics in Universal Algebra. VI, 220 pages. 1972. DM 20,–

Vol. 251: The Theory of Arithmetic Functions. Edited by A. A. Gioia and D. L. Goldsmith VI, 287 pages. 1972. DM 24,–

Vol. 252: D. A. Stone, Stratified Polyhedra. IX, 193 pages. 1972. DM 18,–

Vol. 253: V. Komkov, Optimal Control Theory for the Damping of Vibrations of Simple Elastic Systems. V, 240 pages. 1972. DM 20,–

Vol. 254: C. U. Jensen, Les Foncteurs Dérivés de $\varprojlim$ et leurs Applications en Théorie des Modules. V, 103 pages. 1972. DM 16,–

Vol. 255: Conference in Mathematical Logic – London '70. Edited by W. Hodges. VIII, 351 pages. 1972. DM 26,–

Vol. 256: C. A. Berenstein and M. A. Dostal, Analytically Uniform Spaces and their Applications to Convolution Equations. VII, 130 pages. 1972. DM 16,–

Vol. 257: R. B. Holmes, A Course on Optimization and Best Approximation. VIII, 233 pages. 1972. DM 20,–

Vol. 258: Séminaire de Probabilités VI. Edited by P. A. Meyer. VI, 253 pages. 1972. DM 22,–

Vol. 259: N. Moulis, Structures de Fredholm sur les Variétés Hilbertiennes. V, 123 pages. 1972. DM 16,–

Vol. 260: R. Godement and H. Jacquet, Zeta Functions of Simple Algebras. IX, 188 pages. 1972. DM 18,–

Vol. 261: A. Guichardet, Symmetric Hilbert Spaces and Related Topics. V, 197 pages. 1972. DM 18,–

Vol. 262: H. G. Zimmer, Computational Problems, Methods, and Results in Algebraic Number Theory. V, 103 pages. 1972. DM 16,–

Vol. 263: T. Parthasarathy, Selection Theorems and their Applications. VII, 101 pages. 1972. DM 16,–

Vol. 264: W. Messing, The Crystals Associated to Barsotti-Tate Groups: With Applications to Abelian Schemes. III, 190 pages. 1972. DM 18,–

Vol. 265: N. Saavedra Rivano, Catégories Tannakiennes. II, 418 pages. 1972. DM 26,–

Vol. 266: Conference on Harmonic Analysis. Edited by D. Gulick and R. L. Lipsman. VI, 323 pages. 1972. DM 24,

Vol. 267: Numerische Lösung nichtlinearer partieller Differential- und Integro-Differentialgleichungen. Herausgegeben von R. Ansorge und W. Törnig, VI, 339 Seiten. 1972. DM 26,–

Vol. 268: C. G. Simader, On Dirichlet's Boundary Value Problem. IV, 238 pages. 1972. DM 20,–

Vol. 269: Théorie des Topos et Cohomologie Etale des Schémas. (SGA 4). Dirigé par M. Artin, A. Grothendieck et J. L. Verdier. XIX, 525 pages. 1972. DM 50,–

Vol. 270: Théorie des Topos et Cohomologie Etale des Schémas. Tome 2. (SGA 4). Dirigé par M. Artin, A. Grothendieck et J. L. Verdier. V, 418 pages. 1972. DM 50,–

Vol. 271: J. P. May, The Geometry of Iterated Loop Spaces. IX, 175 pages. 1972. DM 18,–

Vol. 272: K. R. Parthasarathy and K. Schmidt, Positive Definite Kernels, Continuous Tensor Products, and Central Limit Theorems of Probability Theory. VI, 107 pages. 1972. DM 16,–

Vol. 273: U. Seip, Kompakt erzeugte Vektorräume und Analysis. IX, 119 Seiten. 1972. DM 16,–

Vol. 274: Toposes, Algebraic Geometry and Logic. Edited by. F. W. Lawvere. VI, 189 pages. 1972. DM 18,–

Vol. 275: Séminaire Pierre Lelong (Analyse) Année 1970–1971. VI, 181 pages. 1972. DM 18,–

Vol. 276: A. Borel, Représentations de Groupes Localement Compacts. V, 98 pages. 1972. DM 16,–

Vol. 277: Séminaire Banach. Edité par C. Houzel. VII, 229 pages. 1972. DM 20,–

Vol. 278: H. Jacquet, Automorphic Forms on GL(2). Part II. XIII, 142 pages. 1972. DM 16,–

Vol. 279: R. Bott, S. Gitler and I. M. James, Lectures on Algebraic and Differential Topology. V, 174 pages. 1972. DM 18,–

Vol. 280: Conference on the Theory of Ordinary and Partial Differential Equations. Edited by W. N. Everitt and B. D. Sleeman. XV, 367 pages. 1972. DM 26,–

Vol. 281: Coherence in Categories. Edited by S. Mac Lane. VII, 235 pages. 1972. DM 20,–

Vol. 282: W. Klingenberg und P. Flaschel, Riemannsche Hilbertmannigfaltigkeiten. Periodische Geodätische. VII, 211 Seiten. 1972. DM 20,–

Vol. 283: L. Illusie, Complexe Cotangent et Déformations II. VII, 304 pages. 1972. DM 24,–

Vol. 284: P. A. Meyer, Martingales and Stochastic Integrals I. VI, 89 pages. 1972. DM 16,–

Vol. 285: P. de la Harpe, Classical Banach-Lie Algebras and Banach-Lie Groups of Operators in Hilbert Space. III, 160 pages. 1972. DM 16,–

Vol. 286: S. Murakami, On Automorphisms of Siegel Domains. V, 95 pages. 1972. DM 16,–

Vol. 287: Hyperfunctions and Pseudo-Differential Equations. Edited by H. Komatsu. VII, 529 pages. 1973. DM 36,–

Vol. 288: Groupes de Monodromie en Géométrie Algébrique. (SGA 7 I). Dirigé par A. Grothendieck. IX, 523 pages. 1972. DM 50,–

Vol. 289: B. Fuglede, Finely Harmonic Functions. III, 188. 1972. DM 18,–

Vol. 290: D. B. Zagier, Equivariant Pontrjagin Classes and Applications to Orbit Spaces. IX, 130 pages. 1972. DM 16,–

Vol. 291: P. Orlik, Seifert Manifolds. VIII, 155 pages. 1972. DM 16,–

Vol. 292: W. D. Wallis, A. P. Street and J. S. Wallis, Combinatorics: Room Squares, Sum-Free Sets, Hadamard Matrices. V, 508 pages. 1972. DM 50,–

Vol. 293: R. A. DeVore, The Approximation of Continuous Functions by Positive Linear Operators. VIII, 289 pages. 1972. DM 24,–

Vol. 294: Stability of Stochastic Dynamical Systems. Edited by R. F. Curtain. IX, 332 pages. 1972. DM 26,–

Vol. 295: C. Dellacherie, Ensembles Analytiques, Capacités, Mesures de Hausdorff. XII, 123 pages. 1972. DM 16,–

Vol. 296: Probability and Information Theory II. Edited by M. Behara, K. Krickeberg and J. Wolfowitz. V, 223 pages. 1973. DM 20,–

Vol. 297: J. Garnett, Analytic Capacity and Measure. IV, 138 pages. 1972. DM 16,–

Vol. 298: Proceedings of the Second Conference on Compact Transformation Groups. Part 1. XIII, 453 pages. 1972. DM 32,–

Vol. 299: Proceedings of the Second Conference on Compact Transformation Groups. Part 2. XIV, 327 pages. 1972. DM 26,–

Vol. 300: P. Eymard, Moyennes Invariantes et Représentations Unitaires. II. 113 pages. 1972. DM 16,–

Vol. 301: F. Pittnauer, Vorlesungen über asymptotische Reihen. VI, 186 Seiten. 1972. DM 18,–

Vol. 302: M. Demazure, Lectures on p-Divisible Groups. V, 98 pages. 1972. DM 16,–

Vol. 303: Graph Theory and Applications. Edited by Y. Alavi, D. R. Lick and A. T. White. IX, 329 pages. 1972. DM 26,–

Vol. 304: A. K. Bousfield and D. M. Kan, Homotopy Limits, Completions and Localizations. V, 348 pages. 1972. DM 26,–

Vol. 305: Théorie des Topos et Cohomologie Etale des Schémas. Tome 3. (SGA 4). Dirigé par M. Artin, A. Grothendieck et J. L. Verdier. VI, 640 pages. 1973. DM 50,–

Vol. 306: H. Luckhardt, Extensional Gödel Functional Interpretation. VI, 161 pages. 1973. DM 18,–

Vol. 307: J. L. Bretagnolle, S. D. Chatterji et P.-A. Meyer, Ecole d'été de Probabilités: Processus Stochastiques. VI, 198 pages. 1973. DM 20,–

Vol. 308: D. Knutson, λ-Rings and the Representation Theory of the Symmetric Group. IV, 203 pages. 1973. DM 20,–

Vol. 309: D. H. Sattinger, Topics in Stability and Bifurcation Theory. VI, 190 pages. 1973. DM 18,–

Vol. 310: B. Iversen, Generic Local Structure of the Morphisms in Commutative Algebra. IV, 108 pages. 1973. DM 16,–

Vol. 311: Conference on Commutative Algebra. Edited by J. W. Brewer and E. A. Rutter. VII, 251 pages. 1973. DM 22,–

Vol. 312: Symposium on Ordinary Differential Equations. Edited by W. A. Harris, Jr. and Y. Sibuya. VIII, 204 pages. 1973. DM 22,–

Vol. 313: K. Jörgens and J. Weidmann, Spectral Properties of Hamiltonian Operators. III, 140 pages. 1973. DM 16,–

Vol. 314: M. Deuring, Lectures on the Theory of Algebraic Functions of One Variable. VI, 151 pages. 1973. DM 16,–

Vol. 315: K. Bichteler, Integration Theory (with Special Attention to Vector Measures). VI, 357 pages. 1973. DM 26,–

Vol. 316: Symposium on Non-Well-Posed Problems and Logarithmic Convexity. Edited by R. J. Knops. V, 176 pages. 1973. DM 18,–

Vol. 317: Séminaire Bourbaki – vol. 1971/72. Exposés 400–417. IV, 361 pages. 1973. DM 26,–

Vol. 318: Recent Advances in Topological Dynamics. Edited by A. Beck, VIII, 285 pages. 1973. DM 24,–

Vol. 319: Conference on Group Theory. Edited by R. W. Gatterdam and K. W. Weston. V, 188 pages. 1973. DM 18,–

Vol. 320: Modular Functions of One Variable I. Edited by W. Kuyk. V, 195 pages. 1973. DM 18,–

Vol. 321: Séminaire de Probabilités VII. Edité par P. A. Meyer. VI, 322 pages. 1973. DM 26,–

Vol. 322: Nonlinear Problems in the Physical Sciences and Biology. Edited by I. Stakgold, D. D. Joseph and D. H. Sattinger. VIII, 357 pages. 1973. DM 26,–

Vol. 323: J. L. Lions, Perturbations Singulières dans les Problèmes aux Limites et en Contrôle Optimal. XII, 645 pages. 1973. DM 42,–

Vol. 324: K. Kreith, Oscillation Theory. VI, 109 pages. 1973. DM 16,–

Vol. 325: Ch.-Ch. Chou, La Transformation de Fourier Complexe et L'Equation de Convolution. IX, 137 pages. 1973. DM 16,–

Vol. 326: A. Robert, Elliptic Curves. VIII, 264 pages. 1973. DM 22,–

Vol. 327: E. Matlis, 1-Dimensional Cohen-Macaulay Rings. XII, 157 pages. 1973. DM 18,–

Vol. 328: J. R. Büchi and D. Siefkes, The Monadic Second Order Theory of All Countable Ordinals. VI, 217 pages. 1973. DM 20,–

Vol. 329: W. Trebels, Multipliers for (C, α)-Bounded Fourier Expansions in Banach Spaces and Approximation Theory. VII, 103 pages. 1973. DM 16,–

Vol. 330: Proceedings of the Second Japan-USSR Symposium on Probability Theory. Edited by G. Maruyama and Yu. V. Prokhorov. VI, 550 pages. 1973. DM 36,–

Vol. 331: Summer School on Topological Vector Spaces. Edited by L. Waelbroeck. VI, 226 pages. 1973. DM 20,–

Vol. 332: Séminaire Pierre Lelong (Analyse) Année 1971-1972. V, 131 pages. 1973. DM 16,–

Vol. 333: Numerische, insbesondere approximationstheoretische Behandlung von Funktionalgleichungen. Herausgegeben von R. Ansorge und W. Törnig. VI, 296 Seiten. 1973. DM 24,–

Vol. 334: F. Schweiger, The Metrical Theory of Jacobi-Perron Algorithm. V, 111 pages. 1973. DM 16,–

Vol. 335: H. Huck, R. Roitzsch, U. Simon, W. Vortisch, R. Walden, B. Wegner und W. Wendland, Beweismethoden der Differentialgeometrie im Großen. IX, 159 Seiten. 1973. DM 18,–

Vol. 336: L'Analyse Harmonique dans le Domaine Complexe. Edité par E. J. Akutowicz. VIII, 169 pages. 1973. DM 18,–

Vol. 337: Cambridge Summer School in Mathematical Logic. Edited by A. R. D. Mathias and H. Rogers. IX, 660 pages. 1973. DM 42,–

Vol. 338: J. Lindenstrauss and L. Tzafriri, Classical Banach Spaces. IX, 243 pages. 1973. DM 22,–

Vol. 339: G. Kempf, F. Knudsen, D. Mumford and B. Saint-Donat, Toroidal Embeddings I. VIII, 209 pages. 1973. DM 20,–

Vol. 340: Groupes de Monodromie en Géométrie Algébrique. (SGA 7 II). Par P. Deligne et N. Katz. X, 438 pages. 1973. DM 40,–

Vol. 341: Algebraic K-Theory I, Higher K-Theories. Edited by H. Bass. XV, 335 pages. 1973. DM 26,–

Vol. 342: Algebraic K-Theory II, "Classical" Algebraic K-Theory, and Connections with Arithmetic. Edited by H. Bass. XV, 527 pages. 1973. DM 36,–